D1228997

Acquisitions Editor: James Walsh
Production: Publishers' Design and Production Services, Inc.
Cover Design: The Printed Image

CHARLES RIVER MEDIA, INC.
10 Downer Avenue
Hingham, Massachusetts 02043
781-740-0400
781-740-8816 (FAX)
info@charlesriver.com
www.charlesriver.com

This book is printed on acid-free paper.

M. Tim Jones. *BSD Sockets from a Multi-Language Perspective.*
ISBN: 1-58450-268-1

Library of Congress Cataloging-in-Publication Data
Jones, M. Tim.
  BSD Sockets programming from a multi-language perspective / M. Tim Jones.
      p. cm.
  ISBN 1-58450-268-1 (Pbk. with CD-ROM : alk. paper)
  1. Internet programming.   2. Computer networks—Design and
construction.   3. Internetworking (Telecommunication)   4. Programming
anguages (Electronic computers)   I. Title.
  QA76.625.J66 2004
  005.2'76—dc22

                          2003016400
inted in the United States of America
  7 6 5 4 3 2 First Edition

BSD SOCKETS PROGRAMMING FROM A MULTI-LANGUAGE PERSPECTIVE

BSD SOCKETS PROGRAMMING FROM A MULTI-LANGUAGE PERSPECTIVE

M. TIM JONES

CHARLES RIVER MEDIA, INC.
Hingham, Massachusetts

This book would not have been possible without the patience and support of my beautiful wife, Jill, and my three children, Megan, Elise, and Marc. Without Jill's support and encouragement, this book would have remained a draft outline on my laptop. I'm also grateful to my parents, Bud and Celeta, whose gift of a TRS-80 (with 4 KB RAM!) in 1979 began what has been a rewarding and successful career in software development.

Contents

Acknowledgments

This book investigates Sockets programming from the perspective of a number of interesting and novel languages. This book would not be possible without these languages, and, therefore, the authors of these languages must be acknowledged. The C language was conceived and implemented by Dennis Ritchie at Bell Labs. Perl was the brainchild of Larry Wall, who continues to direct its implementation and direction. Tcl was initially developed by John Ousterhout at the University of California at Berkeley. Python was created by Guido Van Rossum at CWI in Amsterdam. Java was conceived by James Gosling at Sun Microsystems. Finally, the Ruby language (my personal favorite object-oriented scripting language) was developed in Japan by Yukihiro Matsumoto (aka "Matz").

While one may think of a book as written by a lonely author sitting up late at night researching and scribbling away on a stack of coffee-stained paper, the development of a book is an effort of many dedicated people. I'm thankful for the folks at Charles River Media who made this book possible, including Jim Walsh, Bryan Davidson, and Meg Dunkerley. Thanks also to Jim Lieb for his extremely helpful reviews.

PART

I

Introduction to Sockets Programming

The first part of this book is devoted to Sockets programming using the BSD Sockets API. The standard BSD API is covered in a detailed fashion—covering not only the functions provided, but also behavioral consequences largely ignored by most developers.

The C language is used to discuss Sockets programming in Part I, primarily because it's a very common language and one that most developers understand. The reader is assumed to have at least a rudimentary knowledge of the C language.

Additional topics covered in Part I of this book include a detailed discussion of socket options, common pitfalls suffered by many developers, and optimizations that can be employed to make Sockets programming more efficient.

Part II of this book covers the Sockets API from a multilanguage perspective, using Part I as the platform from which we move forward. A number of current languages are discussed, including Ruby, Perl, Python, and Tcl.

Finally, in Part III, we examine a number of code patterns that illustrate how to develop specific kinds of network applications within the languages covered in Part II. These patterns include stream, datagram, broadcast, and multicast clients and servers. Protocol specific applications are also discussed, including a simple HTTP server and an SMTP client.

1 Networking Overview

	Application Layer	
	Sockets API (BSD)	
Transport Layer	TCP	UDP
Network Layer	IP	
	Driver	
	Physical Layer	

In this first chapter, fundamental networking concepts are discussed to provide a basis for more advanced discussions in this book. If the reader is already aware of the networking model and types of communication available on the Internet, this chapter may be skipped.

WHAT IS THE INTERNET?

The Internet is quite simply a network of networks. A network can be defined as a collection of computers and specialized devices that support communication between them.

The basic unit of communication on the Internet is the packet. A packet is made up of a payload, which contains the data to be communicated, and a header, which describes the payload and includes information about the sender and intended receiver of the data. To use a simple letter analogy, the packet header is the envelope, which has information describing where the letter originated and where it's going, and the contents of the envelope is the payload—what is being delivered to the recipient.

Packets are originated by hosts, which are those machines that run applications such as Web browsers or e-mail clients. Hosts are also the primary recipients of packets. In addition to hosts, the Internet is also occupied by routers (see Figure 1.1). These specialized devices route packets between networks. For example, a router interconnects a number of independent networks together. When a packet arrives from a network and is destined for another, the router identifies to which particular network the packet should be sent. The router makes these decisions using packet header information; we look more closely at this process in the Introduction to IP Routing section later in this chapter.

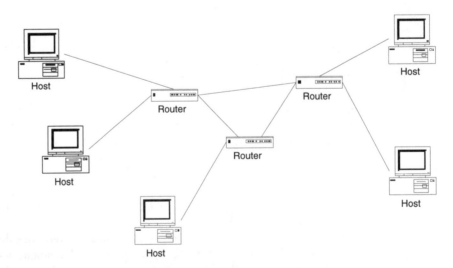

FIGURE 1.1 Fundamental devices on the Internet.

There are many good historical references of the Internet (which conceptually began in 1962 at MIT); some are even written as RFCs (Request for Comments—the Internet protocols' standardization process). A very interesting reference for the history of the Internet is provided in the References section of this chapter under [Leiner et al].

INTERNET MODEL OF COMMUNICATION

The model of communication for the Internet is a layered model in which each layer provides a distinct function (see Figure 1.2).

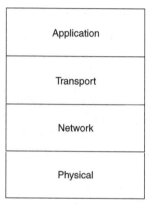

FIGURE 1.2 Layered model
for Internet communication.

At the lowest layer is the Physical layer, which is responsible for communication with the physical medium. This traditionally includes not only the physical interface (such as an Ethernet MAC or serial UART), but also the driver software that abstracts the hardware interface to the software.

Moving up the stack, the next layer is the Network layer. The Network layer provides the means to communicate between hosts on a network. The Network layer includes the addressing information that identifies the source and address hosts of the message.

The next layer is the Transport layer. This layer, from a receive standpoint, handles delivery of the message received from the Network layer to the intended application. From a transmit standpoint, the Transport layer multiplexes outgoing messages for delivery to the Network layer. Note that this is a very simplistic description; we delve into more details in the Transport Layer section later in this chapter.

To segment traffic for multiple applications at the Transport layer, some differentiating mechanism is required. Protocols at the Transport layer include a unique *port* number that uniquely identifies the application from all others. Ports can be dynamically assigned (most common for client applications) or statically assigned (used for server applications). Servers are commonly assigned static ports, also called *well-known* ports. For example, Web servers are assigned port 80, whereas mail servers typically occupy port 25. Client applications are assigned a port dynamically. These dynamic ports are called *ephemeral* ports.

The final layer is the Application layer. Applications, from this perspective, are programs that communicate over the network. This includes applications such as a mail client or a Web server. Note that many applications can run on a single host. The *port* provides the mechanism to demultiplex traffic so that it gets to its intended application on the host.

THE IP PROTOCOL SUITE

The lingua franca of the Internet is the Internet Protocol, otherwise known as IP. IP is the foundation upon which Transport and Application layer protocols are built (see Figure 1.3). The IP packet is recognized throughout the Internet and permits packets to be routed from their source to their destination host. All protocols are encapsulated within IP datagrams, which means that interior nodes of the Internet see only this IP wrapper. Inner protocols are viewed only at the endpoints (except for a few nonstandard scenarios).

In this section, we look at the standard protocols that operate on IP.

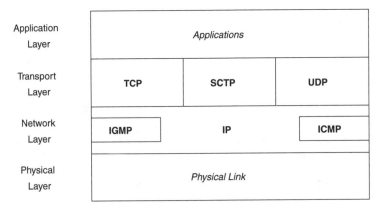

FIGURE 1.3 Layered protocols for IP.

Network Layer

The Network layer provides the ability to transport packets from a source to a destination endpoint. This layer is occupied by the Internet Protocol (IP), Internet Control Message Protocol (ICMP), and the Internet Group Management Protocol (IGMP).

IP and ICMP are standard protocols and found within all IP networking devices. IP is de-fined by a collection of RFCs, including 791, 950, 919, and 922. ICMP is defined by RFC 792 (with updates in RFC 950). Both IP and ICMP are known as STD number 5. IGMP is defined by RFC 1112, but is also covered by STD number 5.

Internet Protocol (IP)

The Internet Protocol (or IP) is the foundation protocol by which all other Internet proto-cols are transported. IP provides the end-to-end transport capability, but it is unreliable and provides only best-effort connectionless delivery. This means that IP datagrams may be lost, duplicated, or even arrive at their destination in a different order than which they were transmitted. This isn't a problem, because Transport layer protocols will provide any relia-bility that may be necessary for the given application. IP provides the very specific end-to-end delivery service.

The IP packet header is shown in Figure 1.4. The important elements to note here are the source address and destination address. These addresses, known as IP addresses, are 32-bit quantities that uniquely identify the source and destination hosts on the Internet. Also of interest is the length field, which permits an IP packet length to range from zero bytes to 65535 octets.

Byte 0	Byte 1	Byte 2	Byte 3
0 1 2 3 4 5 6 7	0 1 2 3 4 5 6 7	0 1 2 3 4 5 6 7	0 1 2 3 4 5 6 7

Version	IHL	Type of Service	Total Length	
Identification			Flgs	Fragment Offset
Time to Live		Protocol	Header Checksum	
Source Address				
Destination Address				
Options				Padding

(20 bytes)

FIGURE 1.4   Internet Protocol packet header.

Internet Control Message Protocol (ICMP)

ICMP is the Internet Control Message Protocol; as its name implies, it provides control messages to report errors to the IP layer and to provide specialized functionality for Appli-cation layer programs. It's important to realize that although ICMP is commonly men-tioned separately from IP (as we're doing here), ICMP is an integral part of IP and utilizes IP as its underlying protocol. ICMP can be thought of as a support protocol for IP, whose most common use is echo request, or ping. ICMP also handles error reporting to notify hosts of errors from routers along the path to the destination, or from the destination host itself.

The ICMP header is shown in Figure 1.5 and, being an IP-based protocol, is preceded by an IP packet header on the wire. Recall from Figure 1.4 that a protocol field is present. This field identifies the protocol header that exists inside the IP header. Within the IP network stack, the IP protocol field is used to route the inner protocol frame to the specific upper-layer transport protocol.

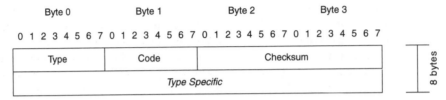

FIGURE 1.5 Internet Control Message Protocol packet header.

ICMP utilizes a very simple header whose type-specific contents depend upon the actual purpose of the packet. Other than ping, ICMP messages can represent router advertisements and solicitations, time stamp transfer, or an indication that the destination listed in a prior IP packet is unreachable.

Internet Group Management Protocol (IGMP)

IGMP, or the Internet Group Management Protocol, is a similar protocol to ICMP. The purpose of IGMP is to advertise to multicast routers on the local network that the interface from which the advertisement was sent is interested in receiving multicast messages. In reality, a host responds to requests sent by routers on the LAN. If the router has no requests for multicast traffic for a given LAN segment, the router will not relay multicast traffic from other segments. Note that for multi-homed interfaces (hosts with more than one interface), an IGMP request must be originated on each interface that desires to converse using multicast. The IGMP packet header is shown in Figure 1.6.

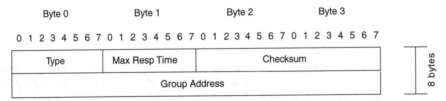

FIGURE 1.6 Internet Group Management Protocol packet header.

IGMP packets, as with ICMP, are encapsulated within an IP datagram. Only two types of messages are possible with IGMP, requests sent by multicast routers and replies sent by hosts.

Transport Layer

The Transport layer provides end-to-end communication services to applications running in the Application layer. Two types of protocols are provided in the Transport layer: datagram-based protocols and stream-based protocols. Stream-based protocols are reliable and guarantee end-to-end, in-order delivery of packets sent. This doesn't mean that the protocol guarantees all packets sent will arrive at the destination. The protocol ensures that no packets are delivered out-of-order. Datagram-based protocols make no guarantees for delivery. Packets may arrive out of order or not at all, with no indication of packet drops. We describe the most common examples of three types of transport protocols, UDP, TCP, and SCTP.

TCP is described in RFC 793 and is known as STD number 7. Additional RFCs such as 1323 describe additions to TCP (in the case of RFC 1323, extensions for high performance). UDP is defined in RFC 791 and is known as STD number 6. SCTP is described in RFC 2960, but is not yet an Internet STD document.

User Datagram Protocol (UDP)

The User Datagram Protocol, or UDP, provides an efficient datagram transport service. UDP offers no guarantees for delivery, nor will it identify packets lost in the network. Despite these limitations, UDP provides a very fast alternative to TCP because no connection is necessary with UDP and no acknowledgment protocol exists. The UDP packet header is shown in Figure 1.7.

Byte 0	Byte 1	Byte 2	Byte 3
0 1 2 3 4 5 6 7	0 1 2 3 4 5 6 7	0 1 2 3 4 5 6 7	0 1 2 3 4 5 6 7

Source Port	Destination Port
Length	Checksum

8 bytes

FIGURE 1.7 User Datagram Protocol packet header.

UDP is represented by a very simple packet header. A checksum field is used to validate the UDP header (and associated payload) and ensure that it wasn't modified en route. The length field defines the length of the payload data provided in the UDP packet. Finally, the source and destination ports are used to differentiate traffic and uniquely represent the source application and the destination application on the recipient host.

Transmission Control Protocol (TCP)

The Transmission Control Protocol, or TCP, is a reliable connection-based protocol that includes flow control and error recovery. TCP operates using the port model (as with UDP). The port defines the particular application for which the packet should be delivered.

From an application perspective, TCP is a stream-based protocol. This means that all data that flows between a sender and receiver is simply a stream of bytes without any indication of boundaries. This differs from UDP, in which data sent by the application is received in the same manner. For example, if an application sends 20 bytes through a UDP datagram, the receiving application receives 20 bytes, even if additional data was sent after the original 20 bytes. With TCP, the 20 bytes sent may be accumulated with other data sent (for the given connection) and, therefore, the application must take care of deconstructing the stream in the event it represents individual packets of data.

TCP requires that a connection between two endpoints be set up prior to data moving between the applications. This allows initialization of the connection at both endpoints for error recovery and flow control. The TCP packet header is shown in Figure 1.8.

Byte 0	Byte 1	Byte 2	Byte 3

0 1 2 3 4 5 6 7 0 1 2 3 4 5 6 7 0 1 2 3 4 5 6 7 0 1 2 3 4 5 6 7

Source Port	Destination Port
Sequence Number	
Acknowledgment Number	

HLen	Reserved	U	A	P	R	S	F	Window

Checksum	Urgent Pointer
Options (variable)	

FIGURE 1.8   Transmission Control Protocol packet header.

TCP provides source and destination ports (to help uniquely identify the source and destination application) and many other fields that are used for other purposes. As TCP is reliable, the sequence number allows the sender to identify a numeric identifier of the first byte of data in the packet (given that each byte sent is uniquely identified by an incrementing number). The receiver replies at various times with an acknowledgment number using the sequence number of the last byte in the packet to tell the sender that this data has been received (or in TCP terms, acknowledged). The sequence and acknowledgment numbers are symmetric; the sequence number represents the data that we send, whereas the acknowledgment number represents our acknowledgment of the data we receive. The peer operates with its own (swapped) versions of this data. A checksum protects not only the TCP header, but also any data contained in the packet.

The flags field (to the right of Reserved in Figure 1.8) allows the packet to alter the behavior of the peer as well as convey additional information. For example, the "S" field represents "Synchronization," and is used to initiate a connection. The "F" field is the "Final" bit and instructs the peer to close its side of the connection.

A final interesting point to note regarding the TCP packet header is the window field. When two endpoints negotiate a connection, they define the size of window. The window is the amount of outstanding data that can be sent by a peer at one time (the amount of data on the wire at any given time). The window can also be thought of as the amount of memory set aside by the host for the connection. When a full window of data is received, but the application awaiting the data has not yet accepted it from the stack, then no room exists for additional data. In this case, the window has closed and the sender must stop sending and await notification from the receiver that space is available. Although this is a simplified version of what happens within TCP, it provides a very powerful mechanism for flow control and adapts to both the speed of the receiving host to process the data and congestion in the network. From an application point of view, large windows improve throughput on long haul, many-hop networks by getting as much data into the pipe as possible.

Stream Control Transmission Protocol (SCTP)

The Stream Control Transmission Protocol, or SCTP, is another reliable Transport layer protocol that is similar to both TCP and UDP, but provides additional powerful features.

Although SCTP is most similar to TCP, it provides message framing similar to UDP. For example, if an application sends 20 bytes, the receiving application will receive those 20 bytes without any additional data that may have been accumulated by the sender. In other words, SCTP operates in a message-oriented fashion (such as UDP), not in a byte-stream fashion (as with TCP).

SCTP also provides for out-of-order delivery of data, provided by neither TCP nor UDP. This feature is designed to solve the head-of-line blocking problem, in which multiple flows may be defined within an SCTP connection, and no subflow may halt the receiving of data of any other. This feature is also called *multistreaming*.

Other important and innovative features of SCTP permit the use of multiple network interfaces on a single host to be aggregated so that the load can be shared across each of the links for a single SCTP connection. TCP and UDP support multiple interfaces, but SCTP provides native support for seamless redundancy and failover.

A network host that provides two or more network interfaces, or assigns more than one IP address to its single interface is known as a multi-homed host.

The SCTP packet header is shown in Figure 1.9. What is shown is the SCTP common header, as the SCTP packet header can be very different based upon the type of packet.

The source and destination ports help define the SCTP association, whereas the verification tag helps uniquely identify the packet as part of the current association rather than a previous embodiment of the association between the same pair of endpoints. The checksum provides a data integrity check on the SCTP packet. This differs from TCP and UDP's 16-bit checksum.

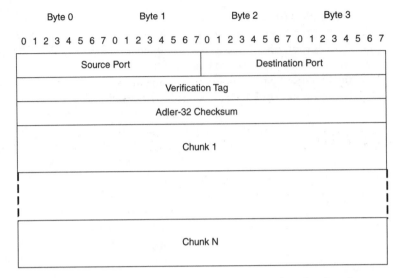

FIGURE 1.9 Stream Control Transmission Protocol packet header.

The final elements of the SCTP packet include what is known as *chunks*. A chunk is a self-descriptive packet within the SCTP packet. The chunk is used for data, initiating SCTP connections, acknowledging data, shutting down a connection, and so on. This differs from the flags method provided in TCP, allowing SCTP to be much more extensible.

Application Layer

The Application layer provides for the application-specific protocols. These protocols utilize the Transport layer protocols to achieve their results. Examples of Application layer protocols include HTTP (HyperText Transport Protocol) and SMTP (Simple Mail Transport Protocol).

The interface between the Application layer and the Transport layer is the subject of this book. The most common interface for the development of Application layer protocols is the BSD (Berkeley Software Distribution) Sockets API (Application Programming Interface). The Sockets API permits the creation of sockets between endpoints on the Internet.

TYPES OF COMMUNICATION

Now that we've looked at the basic protocols in the IP suite, let's look at the various communication models that are possible and how they map to the previously discussed protocols. In the coming chapters, we investigate patterns for creating servers and clients in each of the communication models. Each language chapter also includes example code illustrating each of the models.

Stream

In the stream mode, a conduit between two endpoints is created allowing a stream of bytes to flow bidirectionally. The stream implies that no framing occurs in the communication; the stream protocol may aggregate or split the data as it's sent. Example stream protocols include TCP and SCTP.

Datagram

In datagram mode, messages are sent from a source and a destination, without any connection setup. Messages are transferred between the source and destination, which means framing occurs within the transport protocol. Because no setup is required, datagrams received include not just the payload, but also source information. This allows the receiver to discern the sender, in the event a response is generated. An example datagram protocol is UDP.

Broadcast

The broadcast mode, as the name implies, refers to communication between all entities on a network. Broadcast is based upon the datagram mode, as it's message-oriented and permits transmission without prior setup to all receivers. Broadcast can be performed using UDP.

Multicast

The multicast mode permits the creation of a group of receivers on a network, for which a sender can transmit a message to a single address (the group), but all receivers in the group receive the message. This mode is similar to broadcast, but supports more efficient operation because not all hosts must interact with messages as is the case with broadcast communication. Multicast is also UDP-based.

PACKET PROCESSING IN A NETWORK STACK

Packets within the IP suite of protocols are known as PDUs, or Protocol Data Units. The layered architecture of the Internet protocols permits the conceptually simple process of creating and parsing packets. Consider the illustration in Figure 1.10.

When data is transmitted through a network stack (we'll ignore the communication model for this example), the data passes through the layers of the stack. The application, operating in the Application layer, communicates with the network stack through the Sockets API. The Sockets API, using information provided by the application, determines the connection (the socket) for which the data is intended and passes it to the Transport layer. In the Transport layer, a transport header is constructed and the new packet is passed to the Network layer. The element passed through the stack layers is PDU—each layer changes the PDU in a way specific to the particular protocol. The Network layer adds its packet header

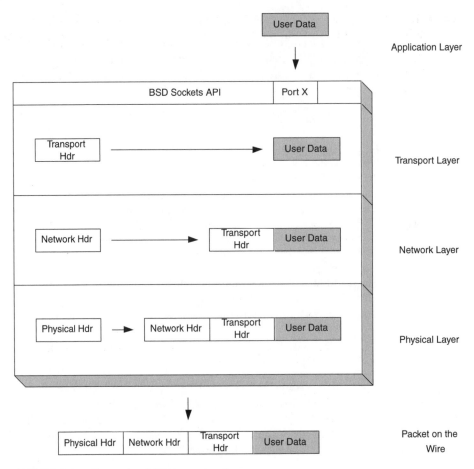

FIGURE 1.10 Example of PDU construction.

and passes it to the Physical layer. After the Physical layer packet header is attached, the packet is "put on the wire" and is on its way to the destination.

The process of creating the PDU is the process of encapsulation. Each layer performs a specific service based upon the particular protocols used in the PDU construction. Packet headers encapsulate the packet passed from the higher layers, changing both the type and purpose of the packet. The Physical layer packet header is used to transport the PDU on a physical medium. The Network layer packet header permits the packet to traverse the Internet from the source to destination host. The Transport layer packet header permits the packet data to be delivered to the specific application.

Consider now the example shown in Figure 1.11. In this example, we illustrate a packet arriving at a host and being delivered to the intended application. The process is now performed in reverse. As the packet is pushed up the layers of the network stack, the packet headers are removed from the PDU to make the relevant packet headers visible to the proper layer.

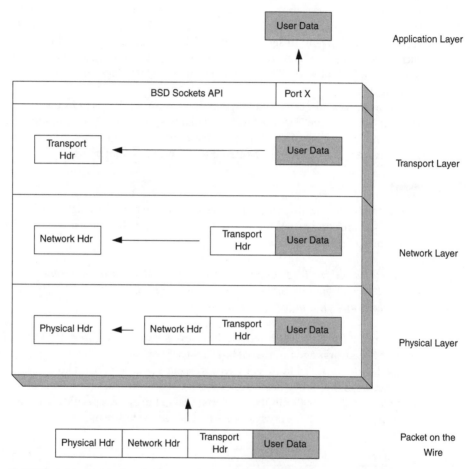

FIGURE 1.11   Example of PDU parsing and delivering payload data to an application.

As the packet is received from the network, the packet header is inspected to determine how to route the packet up the stack. Recall that there may be multiple protocols operating within a given layer (TCP and UDP, for example, in the Transport layer). Each layer inspects the current packet header, identifies how to process the packet, removes the header for the current layer, and routes it up the stack to the necessary protocol. At the Transport layer, the data is queued for the given application (as identified by the protocol and port). The application may then extract the data using functions provided in the Sockets API.

ADDRESSING

Let's now focus on how entities on the Internet are addressed at the Network layer. Internet addresses have two forms. The first is the symbolic form for which most nontechnical users are familiar. An example of this address is *www.mysite.com*. This form is also called a Fully Qualified Domain Name (FQDN). The second form is numeric, and is represented by a 32-bit unsigned number. This is commonly defined in string notation, such as "192.168.1.1", but is represented as a 32-bit word (made up of the 4 bytes in the dotted string notation). The actual 32-bit value is represented in hexadecimal as 0xc0a80101.

Equivalences exist between the symbolic and numeric forms of Internet addresses. The Domain Name System (DNS) provides a global mapping of FQDNs to numeric addresses. Most, if not all, devices on the Internet have access to a resolver that communicates with a DNS server to resolve symbolic names to their numeric counterparts.

IP Addresses

Internet addresses are also called IP addresses because they are a Network layer construct. IP addresses are constructed in two parts, the first part being the network number and the second being the host number.

We focus solely on IPv4 (Internet Protocol version 4) here because it is the current standard for Internet communication. A newer version of the IP protocol called IPv6 may someday replace IPv4, but can coexist with IPv4.

There are five classes of Internet addresses, as shown in Figure 1.12. They differ in the number of networks and number of hosts that are possible in each class. For example, Class A provides for a minimal number of networks, but a large number of hosts. Class B provides more networks, but fewer hosts. Conversely, Class C provides for a large number of networks, each with a small number of hosts. Class D provides specifically for multicasting, which represent groups of hosts. Class E is reserved for future use.

Byte 0	Byte 1	Byte 2	Byte 3

A	0		Network		Host			
B	1	0		Network			Host	
C	1	1	0		Network			Host
D	1	1	1	0		Multicast Group		
E	1	1	1	1				

FIGURE 1.12   Classes of Internet addresses.

An alternative to strict classes of addresses is known as Classless InterDomain Routing (CIDR). In CIDR IP addressing, rather than applying strict classes to the 32-bit IP address, any number of bits can be assigned to a network, with the remaining bits being assigned to the host portion (rather than the strict 8, 16, or 24). This means that the IP address space is more effectively managed. Let's say that a particular site needs 19 IP addresses. This could be accommodated with the Class C domain, but because 8 bits are assigned to the host portion, 235 IP addresses would be wasted. It's easy to see from this example how much space can be wasted. With CIDR, we simply assign the number of bits necessary for the host; in this case (19 IP addresses), it's 5. With a given IP address, such as 192.11.187.1, we apply the network prefix, resulting in 192.11.187.1/27. This means that the upper 27 bits are used to identify the network, with the remaining 5 bits used to identify the host.

With the growth of technologies such as IP masquerading and Network Address Translation (NAT), ideas of public and private addresses are more commonplace. NAT permits a private network to use private addresses and to then translate them to public addresses when communicating over the WAN. The net effect of this technology is minimizing the use of public addresses and, therefore, providing IPv4 with a longer life. This technology is also called "IP Sharing," in which a single public IP address is shared internally by a number of hosts using a number of private IP addresses. The gateway host provides the mapping of requests (translation) to the external world and back.

Subnetting

Within a particular network, classes of addresses can be further refined than what is shown in Figure 1.12. For example, a network of type Class C can support up to 254 hosts. An administrator on that network can utilize the hosts number to define additional networks with fewer numbers of hosts (splitting the bits up between internal networks and internal hosts). Although hosts and other devices are cognizant of the split internally, it is invisible to the external network. This provides for better utilization and separation of subnets within an internal network.

 Although 256 unique hosts can be represented by an 8-bit host identifier, only 254 may be assigned. This is because the zero address is used to represent "this" address (the current address), and the 255 address represents "all" addresses (a local broadcast).

INTRODUCTION TO IP ROUTING

Let's now discuss the routing of IP datagrams on the Internet. Prior to an IP datagram being emitted from a host, it is passed through the IP routing algorithm. This algorithm determines the next hop that the datagram must take in order to get to its destination. When a datagram is emitted from a host, the host has no idea in advance the path that the datagram will take en route to the destination. Therefore, a host worries only about the first hop or, in other words, where the datagram should be forwarded to get to its final destination.

Each device on the Internet includes a routing table. This table includes a set of networks that are available for routing in addition to a default route. The default route defines the destination for the packet if none of the prior routing entries satisfied the datagram's needs.

Simple Routing Example

Let's look at a simple example to illustrate the routing decisions from an endpoint's perspective. Consider the network topology shown in Figure 1.13.

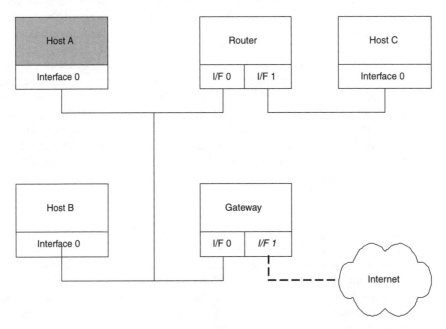

FIGURE 1.13   Sample network topology for routing discussions.

In this sample network, we'll look at routing decisions from the perspective of Host A. We can see from Figure 1.14 that Host A has an IP address of 192.168.1.1 and a network mask of 255.255.255.0. If we mask the IP address by the network mask, the result is the network on which the host resides, 192.168.1.

Note that this masking operation is based on a compact 32-bit representation. The IP address 192.168.1.1 *is hex* 0xC0A80101 *and the network mask is* 0xFFFFFF00. *Using the masking operation (bitwise AND) results in* 0xC0A80100, *or* 192.168.1.0.

Let's now look at the four cases of routing that are possible in this sample network and the decisions that are made by the IP routing algorithm. The IP routing table for Host A is

Host/Interface	IP Address	Netmask
Host A - Interface 0	192.168.1.1	255.255.255.0
Host B - Interface 0	192.168.1.2	255.255.255.0
Router - I/F 0	192.168.1.253	255.255.255.0
Router - I/F 1	192.168.8.254	255.255.255.0
Host C - Interface 0	192.168.8.5	255.255.255.0
Gateway - I/F 0	192.168.1.254	255.255.255.0

FIGURE 1.14   Sample network device addresses.

provided in Figure 1.15. This table defines a destination IP address, network mask, gateway to which packets satisfied by this row should be forwarded, and the textual type of routing entry. In order to route a packet, the destination address is tested using each of the routing entries in the table. If a route is not satisfied for the IP datagram, the default route is selected (if present). Otherwise, the packet is rejected as representing an IP destination that is unreachable.

Destination	Gateway	Netmask	Route Type
192.168.1.1		255.255.255.255	Host Route
192.168.1.0		255.255.255.0	Local Route
192.168.8.0	192.168.1.253	255.255.255.0	Local Route
0.0.0.0	192.168.1.254	0.0.0.0	Default

FIGURE 1.15   Sample IP routing table for Host A.

Consider an application that desires to communicate with another application on the same host (otherwise known as IPC, or InterProcess Communication). In this case, the first row in the routing table (Figure 1.15) satisfies the route. Host A's IP address is 192.168.1.1, and the destination for the IP datagram (given that the destination application is on the same host) is also 192.168.1.1. Masking the destination address with the network mask results in the destination value in the routing table, 192.168.1.1. Therefore, this route is selected and the IP datagram is routed back up the stack toward the recipient application.

Now consider an application on Host A that desires to communicate with an application on Host B, whose IP address is 192.168.1.2. The destination address of the IP datagram is tested with row 2 in our routing table. Because 192.168.1.2 masked with 255.255.255.0 results in 192.168.1.0, this row is satisfied, representing a destination on the local LAN.

The next example illustrates Host A communicating with Host C. In this case, we're dealing with separate networks, so the process changes slightly. We walk through the routing table as discussed before, masking the IP datagram destination with the network mask and comparing with the route entry destination. Upon finding a match (row 3 in Figure 1.15), we now find that a gateway has been defined. In this case, the packet is forwarded to the host defined as the gateway, who will then route the packet forward to its destination.

The Address Resolution Protocol, or ARP, provides low-level address mapping between IP addresses in the Network layer and Ethernet addresses in the data-link layer for a LAN. Each Ethernet interface has a unique 48-bit address. ARP provides the mapping to the IP layer for data-link layer addressing.

Our final example illustrates what happens when a datagram is destined for an IP address that cannot be mapped by any of the local routes in the routing table. In Figure 1.15, we see that the last row specifies a destination of 0.0.0.0 and a network mask of 0.0.0.0. In this case, if we exhaust all prior routing table entries, this will satisfy the destination by default (because masking any number by 0.0.0.0 always results in 0.0.0.0). This entry is also known as the "default route" because it's the route to all other networks.

Local (Interior) Ethernet Routing

One interesting note is the mapping of Ethernet addresses to IP addresses. Recall in our second routing example that Host A wanted to communicate with Host B. In this case, ARP identifies the Ethernet address (also known as a MAC address) for the given IP address. When the IP datagram arrives for transmission, the Ethernet header is attached with the proper Ethernet address. This simple case covers all communication for hosts on the same LAN. However, what happens when the destination host is not on the local network? In this case, the IP destination within the IP header is the actual IP address, but the Ethernet address is the address of the gateway. Therefore, the gateway receives the IP datagram, but notes that the IP address is different than its IP address. The IP datagram is then routed on toward its destination.

Exterior Routing

So, what happens when an IP datagram reaches the gateway defined by the host's routing table? The datagram is further routed using a routing daemon. One of the most commonly found routing daemons is called "gated." This daemon communicates with other routers to identify where IP datagrams should be forwarded. Luckily, from our perspective, all routing is local, so all we need to do is identify the first hop and let the more complicated routing daemons handle routing through the Internet.

NETWORK ENTITIES

Finally, let's review the types of devices found on the Internet and their roles in communication, routing, and hosting applications. See Figure 1.16 for graphical views of each of the devices.

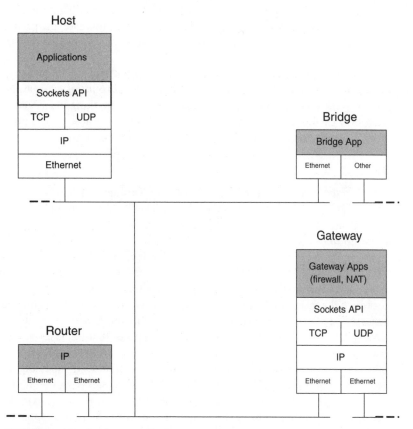

FIGURE 1.16 Devices on the Internet.

Hosts

The host is the basic edge device on the Internet. It hosts a network stack and one or more applications that use the Sockets API to provide or use a service on the Internet. These applications could be client browsers or Web servers (utilizing HTTP).

Hosts can also be considered termination devices because they originate or terminate network connections (the boundaries of communication between two endpoints).

Bridges

The bridge is a specialized device that provides protocol-independent routing of datagrams. We refer to it as specialized because the routing algorithms occur at the MAC or data-link layer. For this reason, the bridge can route both IP datagrams and other types of packets, unrelated to the IP suite of protocols. With the ascent of wireless networks (using the 802.11 IEEE standard), bridge devices that bridge between typical 802.3 networks and 802.11 networks are becoming common.

Routers

A router is a specialized application that may run a particular version of an IP stack to route datagrams between two or more networks. Recall that routing occurs at the IP layer; therefore, it's not necessary for the router to understand anything but IP. As shown in Figure 1.16, only the IP layer is present in the example stack. Because it does not originate or terminate connections, only the IP layer is necessary for the IP routing algorithm.

Gateways

The final device that we discuss is the gateway. This device provides the highest level of IP routing as it commonly inspects datagram packet headers in the transport header. For example, gateways can provide a firewall that is used to restrict access from the external Internet to nodes on the internal network. A gateway can also provide address mapping to hide internal network host addresses or share a single public address with many internal private addresses. This type of functionality is called Network Address Translation, or NAT.

A gateway may also provide higher-level routing algorithms based upon the application for which the frame is destined or originated. For example, a gateway may route Web server requests (HTTP) to one host and mail (SMTP) to another, even though their destination addresses are the same. This type of routing requires that the gateway deeply inspect the packet headers to understand the type of Application layer protocol being transported. Another important use of this type of routing is for system load balancing. For example, the gateway could route HTTP requests for images to one server and requests for HTML files to another. The ability to segregate traffic based upon the request can lead to better system performance (through limited file caching at the server) and can support transparent replication of content to multiple servers behind the gateway.

SUMMARY

In this first chapter, we've investigated a number of networking topics in preparation for detailed discussions of Sockets API programming. We discussed the Internet model of communication, including the layered network stack architecture and the networking protocols that exist at each of the layers. The packet construction and parsing methods of the layered stack architecture were covered in addition to addressing devices and routing datagrams over the Internet. Finally, the different types of devices were discussed, including their roles and simple architecture.

REFERENCES

[Leiner et al] Leiner et al., "A Brief History of the Internet," Internet Society, available online at *http://www.isoc.org/internet/history/brief.shtml*.

RESOURCES

Stewart, Randall R., and Qiaobing Xie, *Stream Control Transmission Protocol (SCTP): A Reference Guide*, Addison-Wesley, 2001.

2 Introduction to Sockets Programming

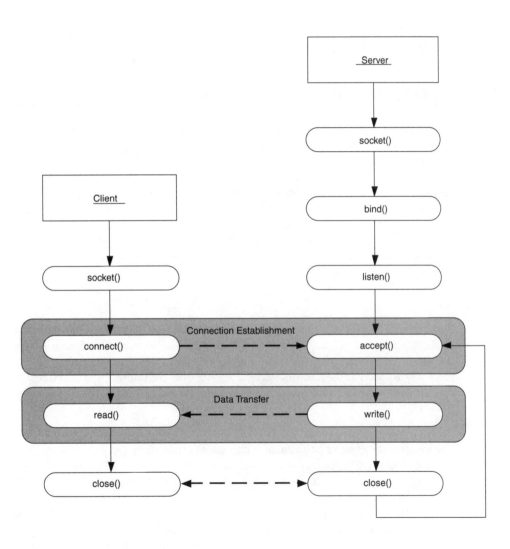

In this chapter, we take a quick tour of Sockets programming. We discuss the basic elements of the Sockets API, addressing, ports, and differences between client and server applications. We complete this quick introduction with a look at a sample client and server application that can communicate over the Internet using the stream model of communication. All code for this chapter can be found on the companion CD-ROM under /software/ch2.

ON THE CD

NOTE

Except for line numbers, boldface is used only for functions that are part of the Socket API in code listings and in the text. It is used to emphasize what is networking specific. This convention is followed in this chapter and throughout the book.

THE SOCKETS PROGRAMMING PARADIGM

The Sockets API, first introduced into the 4.2BSD Unix® system, is the *de facto* standard for network programming. With a very simple API, Sockets provides the ability to build complex communications applications with server and/or client semantics. The Sockets API has been ported to all useful operating systems, including Linux®, commercial kernels, and real-time operating systems, academic operating systems, and even Windows® (though some differences were made here).

In this section, we take a quick tour of the Sockets API using a sample client and server to illustrate the function calls. As with all software examples in Part I of this book, the C language is used. Part II includes software patterns for many other languages, leveraging from the original patterns written in C.

Ports

Recall from Chapter 1 that an application available on the Internet can be defined by two elements, an Internet address (IP address) and a port. The port can be *well-known* (typically defined for servers, so that clients can find it) or *ephemeral* (typically used by clients, dynamically assigned by the network stack). The port is the communication endpoint between two entities on the Internet, connected by a *socket*.

Sockets

The socket is the conduit between applications over the Internet. A stream socket must be logically connected between the two endpoints (using the BSD API), whereas datagram sockets require no connection setup other than the creation and initialization of the local socket endpoints.

After a stream socket is defined and connected, data flows without any additional attention to addressing because a virtual circuit has been defined. With datagram sockets, because no virtual circuit exists, each packet sent must have a recipient address specified to direct the datagram to its destination (with one caveat, which we discuss later in Chapter 6, Advanced Sockets Programming Topics).

It's important to note that a port represents an application, but an application can be represented by multiple ports. Methods for dealing with multiple ports (such as using the `select` *call) are discussed in Chapter 6, Advanced Socket Programming Topics.*

Addressing

If we use standard communication analogies to define the two socket types, we can define a stream connection as a telephone call and a datagram connection as postcard communication. When we make a telephone call, we're required to set up the channel to communicate (in this case, dial the number and await the receiver to pick up the phone). Once the receiver answers, we can communicate on a one-to-one basis in a stream fashion. Datagram communication allows us to define addressing information for each message we send. Therefore, there's no stream, but a number of independent (or related) packets sent separately. Furthering the analogy, if we were provided N-way conferencing on our telephone call, multicast communication would be realized. Finally, broadcast would be much like using a bullhorn to talk to our neighbors.

In either the stream or datagram case, we still need to identify the receiver for our communication. This is done using a *tuple*, which uniquely identifies the receiver. The tuple includes the host address, the port (local identifier), and the protocol, such as

```
{ tcp, 192.168.1.1, 45001 }
```

This tuple is called a half-association. It defines all the necessary elements to uniquely identify an endpoint of a socket. Note that the protocol is included in the tuple, which means that port 45001 for TCP is not the same as port 45001 for another protocol, such as UDP.

To identify a complete socket connection, including both endpoints, we need a 5-tuple association. This defines the protocol and both endpoints for the socket. For example,

```
{ tcp, 192.168.1.1, 45001, 192.168.9.181, 80 }
```

defines a full association representing the hosts and processes for the socket.

SAMPLE SERVER

Let's now take a quick tour of a simple Sockets application. We focus first on the server, a simple Daytime protocol server. This server simply emits the date and time of the current host in string format to the client. This particular server uses stream sockets (provided by the TCP transport protocol). In Part II, we'll look again at this application, and how it can be built using the other communications paradigms (datagram, multicast, and broadcast) in each language.

It should be noted that socket calls are not thread safe. Thread-safe versions do exist, so when operating in a threaded environment, thread calls should be used exclusively.

Source Discussion

The simple Daytime protocol server can be found in Listing 2.1 (line numbers added for the purpose of discussion). We first specify the header files that are necessary for compilation of the source (**lines 1–5**). This gives us access to socket types, symbolics, and a number of necessary host functions. **Lines 7 and 8** define two internal symbolic constants, the maximum buffer size used to send our response, and the port that will be used to identify this server. Note that in this case, port 13 is the well-known port for the Daytime protocol.

Lines 10–15 begin our C main program and define some of our local variables to be used. We discuss each as they're used within the program. **Line 17** shows our first Sockets API call, **socket**. The **socket** call creates a new socket of family AF_INET (for Internet communication) and protocol SOCK_STREAM (the stream protocol, or TCP). This call creates a blank socket that we must now initialize prior to using.

Initialization of our new socket occurs in **lines 19–25**. After clearing out our address structure (servaddr), we define the family as AF_INET and then specify the incoming address and port. For the incoming address, we specify INADDR_ANY, which means that we'll accept incoming connections through any of the interfaces on the host. The port is initialized with our daytime port number. Note the use of **htonl** (host-to-network-long) and **htons** (host-to-network-short). These functions provide the byte swapping to convert host-byte-order to network-byte-order (for either a 4-byte long or 2-byte short). These functions are commonly provided as macros, which may or may not provide any function depending upon whether the host-byte-order is the network-byte-order. Byte ordering is discussed later in this chapter. Finally, we bring our socket and address information together in the **bind** call. This call configures the socket based upon the data defined within the servaddr address structure.

At this point, we have an endpoint (half-association) configured. However, this endpoint is not available for incoming connections. In order to permit connections to be made, a call to the **listen** function is performed (**line 27**). This call permits incoming connections to be made and also defines the number of outstanding connections that are possible, in this case five.

Listing 2.1   Daytime protocol server (server.c).

```
1   #include <sys/socket.h>
2   #include <arpa/inet.h>
3   #include <stdio.h>
4   #include <time.h>
5   #include <unistd.h>
6
7   #define MAX_BUFFER          128
```

```
 8   #define DAYTIME_SERVER_PORT    13
 9
10   int main ( void )
11   {
12     int serverFd, connectionFd;
13     struct sockaddr_in servaddr;
14     char timebuffer[MAX_BUFFER+1];
15     time_t currentTime;
16
17     serverFd = socket(AF_INET, SOCK_STREAM, 0);
18
19     memset(&servaddr, 0, sizeof(servaddr));
20     servaddr.sin_family = AF_INET;
21     servaddr.sin_addr.s_addr = htonl(INADDR_ANY);
22     servaddr.sin_port = htons(DAYTIME_SERVER_PORT);
23
24     bind(serverFd,
25           (struct sockaddr *)&servaddr, sizeof(servaddr));
26
27     listen(serverFd, 5);
28
29     while ( 1 ) {
30
31       connectionFd = accept(serverFd,
32                             (struct sockaddr *)NULL, NULL);
33
34       if (connectionFd >= 0) {
35
36         currentTime = time(NULL);
37         snprintf(timebuffer, MAX_BUFFER, "%s\n",
38                   ctime(&currentTime));
39
40         write(connectionFd, timebuffer, strlen(timebuffer));
41
42         close(connectionFd);
43
44       }
45
46     }
47
48   }
```

At this point, we have a socket server configured and ready for incoming connections. The remainder of the program handles incoming connections and processes them based upon the Daytime protocol. **Line 29** starts an infinite loop for connection handling. In order to now accept incoming connections, we use the **accept** call (**lines 31 and 32**). The

accept call is unique to server applications, particularly stream connections (because no connection setup is performed for datagram connections). We specify our server socket (serverFd) and two NULL pointers. These NULL pointers could represent actual objects, which **accept** would fill in for information on the peer that connected. The **accept** call returns with a new socket (when a client connects), which represents a new preconfigured endpoint. This new socket represents the connection with the peer, which is used for Daytime protocol communication. It's important to note here that our server socket (serverFd) is used solely for accepting new connections, but communicates no user data.

In **line 34**, a quick error check is made to ensure that a proper socket was returned. In the event an error occurred within the **accept**, a value of –1 would be returned.

Passing the socket error check, we build the data to be sent to the client (**lines 36–38**). The current date and time is gathered by the **time** call, which is converted into a string format using the **ctime** call. The call to sprintf copies this new string into our local buffer.

In **line 40**, we send the data to the client using the **write** call. We specify the socket, for the client connection, the data to emit, and the length of the data (computed using strlen).

Finally, we close the new client socket in **line 42**. Note that our server socket (serverFd) remains open and available for new incoming connections. The client socket, per the Daytime protocol, is closed directly after the data is sent. Per the stream protocol, the socket won't be closed right away. Once the data is transmitted, and acknowledged by the peer, the socket is gracefully closed. Our while loop then continues back to **line 31**, where we call **accept** again awaiting a new client connection.

SAMPLE CLIENT

Now, let's look at the client for the previously discussed Daytime protocol server. The C language client is shown in Listing 2.2. For brevity, we omit the include files in this example.

Listing 2.2   Daytime protocol client (client.c).

```
1    int main ( )
2    {
3      int connectionFd, in;
4      struct sockaddr_in servaddr;
5      char timebuffer[MAX_BUFFER+1];
6
7      connectionFd = socket(AF_INET, SOCK_STREAM, 0);
8
9      memset(&servaddr, 0, sizeof(servaddr));
10     servaddr.sin_family = AF_INET;
11     servaddr.sin_port = htons(DAYTIME_SERVER_PORT);
12
13     servaddr.sin_addr.s_addr = inet_addr("127.0.0.1");
14
```

```
15        connect(connectionFd,
16                (struct sockaddr_in *)&servaddr,
17                sizeof(servaddr));
18
19        while ( (in = read(connectionFd, timebuffer,
20                MAX_BUFFER)) > 0) {
21
22          timebuffer[in] = 0;
23          printf("\n%s", timebuffer);
24
25        }
26
27        close(connectionFd);
28
29        return(0);
30    }
```

Source Discussion

Our main program begins on **line 1**, with a set of local variables declared in **lines 3–5**. **Line 7** begins our function with the creation of a new socket (as with the server, a stream socket). We clear out our address structure on **line 9** and then initialize a few of the entries (family and port). Note that in this case, we're defining an address structure for the endpoint to which we want to connect. Recall from the server example, that this same structure was used to bind the name to the socket. This symmetry is important to note as it will be used to link the two endpoints together.

In **line 13**, we define the last element of the address structure, the address of the host to which we're going to connect. We're using the loopback address here (127.0.0.1), which assumes that the client and server are running on the same host. This line could easily reference a different IP address (the address on which the server is actually running) to support connecting over the Internet. The address is converted from a string to a 32-bit value using the **inet_addr** function. This function maps the string "127.0.0.1" to the numeric 0x7f000001, which is the proper mapping for the sockaddr_in structure.

The loopback address is a special address that simply represents the current host. The IP address for the loopback interface is 127.0.0.1.

We create the connection to the server from the client using the **connect** function (**line 13**), specifying connectionFd as our socket endpoint and servaddr as the endpoint to which we want to connect. The **connect** function is specific to stream sockets and performs the connection synchronization and negotiation with the server using TCP. When the **connect** call returns, either an error has occurred, or we're connected to the server. For brevity, we'll assume a successful connection, but a check of the return status can easily determine success or failure of this call.

At **line 19**, we attempt to read from the socket using the **read** call. We specify three parameters, representing our socket endpoint (connectionFd), the buffer for incoming data (timebuffer), and the size of the buffer (maximum amount of data to read, or MAX_BUFFER). The **read** call returns the number of bytes read from the socket. If a value less than 1 is read, an error has occurred. Succeeding this check, we NULL-terminate our buffer in **line 22** and print it in **line 23**. Our read occurs within a while loop, so we try to read again. At this point, the server has closed the socket (**line 42** in Listing 2.1). We'll detect this close (by a return value of 0 to the **read** call) and gracefully exit the while loop.

We close our socket endpoint at **line 27** (using our socket descriptor, connectionFd) and then exit the client at **line 29**.

From the source discussion in the client and server, it's probably become clear that some symmetry exists between the two applications. This topic is explored in the next section.

CLIENT/SERVER SYMMETRY

From our last example, it's clear that for every client that performs a **connect**, a server performs an **accept**. When communication occurs, for every **write**, there's a corresponding **read** of the communicated data (though the granularity of reads and writes may differ depending upon the specific protocol, more on this in Chapter 6, Advanced Sockets Programming Topics). The relationships shown by this symmetry are important because they illustrate how to build client and server Sockets applications.

Consider the illustration shown in Figure 2.1. This figure depicts the Sockets API calls from Listings 2.1 and 2.2 (a stream sockets example).

The server must create a socket (using the **socket** call) and then name it (using **bind**) so that clients can reference it. The **listen** call makes the server visible and allows it to accept connections. The server then must use the **accept** call to accept incoming client connections, which are represented by a newly created socket representing the new endpoint.

For a client to connect to the server, it must also create a socket. Additionally, we must define for the socket what we want to connect to. This is defined in the sockaddr_in structure. Then, by using the **connect** call, we cause the three-way handshake to occur between the client and server, using in this case TCP. Note the relationships between **connect** and **accept**. The **accept** call defines the server's willingness to communicate, whereas the **connect** call defines the client's desire to communicate. For stream connections using TCP, this relationship must always exist.

For the **write** and **read** calls, the same symmetry exists. There's no point in calling the **read** function if the peer never writes any additional data into the socket.

Closing sockets also illustrates some symmetry, but many details lie beneath the surface. When an application performs a close, it defines intent not to send any additional data through the socket. This case is shown in the server example in Listing 2.1. This first close is known as the active close. The peer recognizes this close and performs one of its own. This is the passive close; it completes the closure of the socket because neither endpoint will send or receive any further data through the socket.

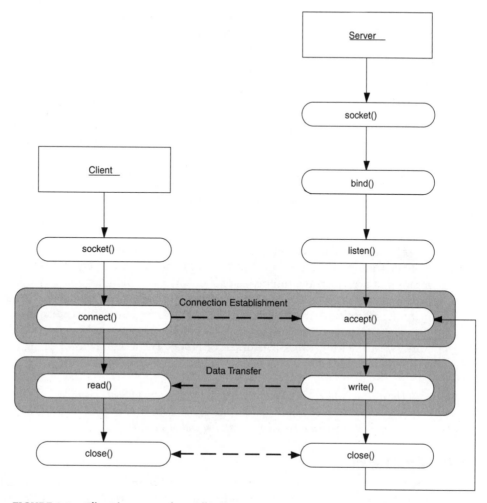

FIGURE 2.1   Client/server socket call symmetry.

Another method for closing the socket communicates different intentions to the peer. We'll discuss the shutdown *method in Chapter 4, Advanced Sockets Functions.*

NETWORK BYTE ORDER

A given computer architecture defines the format by which data is stored within the computer's memory. Two formats exist, which are opposites of one another. The first format, known as *little endian*, stores the contiguous bytes of a type in memory with the least significant byte (0x04) first and the most significant byte last (0x01). The *big endian* format stores contiguous bytes in the opposite order; the most significant byte is stored first and the least significant byte is stored last. See Figure 2.2 for a graphical description.

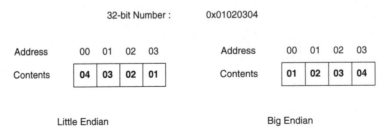

32-bit Number :   0x01020304

| | Address | 00 | 01 | 02 | 03 |
| | Contents | 04 | 03 | 02 | 01 |

| | Address | 00 | 01 | 02 | 03 |
| | Contents | 01 | 02 | 03 | 04 |

Little Endian   Big Endian

FIGURE 2.2   Memory storage of little and big endian formats.

Although the data contained in the messages uses a format predefined by the applications that use it, the values stored in the packet headers are in big endian format. This means that processors that utilize big endian (such as Motorola® processors) require no byte-swapping to occur. Processors that utilize little endian (such as Intel® processors) require all packet fields of more than one byte to be swapped from little endian to big endian. It's a slight advantage, therefore, for a network processor to use the byte ordering of the network (what's called network byte order, or NBO). The lack of byte swapping to convert from host byte order to network byte order can lead to a slight performance advantage.

The original concept of endianness comes from Jonathan Swift's Gulliver's Travels. *In this book, the nations of Liliput and Blefuscu waged war against each another over which end one should cut open a boiled egg—the big end or the little end.*

Swapping Byte Order

The Sockets API provides four functions to swap the byte order of fundamental types. These functions are **ntohs** (network-to-host-short), **htons** (host-to-network-short), **ntohl** (network-to-host-long), and **htonl** (host-to-network-long). These four functions operate on two basic types, shorts (16-bit words) and longs (32-bit words). A function is provided for both converting to network byte order (from host byte order) and converting to host byte order (from network byte order).

It's common practice to use these functions, even if the host architecture uses network byte order (big endian). This permits the software to be easily ported to host architectures that use little endian byte ordering.

SUMMARY

In this short introductory chapter, we've looked at some of the basics of communicating over the Internet using stream sockets and looked at a sample implementation of a simple server and client. The concept of sockets, ports, and addresses were discussed to provide additional framework to understand the relationship between the Sockets API and some features of the underlying protocols. The symmetry of client and server Sockets applications were discussed to better understand the relationships between some of the basic Sockets API functions. Finally, the issue of byte-ordering and computer architecture was identified and the Sockets API functions that can be used to build transparent and portable Sockets applications were discussed.

RESOURCES

Stevens, W. Richard, *UNIX Network Programming*, 2nd Edition, Volume 1, Prentice Hall PTR, 1998.

3 Standard BSD API Functions

	Client	Server	Stream / Datagram
socket	●	●	S/D
bind	●	●	S/D
listen		●	S
accept		●	S
connect	●		S/D
recv/recvfrom	●	●	S/D
send/sendto	●	●	S/D
close	●	●	S/D
htons/ntohs	●	●	
htonl/ntohl	●	●	
inet_addr	●	●	
inet_ntoa	●	●	
inet_aton	●	●	

In this chapter, we take our first look at the details of the BSD Sockets API. This discussion is based on the Linux API, but this implementation follows the standard closely, so it will apply to other implementations as well. We focus on the basic Sockets API primitives first, and then in Chapter 4, Advanced Sockets Functions, we look at the more advanced functions. The Sockets functions that are discussed in this chapter are shown in Table 3.1.

TABLE 3.1   Sockets API functions discussed in Chapter 3

Function	Description
socket	Creates a new socket endpoint
bind	Binds a local address to the socket
listen	Permits incoming connections for a server socket
accept	Accepts a new client connection for a server socket
connect	Connects to a server from a client socket
recv	Receives data through a socket
recvfrom	Receives data through a datagram socket
send	Sends data through a socket
sendto	Sends data through a datagram socket
close	Closes a socket endpoint
htons	Converts short to network byte order
ntohs	Converts short to host byte order
htonl	Converts long to network byte order
ntohl	Converts long to host byte order
inet_addr	Converts a dotted-notation string to a 32-bit network address
inet_ntoa	Converts a 32-bit network address to a dotted-notation string
inet_aton	Converts a dotted-notation string to a 32-bit network address

The following subsections illustrate the API calls, and discuss any special characteristics of their application and common problems that are associated in their use.

In this chapter, we demonstrate the calls as they appear in standard Linux distributions (which follows the BSD API). For embedded Sockets API implementations, the calls and include files may differ.

socket **FUNCTION**

The purpose of the **socket** function is to create a new socket endpoint for communication. Calling the **socket** function to retrieve a new socket is the first step in building a Sockets application. Only one other function returns a new socket, the **accept** function, which is discussed in the accept Function section later in this chapter.

The **socket** function has the following prototype:

```
#include <sys/types.h>
#include <sys/socket.h>

int socket( int domain, int type, int protocol );
```

The domain argument specifies the protocol family that is desired for the new socket. In this book, we focus solely on the IPv4 protocol, so we see the use of the AF_INET constant (Address Family Internet) in all examples. Other possible arguments include AF_INET6 for IPv6 and AF_ROUTE for the creation of a routing socket (used to add or delete rows in the routing table).

The type argument specifies the semantics of communication within the context of the previously defined domain. Some of the common types are shown in Table 3.2, though SOCK_SEQPACKET and SOCK_RDM are available only in specialized stacks.

TABLE 3.2   Socket types used as the type argument for socket

Socket Type	Description
SOCK_STREAM	Stream socket (connection-based, reliable)
SOCK_DGRAM	Datagram socket (connectionless, unreliable)
SOCK_SEQPACKET	Sequenced stream socket (connection-based, reliable)
SOCK_RAW	Raw socket (raw socket access)
SOCK_RDM	Reliable datagram socket (unordered, reliable)

Finally, the protocol argument identifies a particular protocol to use within the domain and type previously specified. Many times, this argument is specified as zero, because only one protocol exists for the given domain and type. When using specialized protocols (such as SCTP), we'll see this argument used. Examples of the **socket** function for a number of different protocols are providing in Listing 3.1.

Listing 3.1   Example usage of the **socket** function.

```
tcpSocket = socket( AF_INET, SOCK_STREAM, 0 );
tcpSocket = socket( AF_INET, SOCK_STREAM, IPPROTO_TCP );

udpSocket = socket( AF_INET, SOCK_DGRAM, 0 );
udpSocket = socket( AF_INET, SOCK_DGRAM, IPPROTO_UDP );

ipSocket = socket( AF_INET, SOCK_RAW, 0 );
ipSocket = socket( AF_INET, SOCK_RAW, IPPROTO_IP );

sctpSocket = socket( AF_INET, SOCK_STREAM, IPPROTO_SCTP );
sctpSocket = socket( AF_INET, SOCK_SEQPACKET, IPPROTO_SCTP );
sctpSocket = socket( AF_INET, SOCK_DGRAM, IPPROTO_SCTP );

rdmSocket = socket( AF_INET, SOCK_RDM, 0 );
```

Therefore, the **socket** function not only creates a new socket endpoint, but it also defines the semantics of communication that will be used through the socket. The return value of the **socket** function is a simple integer, though this may be different depending upon your particular Sockets implementation and language. If an error occurs within the **socket** function, a negative value is returned indicating this.

From Listing 3.1, we can see the creation of a number of different types of sockets. A TCP stream socket can be created either as the default of the SOCK_STREAM type, or by explicitly defining the protocol as IPPROTO_TCP. Creating a UDP datagram socket and raw IP socket is performed similarly. The SCTP socket is created depending on the style of socket requested. SCTP can operate in a datagram-style model or a stream-style model [Stewart01]. Finally, a reliable datagram socket is illustrated last with SOCK_RDM.

TIP

*One interesting difference between stream and datagram sockets is what's known as framing. When data is sent through a datagram socket, the receiver always receives the data in the units that were provided by the sender. For example, if the sender sent 100 bytes and then sent another 50 bytes, the receiver would receive two datagrams of 100 bytes and 50 bytes (through two separate **read** calls). Stream sockets operate differently, in that the boundaries defined by the sender through the **write** calls are not preserved. In the stream socket, using the previous example, the receiver might very well end up reading 150 bytes. This concept is important, and makes datagram sockets useful for message-based communication.*

bind **FUNCTION**

The **bind** function binds a local name to a newly created socket. A name within the AF_INET context is an IP address (that represents an interface) and a port number. The **bind** function has the following prototype:

```
#include <sys/types.h>
#include <sys/socket.h>

int bind( int sock, struct sockaddr *addr, int addrlen );
```

Upon successful completion of the **bind** function (identified by a return value of 0), the socket identified by argument sock will be bound to the address defined in addr. The sock argument is a previously created socket returned from the **socket** function. The addr argument is a special structure that defines the interface address and port number (collectively called an address). Although the **bind** function utilizes the sockaddr structure, for AF_INET sockets, the sockaddr_in structure is used instead. This structure has the following layout as is shown in Listing 3.2.

Listing 3.2   Format of the sockaddr_in address structure.

```
struct sockaddr_in {
    int16_t  sin_family;
    uint16_t sin_port;
    struct in_addr sin_addr;
    char     sin_zero[8];
};

struct in_addr {
    uint32_t s_addr;
};
```

For Internet communication using the IPv4 protocol suite, we'll use AF_INET solely for sin_family. Field sin_port is used to define our specified port number in network byte order. Therefore, we must use **htons** to load the port and **ntohs** to read it from this structure. Field sin_addr is, through s_addr, a 32-bit field that represents an IPv4 Internet address. Recall that IPv4 addresses are four-byte addresses. We'll see quite often that the sin_addr is set to INADDR_ANY, which is the wildcard. When we're accepting connections (server socket), this wildcard says we can accept connections from any available interface on the host. For client sockets, the rules differ.

Let's now look at a quick example of addressing for both a server and client socket. In this example, we create a stream socket (servsock), create a socket address, and initialize it (servaddr), and then finally bind the address to the socket, as shown in Listing 3.3.

Listing 3.3   Example of the **bind** function for a server socket.

```
int servsock;
struct sockaddr_in servaddr;

servsock = socket( AF_INET, SOCK_STREAM, 0);
```

```
memset( &servaddr, 0, sizeof(servaddr) );
servaddr.sin_family = AF_INET;
servaddr.sin_port = htons( 25 );
servaddr.sin_addr.s_addr = inet_addr( INADDR_ANY );

bind( servsock, (struct sockaddr_in *)&servaddr, sizeof(servaddr) );
```

Note that if the sin_port had been defined as zero, the **bind** function would automatically assign an ephemeral (dynamic) port to it. This is a common method for client sockets, but server sockets typically require a known port number and, therefore, define one statically.

Let's now look at what the **bind** function actually does from an interface perspective for both a client and a server. In Figure 3.1, we see an example in which we bind our server to a specific interface using the **bind** function. In the source code shown in the listing contained in Figure 3.1, what is interesting is the use of an IP address to define the address structure's sin_addr element. An IP address is specified that is associated with a particular interface (bottom right of Figure 3.1). Once the bind is performed with this address, incoming connections will occur only through this interface. Clients attempting to connect to the server application in Figure 3.1 through the alternate interface (defined by "10.0.0.1") will be refused.

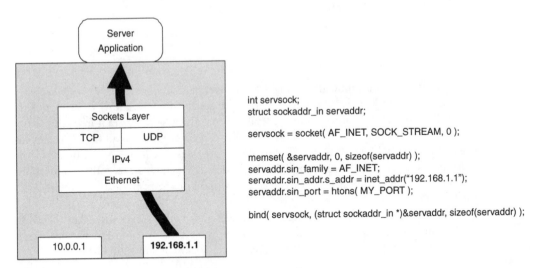

FIGURE 3.1 Binding a specific address to a server socket.

In the next example, Figure 3.2, we see another server **bind** function example. This example is the more traditional use of the **bind** function in which all interfaces are permitted for incoming connections to the server application. This functionality is configured using the INADDR_ANY symbolic constant, which represents the wildcard address (all available interfaces). In this example, clients can connect to the server application through either interface; neither will refuse connections.

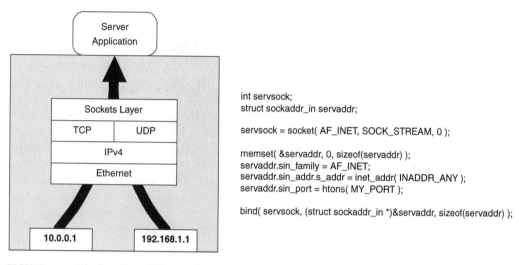

```
int servsock;
struct sockaddr_in servaddr;

servsock = socket( AF_INET, SOCK_STREAM, 0 );

memset( &servaddr, 0, sizeof(servaddr) );
servaddr.sin_family = AF_INET;
servaddr.sin_addr.s_addr = inet_addr( INADDR_ANY );
servaddr.sin_port = htons( MY_PORT );

bind( servsock, (struct sockaddr_in *)&servaddr, sizeof(servaddr) );
```

FIGURE 3.2   Binding the wildcard address to a server socket.

Finally, let's look at a client example. Figure 3.3 shows an example of **bind** within the context of a client socket. This differs from the server example because instead of determining from which interface a connection can arrive, the client example determines through which interface a client may connect. Therefore, when a client performs the **connect** function (discussed later in this chapter), the outgoing socket will utilize only the "10.0.0.1" interface.

A valid question for binding an address to a client socket is why restrict the outgoing connection to a single interface? One application is the balancing of sockets over two or more interfaces. As each socket is created, a different interface address is bound to the client socket. In this way, the number of sockets created are bound to a separate interface and, therefore, are shared among the available interfaces.

One final item to note is the effect of calling the **bind** function with incomplete information in the address structure. If the port number is set to zero in the sockaddr_in structure (sin_port), the **bind** function automatically assigns an ephemeral port to the socket. If the sin_addr.s_addr field is set to zero, then this has the same effect as the wildcard address (INADDR_ANY is zero on most systems). For a server socket, incoming connections are permitted from any interface. For a client socket, outgoing connections are permitted through any interface. Therefore, if a developer desired to provide a static port to a socket, the **bind** function with a sin_addr.s_addr of zero would provide this functionality.

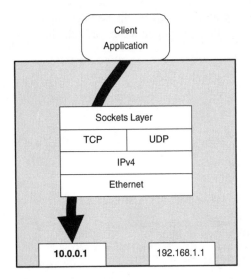

```
int clisock;
struct sockaddr_in cliaddr;

clisock = socket( AF_INET, SOCK_STREAM, 0 );

memset( &cliaddr, 0, sizeof(cliaddr) );
cliaddr.sin_family = AF_INET;
cliaddr.sin_addr.s_addr = inet_addr( "10.0.0.1" );
cliaddr.sin_port = htons( 0 );

bind( clisock, (struct sockaddr_in *)&cliaddr, sizeof(cliaddr) );
```

FIGURE 3.3   Binding an address to a client socket.

listen FUNCTION

The **listen** function is used for stream sockets to mark a socket as willing to accept incoming connections. This function may only be called for sockets initially created of type SOCK_STREAM or SOCK_SEQPACKET. The **listen** function has the following prototype:

```
#include <sys/socket.h>

int listen( int sock, int backlog );
```

The sock argument is a previously created socket (of the previously defined type) and backlog represents the size of the queue that will hold accepted connections.

NOTE

Some systems define the backlog as the size of the queue for incomplete connections, whereas others define it as the size for completed connections (still others split backlog evenly for two queues). An incomplete connection is one that has not yet completed the TCP three-way handshake (for SOCK_STREAM connections). Complete connections are those that have completed the TCP three-way handshake. The TCP three-way handshake is a synchronization step in which a client sends a SYN packet, the server responds with a SYN/ACK packet, and then the client responds with an ACK. Also performed during this state is the communication of parameters that are to be used for the connection (such as time stamp, maximum size of segments, and so on).

The **listen** function is required to be performed before an **accept** function is possible on a server socket (also known as a passive socket in this state). What's important to note is that no data is ever communicated through a passive socket; all that is permitted is the acceptance of new client sockets.

The backlog parameter is often defined as the number 5 in many applications, but this value ultimately depends upon the number of incoming connections that are desired to be queued for accept at any given time. Consider an HTTP server that specified a value of 5. With a large load (thousands of connections per second), many of these connections would be refused by the server, as there would be no room on the backlog queue awaiting accepts by the server. For this reason, many HTTP servers specify very large values to the **listen** function to ensure a sufficient queue for connection accept.

accept **FUNCTION**

The **accept** function is used by a server application (using a socket that previously called **listen**) to accept incoming client connections. These client connections are dequeued from the accepted connection queue (see Figure 3.4). If the queue is empty, the **accept** function blocks, unless the server socket was previously defined as nonblocking. The **accept** function has the following prototype:

```
#include <sys/types.h>
#include <sys/socket.h>

int accept(int sock, struct sockaddr *addr, socklen_t *addrlen);
```

When a new connection is queued in the accepted connection queue, the **accept** function unblocks and returns the new socket to the caller. Sockets on the accepted connection queue are returned in First In, First Out (FIFO) order.

The calling application provides a server socket for which the **listen** function was previously invoked. The caller also provides a sockaddr_in structure (because we're operating within the AF_INET domain) and an int pointer (socklen_t) that identifies the size of the returned addr structure. The address structure returned within the **accept** function defines the host and port of the remote client that was accepted by the server.

Now, let's look at an example of the **accept** function (see Listing 3.4). This application demonstrates the **accept** function, including all of the necessary elements that must precede it (such as **listen** and **bind**). When the **accept** function in Listing 3.4 returns, a new client socket will be contained within connectionFd, and the remote host and port will be contained within the address structure cliaddr.

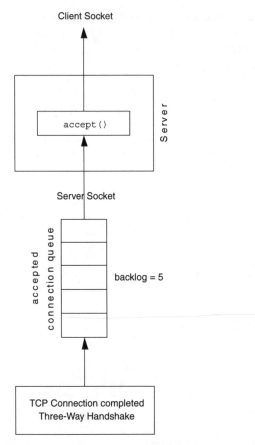

FIGURE 3.4 Graphical depiction of the **accept** function.

Listing 3.4 Sample code example for the **accept** function.

```
int serverFd, connectionFd, clilen;
struct sockaddr_in servaddr, cliaddr;

serverFd = socket( AF_INET, SOCK_STREAM, 0 );

memset( &servaddr, 0, sizeof(servaddr) );
servaddr.sin_family = AF_INET;
servaddr.sin_addr.s_addr = htonl( INADDR_ANY );
servaddr.sin_port = htons( MY_PORT );
```

```
bind( serverFd, (struct sockaddr *)&servaddr, sizeof(servaddr) );

listen( serverFd, 5 );

clilen = 0;
connectionFd = accept( serverFd,
                       (struct sockaddr_in *)&cliaddr, &clilen );
```

If the server did not require information about the client, a NULL could be passed for the second and third parameters to force the **accept** function to avoid passing them back. For example:

```
connectionFd = accept( serverFd,
                       (struct sockaddr_in *)NULL, NULL );
```

Note that if the server application required this information, it could use the **getpeername** function on the connectionFd to identify the remote host and port. We look at the **getpeername** function in Chapter 4, Advanced Sockets Functions.

connect **FUNCTION**

The **connect** function is a client-specific function that connects a client socket to a server socket. With a previously created socket, the **connect** function is called with the address structure identifying the host and port to which to connect. The **connect** function prototype appears as:

```
#include <sys/types.h>
#include <sys/socket.h>

int connect( int sock,
             struct sockaddr *name, socklen_t namelen );
```

The purpose of the **connect** function differs based upon the type of socket being used. With a SOCK_STREAM socket (TCP), the **connect** function attempts to create a stream connection to the defined peer in the name address. With a SOCK_DGRAM socket (UDP), the **connect** function simply associates the local socket with a peer datagram socket. Recall that UDP is unconnected, so the primary function is to permit datagram sockets to call the **send** function instead of **sendto** (which requires the peer address information in addition to the message to send). Because SOCK_STREAM sockets are connection-oriented, a **connect** may be performed once before it is closed. Because SOCK_DGRAM sockets are unconnected, the datagram socket may call **connect** multiple times to associate itself with a new peer datagram socket.

Listing 3.5 provides an example of the **connect** function usage for a stream socket. After the stream socket is created, the peer address to which we'll connect is initialized (servaddr). Using the **connect** function, the stream socket (connectionFd) is connected to the peer. The **connect** function returns zero on success, representing a successful connect to the peer. Otherwise, a –1 is returned representing an error. On Unix systems, the errno variable can be checked for the error. Other TCP/IP stacks may have other variables or functions for checking error status.

Listing 3.5   Sample code for the **connect** function in a stream socket setting.

```
int connectionFd, in;
struct sockaddr_in servaddr;
char timebuffer[MAX_BUFFER+1];

connectionFd = socket( AF_INET, SOCK_STREAM, 0 );

memset( &servaddr, 0, sizeof(servaddr) );
servaddr.sin_family = AF_INET;
servaddr.sin_port = htons( 25 );

servaddr.sin_addr.s_addr = inet_addr( "10.0.0.200" );

connect( connectionFd, (struct sockaddr_in *)&servaddr,
         sizeof(servaddr) );
```

In Figure 3.5, we see a progression of connection state through a call with the **connect** function. In step (**a**), a socket is created at the client (left-hand side). The server in this step has already created a socket (shown initialized with port 25), and is available for connections via the **accept** function (right-hand side). At this point, no connectivity exists between the client and server applications.

Continuing at step (**b**), the **connect** function has been performed and the three-way handshake begins. Once the three-way handshake is complete, we continue to step (**c**) in which the **connect** function returns at the client and the **accept** function returns at the server. These two socket endpoints are now connected (otherwise known as an established connection) in step (**c**) and data communication may occur.

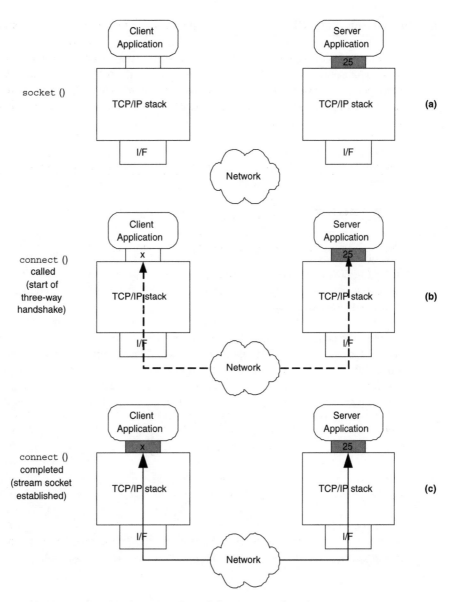

FIGURE 3.5 Graphical progression of the **connect** function.

recv/recvfrom **FUNCTION**

The **recv** and **recvfrom** functions are used to retrieve data that has been queued for a given socket. The semantics of the functions are based upon the type of socket being used (stream or datagram), in addition to a set of flags that can be provided to the functions. The prototypes for the **recv** and **recvfrom** functions are:

```
#include <sys/types.h>
#include <sys/socket.h>

int recv( int sock, char *buf, int len, int flag );

int recvfrom( int sock, const char *buf, int len, int flag,
              struct sockaddr *addr, socklen_t *len );
```

The **recv** function is used by sockets in a connected state. This implies sockets of type SOCK_STREAM as well as SOCK_DGRAM sockets that have previously performed a **connect** function to associate the remote peer to the local socket. The **recvfrom** function can be used by stream and datagram sockets.

Let's now take a deeper look at the **recv**. The **recv** function returns either the number of bytes received from the socket, or a −1 if an error occurred. The socket from which data is desired is passed as the first argument (sock), followed by a buffer argument called buf into which the **recv** function will copy the socket data. The amount of data requested (which must be at least the size in bytes of the preallocated buf argument) is provided next as the len argument. Finally, the flag argument is provided, which can be used to alter the behavior of the **recv** call (see Table 3.3).

TABLE 3.3 Most common flags for the recv/recvfrom functions

Flag	Description
MSG_OOB	Requests out-of-band data from the socket
MSG_PEEK	Returns data, but doesn't consume
MSG_WAITALL	Requests blocking call until all requested data is returned

MSG_OOB refers to the request of out-of-band data through the socket. Out-of-band data, otherwise known as expedited data, permits an application to communicate data with higher priority to the peer. MSG_PEEK allows an application to preview the data that's available in the socket buffer to read, without actually consuming it at the Sockets layer. This means that an application must at some point read the data without specifying the MSG_PEEK flag to remove it from the Sockets buffer. Finally, the MSG_WAITALL flag tells the Sockets layer to block on the **recv** and **recvfrom** call until the specified number of bytes is available to read.

Let's look at a couple of examples of the **recv** and **recvfrom** calls. Either call can be used in a stream or datagram setting, though the **recvfrom** call is rarely used for stream sockets. Because the **recvfrom** call provides peer socket information, and stream sockets are tied to a specific peer, the **recvfrom** call is not useful in this setting. If a stream socket needs the peer information, the **getpeername** function can be used instead.

The first example illustrates the **recv** call in a stream setting. We'll omit some of the intermediate functions to concentrate solely on **recv** (see Listing 3.6). After the following **recv** call completes, the ret variable contains the number of bytes returned or an error status (−1). The number of bytes returned is contained in the buffer character array. A maximum of MAX_BUF are returned, though a smaller number of bytes may be returned.

Listing 3.6 Example of the recv function in a stream setting.

```
int sock, ret;
struct sockaddr_in sa;
char buffer[MAX_BUF+1];

sock = socket( AF_INET, SOCK_STREAM, 0 );

...

ret = recv( sock, buffer, MAX_BUF, 0 );

printf( "%d bytes returned\n", ret );
```

Now let's turn our attention to the **recvfrom** call (shown in Listing 3.7). Prior to calling **recvfrom**, we must initialize the palen variable with the size of the address structure. The **recvfrom** call updates the palen variable internally (note the pass by reference) to identify the size of the resulting pa structure. We're operating in the AF_INET domain, so this size will not change before or after the call to **recvfrom**. After **recvfrom** has completed, the size of the message is returned into the ret variable. Our datagram socket is provided in sock along with the message in the buffer character array (up to a maximum size MAX_BUF). No flags are passed in this example. Finally, the address structure is passed by reference (pa) along with the structure length (palen).

Listing 3.7 Example of the recvfrom function in a datagram setting.

```
int sock, ret, palen;
struct sockaddr_in sa, pa;
char buffer[MAX_BUF+1];

sock = socket( AF_INET, SOCK_DGRAM, 0 );
```

```
...

palen = sizeof( pa );

ret = recvfrom( sock, buffer, MAX_BUF, 0,
                (struct sockaddr_in *)&pa, &palen );

printf( "Returned %d bytes\n", ret );
printf( "Peer Address %s\n", inet_ntoa( pa.sin_addr ) );
printf( "Peer port %d\n", pa.sin_port );
```

To demonstrate the returns, the three printf calls at the end of Listing 3.7 illustrate the result of the **recvfrom** call. The size of the message is emitted first (ret), followed by the IP address of the peer (sin_addr), reconstructed into a string using the **inet_ntoa** function (to be discussed later in this chapter). Finally, the port number of the peer socket is emitted from the sin_port field of the socket address structure.

send/sendto **FUNCTION**

The **send** and **sendto** functions are used to send data to a given peer socket through a local socket. Like the **recv** functions, the semantics of the functions are based upon the type of socket being used (stream or datagram), in addition to a set of flags that can be provided to the functions. The prototypes for the **send** and **sendto** functions are:

```
#include <sys/types.h>
#include <sys/socket.h>

int send( int sock, const void *buf, int len,
          unsigned int flags );

int sendto( int sock, const void *buf, int len,
            unsigned int flags,
            struct sockaddr *addr, socklen_t len );
```

The **send** function is used by sockets in a connected state. This implies sockets of type SOCK_STREAM as well as SOCK_DGRAM sockets that have previously performed a **connect** function to associate the remote peer to the local socket. The **sendto** function can be used by stream and datagram sockets in most implementations. On implementations that support the **sendto** function on a stream socket, the addr and len reference parameters are ignored.

Let's take a more in-depth look at these functions. Both the **send** and **sendto** functions take as arguments the socket through which the data is sent (sock), the buffer containing the data to send (buf), the length of the data contained in the buffer (len), and a flags bit field. The **sendto** function also requires two arguments that define the recipient of the data. The addr argument specifies the socket address (IP address and port within the AF_INET address family) and the len argument passes in the size of the sockaddr_in structure.

The flags field of the **send** and **sendto** functions alters the behavior of the calls (see Table 3.4). The MSG_OOB flag defines that the caller is providing out-of-band data within the buffer. As defined in [Stevens98], the caller may provide only one byte of out-of-band data. The receiver can read this out-of-band data using the same flag within the **recv/recvfrom** function.

The MSG_DONTWAIT flag makes this call nonblocking. This means that the call doesn't block if it's unable to queue all of the user buffer data in the socket send buffer. In this case, the number of bytes consumed in the user buffer is returned. The socket could also be made nonblocking for all calls, which differs depending upon the particular API being used. Unix-like systems utilize the fcntl system call for this purpose, whereas some stacks provide a socket option for this purpose.

Finally, the MSG_DONTROUTE flag disables routing algorithms on the local host for this emission. Typically, the destination address is checked to see if it's on the subnet, and if not, the routing algorithms identify how to route the packet to the destination (first hop). With this flag set, no routing is performed and the resulting packet is emitted as if the destination was on the local subnet.

TABLE 3.4 Most common flags for the send/sendto functions

Flag	Description
MSG_OOB	Requests transmission of out-of-band data through the socket
MSG_DONTWAIT	Forces this call to be nonblocking
MSG_DONTROUTE	Bypasses routing tables

The **send** and **sendto** functions return either the number of bytes queued for transmission in the socket, or a −1 if a local error was detected. Note the distinction between sent and queued for transmission. Upon return of the **send** (and **sendto**) call, the bytes passed down in buf are copied into the local socket send buffer. The return value represents the number of bytes that were placed into the socket buffer. If the socket is nonblocking, this may be less than what was attempted, which means that the caller may need to call the **send** function again to queue the remaining data for transmission. Within the Transport layer, the send buffer is consumed and sent in a manner consistent with the particular transport protocol. Another item of interest is that local errors are detected, but there is no immediate indication that the peer socket received the data.

Let's now look at a couple of examples. The first example illustrates the **send** function in a stream socket setting (Listing 3.8) and the second example demonstrates the **sendto** function in a datagram Sockets application (Listing 3.9).

Listing 3.8   Example of the `send` function with a stream socket.

```
int sock, ret;
struct sockaddr_in sa;
char buf[MAX_BUFFER+1];

sock = socket( AF_INET, SOCK_STREAM, 0 );

memset( &sa, 0, sizeof(sa) );
sa.sin_family = AF_INET;
sa.sin_port = htons( SERVER_PORT );

sa.sin_addr.s_addr = inet_addr( "192.168.1.1" );
connect( sock, (struct sockaddr_in *)&sa, sizeof(sa) );

strcpy( buf, "Hello\n" );
ret = send( sock, buf, strlen(buf), 0 );
```

The stream socket example (Listing 3.8) demonstrates the typical setup of a stream socket, along with defining a server address (through a sockaddr_in structure) and then connecting to it. Once connected, the **send** function is utilized to send a simple string message to the peer socket. To send, we specify our socket (sock), the data to send to the peer (buf) along with its length (computed using the **strlen** function). No flags are specified in this example (fourth argument to **send**, 0).

Listing 3.9   Example of the **sendto** function with a datagram socket.

```
int sock, ret;
struct sockaddr_in sa;
char buf[MAX_BUFFER+1];

sock = socket( AF_INET, SOCK_DGRAM, 0 );

memset( &sa, 0, sizeof(sa) );
sa.sin_family = AF_INET;
sa.sin_port = htons( SERVER_PORT );

sa.sin_addr.s_addr = inet_addr( "192.168.1.1" );

strcpy( buf, "Hello\n" );
ret = sendto( sock, buf, strlen(buf), 0,
              (struct sockaddr_in *)&sa,
               sizeof(sa) );
```

The datagram socket example (Listing 3.9) demonstrates a typical setup of a datagram socket. After the datagram socket is created, we create the address structure (sa), which represents the peer of our socket (the recipient of the datagrams that we'll generate). We load our message into buf (using the string copy function, **strcpy**). Using **sendto**, we specify the same parameters as before for the **send** function (socket descriptor, buffer, length of buffer, and flags), and then provide the recipient address. This comes in the form of the previously created sockaddr_in structure, along with the size of the structure using the sizeof primitive.

close **FUNCTION**

The **close** function is used to mark the socket as closed so that no further activity may occur on it. This activity specifically refers to Application layer activities, as the Sockets layer will continue to try to send any remaining data in the send queue. Therefore, once the socket is closed, an application may no longer call any Sockets API functions using the socket as an argument. The **close** function has the following prototype:

```
int close( int sock );
```

The sock argument refers to a previously created socket through the **socket** function. Success is indicated by return of zero, while a –1 return represents an error. It's important to close unused sockets because they tie up resources not only on the local host, but also on the remote peer for a stream-based socket.

After the **close** function is performed on a stream socket and no data remains to be sent to the peer socket, the socket will begin the connection termination sequence to close both ends of the connection.

htons/ntohs/htonl/ntohl **FUNCTIONS**

The host and network byte order conversion functions are used to convert short and long values between host and network byte order. The function prototypes are:

```
#include <netinet/in.h>

unsigned int htonl( unsigned int hostlong );

unsigned short int htons ( unsigned short hostshort );

unsigned int ntohl( unsigned int netlong );

unsigned short int ntohs( unsigned short netshort );
```

In the functions, "h" represents host and "n" represents network. Therefore, **htonl** represents host-to-network-long. Developers are required to use these functions for portability reasons. If an application runs on a processor that is big endian, then the functions are optimized away (as they perform no function), but the functions should still be included so that the application can be ported to a little-endian architecture without any issues.

Developers commonly use these functions when specifying the port number (for the sin_port field of sockaddr_in), as well as raw IPv4 address structure manipulation. In the next section, address functions are shown, which perform these functions internally so that the addresses are in the proper byte order for the given architecture.

inet_addr/inet_aton/inet_ntoa FUNCTIONS

These miscellaneous conversion functions provide the ability to convert IP address from string form (dotted-notation) to binary network format (and vice versa). All of these functions are prefixed with inet, which refers to the fact that they operate on Internet addresses (IPv4).

```
#include <netinet/in.h>
#include <arpa/inet.h>

unsigned long inet_addr( const char *addr );

char *inet_ntoa( struct in_addr in );

int inet_aton( const char *addr, struct in_addr *inp );
```

The most common of the address conversion functions is **inet_addr**. This function takes an IP address string in dotted-notation ("<N>.<N>.<N>.<N>") and converts it into a binary format address (unsigned int) in network byte order. The **inet_ntoa** function takes an in_addr structure (contained within the sockaddr_in structure) and returns a character pointer of a string in dotted-IP notation. Finally, the **inet_aton** function provides conversion of a dotted-notation IP string into binary format address in network byte order. This function differs in format from the **inet_addr** in that a return value is available for error checking. Let's now look at examples of each of these functions (see Listing 3.10).

Listing 3.10   Examples of the socket address functions.

```
/* inet_addr() example */
unsigned int addr;
const char *straddr = "192.168.1.1";

addr = inet_addr( straddr );

/* inet_ntoa() example */
char *str;
struct sockaddr_in sa1;
```

```
sa1.sin_addr.s_addr = inet_addr( "192.168.1.2" );
str = inet_ntoa( sa1.sin_addr );
printf( "%s\n", str );

/* inet_aton example */
char str[30];
int ret;
struct sockaddr_in sa;

strcpy( str, "192.168.1.3" );

ret = inet_aton( str, &sa.sin_addr );

printf( "%08x\n", sa.sin_addr.s_addr );
```

SUMMARY

In this chapter, we investigated the basic Sockets API functions that permit the construction of simple stream and datagram servers and clients. Although the Linux API was provided as the example Sockets API, it follows the BSD API very closely. Third-party Sockets implementations may differ slightly in naming and types used, but in most cases, the spirit of the API remains. In the next section, we continue our investigation of the Sockets API, looking at the more advanced functions.

REFERENCES

[Stevens98] Stevens, W. Richard, *UNIX Network Programming*, 2nd edition, Volume 1, Prentice Hall PTR, 1998.

[Stewart01] Stewart, Randall R., and Qiaobing Xie, *Stream Control Transmission Protocol (SCTP): A Reference Guide*, Addison-Wesley, 2001.

4 Advanced Sockets Functions

	socket-based	no socket needed
select	●	
poll	●	
getpeername	●	
getsockname	●	
getsockopt	●	
setsockopt	●	
shutdown	●	
gethostname		●
sethostname		●
gethostbyaddr		●
gethostbyname		●
getservbyname		●
getservbyport		●

In this chapter, we continue looking at the details of the Sockets API, in this case some advanced functions and some other functions whose usage is rare. This discussion, as in the prior chapter, is based on the Linux API. The prototypes and usage apply to other implementations as well. The Sockets functions that are discussed in this chapter are shown in Table 4.1. All code for this chapter can be found on the companion CD-ROM under /software/ch4.

TABLE 4.1   Sockets functions discussed in Chapter 4

Function	Description
select	Provides event notification of socket events
getsockname	Returns the address of the local socket
getpeername	Returns the address of the remote socket
getsockopt	Returns the value of a socket option for a given socket
setsockopt	Manipulates a socket option for a given socket
shutdown	Closes a socket with more control
gethostname	Retrieves the name of the host on which this stack operates
sethostname	Sets the name of the host on which this stack operates
gethostbyaddr	Returns the FQDN for a given IP address
gethostbyname	Returns the IP address for a given FQDN
getservbyname	Returns the port number for a given service string
getservbyport	Returns the service string for a given port number

The following subsections illustrate the API calls, and discuss any special characteristics of their application and common problems that are associated in their use.

select **FUNCTION**

The **select** function provides a notification service for a collection of events on one or more sockets. Given a set of socket descriptors, the **select** function can be used to determine if data is available to read, space is available in the send buffer for write, or if an error occurred (otherwise known as an exception). Additionally, the **select** function can be configured to return after some amount of time if no events have occurred.

The **select** function has the following prototype:

```
#include <sys/select.h>
#include <sys/time.h>

int select( int num, fd_set *readsds, fd_set *writesds,
            fd_set *exceptfds, struct timeval *timeout );
```

The num argument here refers to the number of socket descriptors that are present in the fd_set structures (plus one). The fd_set structure is an array of bit fields whose manipulation is masked by a set of functions (we look at these in the following paragraphs). Three of the fd_set structures are shown, one for read events, one for write events, and one for exceptions. The final argument, timeout, is used to define a timeout value for the **select** function. We look at this structure in more detail shortly. The return value for **select** is shown in Table 4.2.

TABLE 4.2   Return values for the **select** function

Value	Description
> 0	Number of descriptors that are contained within the descriptor sets
0	Expiration of the timeout argument
-1	An error occurred in the **select** function

The fd_set manipulation functions are discussed next. These functions permit a developer to enable, set, or clear a socket for a given event, test whether an event occurred for a socket, and clear the events within a given fd_set structure.

```
FD_CLR( int socket, fd_set *set );
FD_SET( int socket, fd_set *set );
FD_ISSET( int socket, fd_set *set );
FD_ZERO( fd_set *set );
```

The **FD_CLR** function is used to remove a socket descriptor from a set, whereas the **FD_SET** function adds a socket descriptor to the set for notification. The **FD_ISSET** function is used to test whether the particular descriptor was in the response set, and is used after the **select** function returns. Finally, the **FD_ZERO** function is used to clear out the set structure. This function is commonly used before the socket descriptors of interest are set using the **FD_SET** function.

Using the **select** function is quite simple, though **select** is commonly not used in high performance applications due to its inherent inefficiency. Applications that use **select** typically follow a similar model. First, the descriptor sets are defined for the events for which notification is needed. Note that not all of the sets must be defined; if the application was interested only in read events, only this set need be specified. For each event set of interest, the set is cleared using **FD_ZERO**. Next, for each of the socket descriptors to be monitored, the **FD_SET** function is called for the particular event set. Note that a socket descriptor may be specified in more than one event set (for example, if read and exception events were desired for the socket). If a timeout is needed, the timeout parameter is also defined. Upon return, the **FD_ISSET** function is used to determine which sockets reported which events and then take whatever action is necessary.

Let's now look at a simple example of the **select** function. This example illustrates notification on a read and exception event for a given socket (see Listing 4.1).

Listing 4.1   Simple example of the **select** function for a read event.

```
fd_set read_set, exc_set;
int sock;
int ret;

...

/* Set up our read events */
FD_ZERO( &read_set );
FD_SET( sock, &read_set );

/* Set up our exception events */
FD_ZERO( &exc_set );
FD_SET( sock, &exc_set );

ret = select( sock+1, &read_set, NULL, &exc_set, NULL );

if (ret > 0) {

  if       (FD_ISSET( sock, &read_set )) {

    /* Read event for socket sock */

  } else if (FD_ISSET( sock, &exc_set )) {

    /* Exception event for socket sock */

  }

} else {

  /* An error occurred */

}
```

In the very simple example from Listing 4.1, we enable a read event and an exception event for our socket (represented by sock) and then call **select**. Note that we specify the first argument as sock+1, because this must be exactly one greater than the largest socket descriptor that was loaded in the socket descriptor sets. Because we're looking for read events and exception events, we pass in our read set structure as argument 2, and then our exception set as argument 4 (both as references, because the **select** function will modify these sets). We're not looking for write events, so we pass a NULL for this argument of **select**. We also have no timeout constraints for the **select** function, so we pass a NULL for timeout as well. Upon return from **select**, we check the return value, and if greater than zero, we know that one or more events were triggered. We then use the **FD_ISSET** function to test which event actually fired. Otherwise, a return value less than zero indicates an error, and should be treated accordingly (commonly, a bad socket was passed in to one of the sets).

Let's now look at another example that utilizes the timeout feature of **select**. In this example, we'll try to read from a socket for 10 seconds, and if no data is read within that time, we'll time out the **select** call and perform some other action (see Listing 4.2).

Listing 4.2   Simple **select** example for read events with a timeout.

```
fd_set read_set;
struct timeval to;
int sock, ret;

...

/* Set up our read event */
FD_ZERO( &read_set );
FD_SET( sock, &read_set );

/* Set up our timeout */
to.tv_sec = 10;
to.tv_usec = 0;

ret = select( sock+1, &read_set, NULL, NULL, &to);

if (ret == 0) {

  /* Timeout occurred */

} else if (ret > 0) {

  if (FD_ISSET( sock, &read_set )) {

    /* Read event for socket sock */

  }
```

```
} else {

  /* An error occurred */

}
```

The difference to note in Listing 4.2 is that we create a new variable called to, which is a structure of type timeval. This contains a tv_sec field to represent time in seconds, and a tv_usec field for microsecond time values. These are loaded, in this example, for 10 seconds. We pass the timeout argument as a reference in this example, and then check for a zero return after **select**, which indicates a timeout occurred. What happens to timeout values when a read event occurs before the timeout value is reached? Linux systems place the remaining time left in the to argument, which could then be used to pass back into the **select** function to keep regular timeout intervals. This is sloppy time-wise, but better than restarting the timeout with each call to **select**.

Another common use of the **select** function is that of time delay. Because the event sets can all be set to NULL, the application can call **select** as follows:

```
/* Set up our timeout */
to.tv_sec = 0;
to.tv_usec = 100000;

ret = select( 0, NULL, NULL, NULL, &to);

if (ret == 0) {

  /* Timeout */

}
```

In this example, the **select** call will return after approximately 100 milliseconds. In Linux, any file descriptor can be set in the event sets (socket descriptors or file descriptors). Therefore, the application can multiplex across not only event types, but also medium types. The Windows operating system provides only support for sockets within the context of **select**.

In addition to looking for read, write, and exception events, the **select** call can be used for other types of socket events. For example, a read event is generated on a socket if the peer closed its end of the socket connection. An application can also detect whether an **accept** call needs to be performed on a listening socket (to avoid blocking on the **accept** call). In this case, a read event on a listening (server) socket indicates that one or more client connections are awaiting **accept**. Write events can detect peer closures in addition to buffer space available for the socket write buffer. Another use is to detect the completion of a socket **connect** from a client to a server.

getsockname **FUNCTION**

The purpose of the **getsockname** function is to return the local address of a given socket. Local address refers here to the local IP address and local port number. The prototype for the **getsockname** function is:

```
int getsockname( int sock,
                 struct sockaddr *addr, socklen_t *len );
```

The caller provides an existing socket (either connected or bound), an address structure, and the length of the address structure. Because we're focusing on AF_INET type sockets, we'll use the sockaddr_in structure exclusively. The **getsockname** function returns either 0 on success or –1 if an error occurred (commonly, a bad socket was passed in). An example of this function in use is provided in Listing 4.3.

Listing 4.3   Using getsockname to gather the local address information for a socket.

```
int socket;
struct sockaddr_in localaddr;
int la_len, ret;

...

la_len = sizeof( localaddr );

ret = getsockname( socket,
                   (struct sockaddr_in *)&localaddr,
                   &la_len );

if (ret == 0) {

  printf( "Local Address is : %s\n",
            inet_ntoa( localaddr.sin_addr ) );

  printf( "Local Port is : %d\n", localaddr.sin_port );

}
```

An interesting use of the **getsockname** function is to identify the interface used by a given socket in a multiport design. For example, if a host contains a number of interfaces, how the socket will connect through those interfaces depends upon the peer address, routing tables, and other factors. The **getsockname** function can be used to identify which of the local interfaces (localaddr.sin_addr, in this case) the socket has used as its egress.

getpeername **FUNCTION**

The purpose of the **getpeername** function is to return the peer address of a given socket (also known as the foreign address). Peer address refers here to the peer IP address and peer port number to which the local socket is connected. The prototype for the **getpeername** function is:

```
int getpeername( int sock,
                 struct sockaddr *addr, socklen_t *len );
```

The caller provides an existing socket (in an established state), an address structure, and the length of the address structure. Like **getsockname**, we're focusing on AF_INET type sockets, we'll use the sockaddr_in structure exclusively. The **getpeername** function returns either 0 on success or –1 if an error occurred (commonly, a bad socket was passed in or the socket was not yet connected). An example of this function in use is provided in Listing 4.4.

Listing 4.4 Using getpeername to gather the peer address information for a socket.

```
int socket;
struct sockaddr_in peeraddr;
int la_len, ret;

...

la_len = sizeof( peeraddr );

ret = getpeername( socket,
                   (struct sockaddr_in *)&peeraddr,
                   &la_len );

if (ret == 0) {

    printf( "Peer Address is : %s\n",
            inet_ntoa( peeraddr.sin_addr ) );

    printf( "Peer Port is : %d\n", peeraddr.sin_port );

}
```

The **getpeername** function is commonly used by servers, as clients would already know this information (because they define to whom they will connect). A server accepts a client connection, and can, at this point, through the **accept** call also collect the peer address information. If the peer socket information is not collected here, the **getpeername** function can be used to identify the identity of the peer socket.

getsockopt/setsockopt FUNCTIONS

Unix systems provide a number of ways to change the behavior of a socket. These include not only **setsockopt** but also other Unix-specific functions such as ioctl. We focus here on the socket option functions and how they're used. In the next chapter (Chapter 5, Socket Options), we look at the specific socket options in more detail.

The prototypes for the **getsockopt** and **setsockopt** functions are defined as:

```
#includes <sys/types.h>
#include <sys/socket.h>

int getsockopt( int sock, int level, int optname,
                void *optval, socklen_t *optlen );

int setsockopt( int sock, int level, int optname,
                void *optval, socklen_t optlen );
```

Each function accepts an existing socket as the first argument. It's important to note that some options must be set prior to connection establishment, whereas others can be performed at any time. We investigate this further in Chapter 5, Socket Options. The second parameter level defines the level of the socket option. The level refers to the layer of the networking stack for which this socket option will take effect. The available levels are shown in Table 4.3.

TABLE 4.3 Standard socket option levels

Level	Description	Option Prefix
SOL_SOCKET	Sockets layer	SO_
IPPROTO_TCP	TCP Transport layer	TCP_
IPPROTO_IP	IP Network layer	IP_

The next argument, optname, defines the particular option that we desire to read or write. A variety of socket option names exist split among the various levels defined in Table 4.3 (more on these in Chapter 5, Socket Options). The optval is a pointer to a variable that defines a value that we want to write (or location for the option value read from the socket). For both read and write, this will be a void pointer because there are a number of different types that can be specified. Finally, optlen defines the length of the variable passed in optval. Because this can be a number of different types of scalars or structures, the optlen parameter provides the size of optval, and in the case of **getsockopt**, returns the resulting size of optval (as a value-result parameter).

Let's look at a quick example of both **getsockopt** and **setsockopt**. In this example, we're going to read the size of the send socket buffer (socket option SO_SNDBUF), and then set it to twice that value (see Listing 4.5). The SO_SNDBUF is a Sockets layer option within level SOL_SOCKET.

Listing 4.5   Getting and then setting the send buffer size for a socket.

```
int sock, ret, size, len;

sock = socket( AF_INET, SOCK_STREAM, 0 );

len = sizeof( size );

ret = getsockopt( sock, SOL_SOCKET, SO_SNDBUF,
                  (void *)&size, (socklen_t *)&len );

size = size * 2;

ret = setsockopt( sock, SOL_SOCKET, SO_SNDBUF,
                  (void *)&size, sizeof( size ) );

...
```

Both **getsockopt** and **setsockopt** return 0 on success and –1 on error. Socket options are an important element of Sockets programming. They can help to improve performance, increase reliability, or simply change the standard behavior of a socket. Chapter 5, Socket Options, discusses a large number of available socket options and their use in Sockets API applications.

shutdown **FUNCTION**

The **shutdown** function is an advanced form of the **close** function (from Chapter 3, Standard BSD API Functions) that provides greater control over the process of closing down a socket connection. In most cases, a **close** will suffice to end a connection, but there are cases to consider for **shutdown**. The prototype for the **shutdown** function is defined as:

```
#include <sys/socket.h>

int shutdown( int sock, int how );
```

The **shutdown** function operates on a currently connected socket, as defined by sock. The how parameter is used to determine in what manner the **shutdown** function should operate. Three values are possible for the how parameter, as shown in Table 4.4.

TABLE 4.4   Range of values for the how parameter of `shutdown`

Value	Description
0	Subsequent receives are not permitted
1	Subsequent writes are not permitted
2	Subsequent receives and writes are not permitted (**close**)

The **shutdown** function historically was a way to close a socket regardless of the number of users of the socket [Stevens98].

If the **shutdown** function is called with a how of 0, the application may no longer read from the socket, but data is still permitted to be sent. Received data is discarded, though the sender may continue to send while the peer application is unaware of this situation. A **shutdown** with a how of 1 means that the socket can still be read, but writes are not permitted (discarded). Any data in the socket's send buffer will continue to be sent. This scenario (writes not permitted) is called a TCP half-close. Finally, a how of 2 is similar to a **close** because both reading and writing are no longer permitted through the socket.

GETHOSTNAME/SETHOSTNAME **FUNCTIONS**

The **gethostname** function is a miscellaneous function that permits the application to identify the name of the host on which it executes. The function prototype for **gethostname** is defined as:

```
#include <unistd.h>

int gethostname( char *name, size_t len );
```

The caller provides a preallocated buffer pointer (of at least size 255) that the **gethostname** function uses to store the host name. The size of the buffer is provided as the len argument. Sample usage of the **gethostname** function is illustrated in Listing 4.6.

Listing 4.6   Sample usage of the `gethostname` function.

```
#define MAX_HOSTNAME        255

char hostbuffer[MAX_HOSTNAME+1]
int ret;

ret = gethostname( hostbuffer, MAX_HOSTNAME );
```

```
if (ret == 0) {

  printf( "Host name is %s\n", hostbuffer );

}
```

The result of **gethostname** is a NULL-terminated string representing the name of the host. The host name is commonly truncated if the application provides a buffer of insufficient size.

The application can also set the name of the current host using the **sethostname** function:

```
#include <unistd.h>

int sethostname( const char *name, size_t len );
```

A simple example of **sethostname** is shown in Listing 4.7.

Listing 4.7 Sample usage of the **sethostname** function.

```
#define MAX_HOSTNAME     255

char hostbuffer[MAX_HOSTNAME+1]
int ret;

strcpy( hostbuffer, "Elise" );

ret = sethostname( hostbuffer, strlen(hostbuffer) );

if (ret == 0) {

  printf( "Host name is now %s\n", hostbuffer );

}
```

Both **gethostname** and **sethostname** return 0 on success and –1 if an error occurs.

gethostbyaddr FUNCTION

The **gethostbyaddr** function is used to identify the fully qualified domain name of a host given its IP address. In order to take advantage of this functionality, the host on which the application operates must support a DNS resolver. This DNS client communicates with a DNS server to resolve host names to IP addresses and vice versa. The prototype for the **gethostbyaddr** function is:

```
#include <netdb.h>

struct hostent *gethostbyaddr( const char *addr,
                                      int len, int type);
```

We pass in our character string IP address as addr and its length as len. The final type argument refers to the address family, which will be AF_INET for all examples in this book (IP-based protocols). The **gethostbyaddr** function returns a pointer to a hostent structure that will contain the relevant naming information. If an error occurs, the return value will be NULL, otherwise it's a valid pointer to a hostent structure. The hostent structure has the following format:

```
struct hostent {
    char    *h_name;           /* official name of host */
    char    **h_aliases;       /* alias list            */
    int     h_addrtype;        /* host address type     */
    int     h_length;          /* length of address     */
    char    **h_addr_list;     /* list of addresses     */
};

#define h_addr h_addr_list[0]  /* first address, NBO    */
```

Field h_name refers to the official name of the host (compared to the aliases that can also be used for the host, defined as the list h_aliases). The h_addrtype will always be AF_INET. Because a host may be accessible by more than one address, a list of addresses is provided in h_addr_list. Finally, the h_addr macro is provided for backward compatibility reasons (an old member of this structure) and refers to the first element of the h_addr_list. We look at each of these fields in the discussion of **gethostbyname**.

A simple example of the **gethostbyaddr** function is shown in Listing 4.8. In this example, we'll emit the fully qualified domain name given an IP address.

Listing 4.8 Sample usage of the gethostbyaddr function.

```
struct in_addr in;
struct hostent *hp;

inet_aton( "192.168.1.1", &in );

if ( hp = gethostbyaddr( (char *)&in.s_addr,
                            sizeof(in.s_addr), AF_INET )) {

  printf( "Host name is %s\n", hp->h_name );

}
```

For demonstration purposes, we fill an in_addr structure with a binary representation of our string IP address using the **inet_aton** function. The **gethostbyaddr** function is called next with the s_addr field of the in_addr structure, storing the response in the hostent structure pointer. The h_name field of the hostent structure refers to the host name, and is emitted in this example by printf.

gethostbyname **FUNCTION**

The **gethostbyname** function is used to identify the IP address given a fully qualified domain name. As with **gethostbyaddr**, a local resolver client must be available in order to communicate with the DNS server. The prototype for the **gethostbyname** function is:

```
#include <netdb.h>

struct hostent *gethostbyname( const char *name );
```

The name argument refers to the fully qualified domain name for which we desire an IP address translation. The response hostent structure reference contains all of the relevant information, or NULL is returned to indicate that an error occurred.

An example of the **gethostbyname** function is shown in Listing 4.9. In this example, we'll specify a fully qualified domain name and emit all of the relevant information about the name (to illustrate the usage of the hostent structure fields).

Listing 4.9   Sample usage of the gethostbyname function.

```
struct hostent *hp;
int i;

hp = gethostbyname( "www.microsoft.com" );

if (hp) {

  printf( "h_name is %s\n", hp->h_name );
  printf( "h_addrtype is %d\n", hp->h_addrtype );

  i=0;
  printf( "Aliases:\n" );
  while (1) {

    if ( hp->h_aliases[i] ) {

      printf( "h_aliases[%d] = %s\n", i, hp->h_aliases[i] );
      i++;
```

```
      } else break;

  }

  i=0;
  printf( "Addresses:\n" );
  while (1) {

    if ( hp->h_addr_list[i] ) {

      struct in_addr theAddr;
      memcpy( &theAddr.s_addr,
              hp->h_addr_list[i], sizeof(theAddr.s_addr) );
      printf( "  h_addr_list[%d] = %s\n",
              i, inet_ntoa( theAddr ) );
      i++;

    } else break;

  }

}
```

The function **gethostbyname** is the most commonly used of the **gethostbyname** and **gethostbyaddr** pair. A common code pattern for resolving fully qualified domain names to IP addresses is shown in Listing 4.10.

Listing 4.10   Typical usage of the `gethostbyname` function with `inet_addr` (`resolve.c`).

```
#include <netinet/in.h>
#include <arpa/inet.h>
#include <netdb.h>

int resolve_name( struct sockaddr_in *addr, char *hostname )
{
  addr->sin_family = AF_INET;

  /* See if host name actually is a string IP address */
  addr->sin_addr.s_addr = inet_addr( hostname );

  /* If host name wasn't a string notation IP address,
   * use the gethostbyname function to resolve it.
   */
  if ( addr->sin_addr.s_addr = 0xffffffff ) {
```

```
struct hostent *hp;

hp = (struct hostent *)gethostbyname( hostname );

/* Don't know what hostname is... */
if (hp == NULL) return -1;
else {

  memcpy( (void *)&addr->sin_addr,
          (void *)hp->h_addr_list[0],
          sizeof( addr->sin_addr) );

  }

}

return 0;
}
```

In this example, the code first tries the **inet_addr** function. If the host name variable happens to be a string IP address in dotted notation, then **inet_addr** will convert this into a binary notation address in network byte order. Otherwise, 0xffffffff is returned to indicate the error. Upon detecting this, we assume that the address is a fully qualified domain name and rely on **gethostbyname** to resolve the name to an address. After successful return from **gethostbyname** (a valid pointer to a hostent structure), we copy the first address of the h_addr_list to the address portion of the sockaddr_in structure. Sample usage of this function is shown in Listing 4.11.

Listing 4.11 Sample usage of the resolve_name function.

```
struct sockaddr_in addr;
int ret;

ret = resolve_name( &addr, "www.microsoft.com" );

if (ret == 0) {

  printf( "address is %s\n", inet_ntoa( addr.sin_addr ) );

}
```

In Listing 4.11, we pass in our sockaddr_in structure with the name for which we want to resolve. Upon return (successful resolution indicated by a return value of zero), we convert the binary address into string dotted-notation using inet_ntoa and emit it using printf.

getservbyname **FUNCTION**

Within the IP suite of protocols, a service sits on a unique port for a given Transport layer protocol. For example, an SMTP server resides on port 25 within the TCP layer. An SNMP agent resides on port 161 within the UDP layer. The **getservbyname** function takes a service name and protocol and yields among other things, the port to be used to either register the service or find the service. The **getservbyname** function has the following prototype:

```
#include <netdb.h>

struct servent *getservbyname( const char *name,
                               const char *proto );
```

The caller provides a service name, such as "smtp" or "http" and a protocol such as "tcp" or "udp." The function returns a pointer to a servent structure representing the requested protocol, or NULL representing either an error or service not found. The servent structure has the following format:

```
struct servent {
    char *s_name;       /* The official service name */
    char **s_aliases;   /* List of alias names       */
    int  s_port;        /* Port number               */
    char s_proto;       /* Required Protocol         */
};
```

The s_name field represents the official service name (such as "smtp"), whereas s_alias is a list of other names for the service. For example, service "smtp" has an alias "mail." Field s_port is the port number for the service; in the case of "smtp," the port number is 25 (the well-known reserved port for SMTP). Finally, the s_proto field is the protocol that is required by the particular service (in this case, "tcp").

Let's now look at a simple example of the **getservbyname** function. In Listing 4.12 is the simple case of an application identifying the port number of a given service. We provide the **getservbyname** function our service name ("http") and the transport protocol over which it runs ("tcp"), and then store the result in sp. The service name (which may differ from the service name we passed in) is stored in sp->s_name, whereas the port number is stored in sp->s_port. The port number is stored in network byte order, so we use the **ntohs** function to convert this into host byte order for proper display.

Listing 4.12   Simple example of getservbyname.

```
#include <netdb.h>

struct servent *sp;
```

```
...

sp = getservbyname( "http", "tcp" );

if (sp) {

  printf( "Service %s is at port %d\n",
          sp->s_name, ntohs( sp->s_port ) );

}
```

More information is provided by **getservbyname** (as is illustrated by the contents of the servent structure). Listing 4.13 illustrates the use of the other fields within the servent structure. The protocol field (s_proto) identifies the protocol required for the particular service. The s_aliases list defines some other service names that can be used. If the application uses an alias in the **getservbyname** call, the official name still appears in the s_name field of the servent structure.

Listing 4.13   Complete example of getservbyname.

```
#include <netdb.h>

struct servent *sp;
int i;

sp = getservbyname( "smtp", "tcp" );

if (sp) {

  printf( "s_name  = %s\n", sp->s_name );
  printf( "s_port  = %d\n", ntohs(sp->s_port) );
  printf( "s_proto = %s\n", sp->s_proto );

  i = 0;
  printf( "Aliases:\n" );
  while (1) {

    if ( sp->s_aliases[i] ) {
      printf(" s_aliases[%d] = %s\n", i, sp->s_aliases[i] );
      i++;
    } else break;

  }

}
```

An item to note with **getservbyname** is that although the application may know the service name of interest, it may not know the protocol over which it operates. In this case, the protocol argument of **getservbyname** may be passed as NULL. The application should then consult the s_proto field to know how to construct the socket for creating or connecting to the service. For example:

```
sp = getservbyname( "mail", NULL );
```

results in the same information as the prior example in Listing 4.13. Note that in addition to not specifying the protocol parameter, we also specify an alias for the "smtp" service.

getservbyport **FUNCTION**

The **getservbyport** function can be used to identify the service and its characteristics given a port number. The result of the **getservbyport** function is identical to that of **getservbyname**, but how these results are achieved differs. The prototype for the **getservbyport** function is defined as:

```
#include <netdb.h>

struct servent *getservbyport( int port, const char *proto );
```

The port argument is the port number for the desired service in network byte order. The proto argument (protocol) specifies the particular Transport layer protocol used by the service. As with **getservbyname**, this field is optional. An example of the **getservbyport** function is shown in Listing 4.14.

Listing 4.14 Example of the getservbyport function.

```
#include <netdb.h>

struct servent *sp;

...

sp = getservbyport( htons(80), NULL );

if (sp) {

  printf( "Service %s is at port %d\n",
          sp->s_name, ntohs( sp->s_port ) );

}
```

The database that's used by the **getservbyname** and **getservbyport** functions is commonly found in Unix systems in the /etc/services file. This file (a sample of which is shown in Listing 4.15) provides a line per service and includes the official service name, the port number, the transport protocol, and a set of aliases (note the mapping back to the servent structure). The location of this file may differ depending upon the operating system, and may be hard coded into the source of an embedded systems stack.

Listing 4.15   Sample of a Unix /etc/services file.

```
#Service   Port/Proto     Aliases
discard      9/tcp       sink null
discard      9/udp       sink null
daytime     13/tcp
daytime     13/udp
ftp         21/tcp
smtp        25/tcp       mail
www         80/tcp       http
www         80/udp
```

The value of the service functions is that they allow applications to be built without having to hard code the service port numbers (and protocols) directly into the source. This permits more flexible applications and simpler maintenance.

SUMMARY

This chapter investigated some of the more advanced and less commonly used functions associated with Sockets programming. In addition to multiplexing communication with **select**, we investigated name resolution and service identification. The notion of sockets options was briefly discussed in this chapter. The next chapter provides a detailed look at the various sockets options and how they can be used to change standard socket behavior.

REFERENCES

[Stevens98] Stevens, W. Richard, *UNIX Network Programming*, 2nd edition, Volume 1, Prentice Hall PTR, 1998.

5 Socket Options

Sockets Layer	`SOL_SOCKET`
TCP Layer	`IPPROTO_TCP`
IP Layer	`IPPROTO_IP`

In this chapter, we investigate the socket options that are available for each of the layers of the TCP/IP stack. These options can affect the behavior of the Sockets layer, Transport layer, and IP layer. The most common use of socket options is to increase the performance of a connection, but they're also widely used in other scenarios. We look at the plethora of options available, and provide examples for each. All code for this chapter can be found on the companion CD-ROM at /software/ch5.

SOCKET OPTIONS API

Getting and setting options for a given socket are performed through two functions, **getsockopt** and **setsockopt**. These functions provide a single interface for getting and setting a variety of options using a number of different structures. The socket option prototypes are defined as:

```
#include <sys/types.h>
#include <sys/socket.h>

int getsockopt( int sock, int level, int optname,
                void *optval, socklen_t *optlen );

int setsockopt( int sock, int level, int optname,
                void *optval, socklen_t optlen );
```

All socket options require that the application specify the socket for which the option is to be applied; this is argument one of the call (sock). The level refers to the layer of protocol to which this option will be applied (see Table 5.1 for a list of the protocols and the symbolic constants used by the API). The option name is defined by optname. This is the particular option to be used. Numerous options exist and they are divided by the protocol of interest. The optval argument specifies the value to be set or the location to store the option in a GET request. Finally, the optlen defines the length of the option structure. As there are a number of different structures that can be used to set or get options, this parameter defines the length to avoid the call from overrunning the buffer.

TABLE 5.1 Level argument for the **setsockopt/getsockopt** functions

Level	Description	Option Prefix
SOL_SOCKET	Sockets layer	SO_
IPPROTO_TCP	TCP Transport layer	TCP_
IPPROTO_IP	IP Network layer	IP_

Within each of the option levels, a number of options can be manipulated. The options are split by level because they affect the operation of the stack at the indicated level. The following sections investigate the socket options that are available and illustrate how to manipulate them.

Sockets Layer Options

The Sockets layer options are those defined within the context of level SOL_SOCKET and focus on the Sockets API layer. The typical options for the Sockets layer are defined in Figure 5.1.

Option Name	Description	get/set	value
SO_BROADCAST	Permits transmit of broadcast datagrams	g/s	int
SO_DEBUG	Enables debug logging	g/s	int
SO_DONTROUTE	Enables bypass of routing tables	g/s	int
SO_ERROR	Retrieve the current socket error	g	int
SO_LINGER	Enables linger on close if data present	g/s	struct linger
SO_KEEPALIVE	Enables TCP Keepalive probes	g/s	int
SO_RCVBUF	Modifies the size of the socket receive buffer	g/s	int
SO_SNDBUF	Modifies the size of the socket send buffer	g/s	int
SO_RCVLOWAT	Sets the minimum byte count for input	g/s	int
SO_SNDLOWAT	Sets the minimum byte count for output	g/s	int
SO_SNDTIMEO	Sets the timeout value for output	g/s	struct timeval
SO_RCVTIMEO	Sets the timeout value for input	g/s	struct timeval
SO_REUSEADDR	Enables local address reuse	g/s	int
SO_TYPE	Retrieves the socket type	g	int

FIGURE 5.1 Sockets layer options.

The get/set column in Figure 5.1 defines whether the option can be retrieved, set, or both. The value column defines what is expected to retrieve or set the option. Let's now look at examples of each of the options and better understand what effect they have.

SO_BROADCAST Option

The purpose of SO_BROADCAST is to permit a socket to send datagrams to a broadcast address. In order to send a broadcast datagram, the application must specify the destination of the datagram as the broadcast address. If the SO_BROADCAST socket option is not enabled, the broadcast datagrams will be dropped. If set, the datagrams are permitted to be sent. An example of using the SO_BROADCAST option and then sending a broadcast datagram is shown in Listing 5.1.

Listing 5.1   SO_BROADCAST and sending a broadcast datagram.

```
int    sock, cnt, addrLen, on=1;
struct sockaddr_in addr;
char   buffer[512];

sock = socket( AF_INET, SOCK_DGRAM, 0 );

/* Permit sending broadcast datagrams */
setsockopt( sock, SOL_SOCKET, SO_BROADCAST,
               &on, sizeof(on) );

memset(&addr, 0, sizeof(addr));
addr.sin_family = AF_INET;
addr.sin_port = htons(BCAST_PORT);
addr.sin_addr.s_addr = inet_addr("255.255.255.255");
addrLen = sizeof(addr);

...

/* Send a broadcast datagram */
cnt = sendto(sock, buffer, strlen(buffer), 0,
               (struct sockaddr_in *)&addr, addrLen);
```

SO_DEBUG Option

The SO_DEBUG socket option enables internally logging of interesting events within the TCP layer of a stack within a circular buffer. This logging is commonly compiled away, through conditional compilation, so in a production environment the data is rarely available. If the stack has been compiled with debug logging enabled, enabling this option allows the stack to log this data for later collection. Enabling the option is performed as shown in the code fragment in Listing 5.2.

Listing 5.2   Enabling TCP layer debugging with SO_DEBUG.

```
int    sock, on=1;

sock = socket( AF_INET, SOCK_DGRAM, 0 );

/* Enable TCP layer debugging */
setsockopt( sock, SOL_SOCKET, SO_DEBUG, &on, sizeof(on) );
```

The method for retrieving the logged data is different for each stack implementation and the relevant documentation should be consulted.

SO_DONTROUTE Option

The SO_DONTROUTE option is used to disable the underlying routing algorithms for a given socket. Before a datagram is emitted onto the physical medium, a set of algorithms is employed to determine where the datagram should be directed. In some cases, the default gateway is used to route datagrams on to their destination. If SO_DONTROUTE is set, the datagram is given to the interface that matches the network portion of the destination address and the route is never used.

The SO_DONTROUTE option is a simple integer option, and is retrieved and set as shown in Listing 5.3.

Listing 5.3   Manipulating the SO_DONTROUTE socket option.

```
int sock;
int val, len, ret;

sock = socket( AF_INET, SOCK_DGRAM, 0 );

...

len = sizeof( val );
ret = getsockopt( sock, SOL_SOCKET, SO_DONTROUTE,
                     (void *)&val, &len );
printf(" so_dontroute = %d\n", val );

val = 1;
ret = setsockopt( sock, SOL_SOCKET, SO_DONTROUTE,
                      (void *)&val, sizeof(int) );
```

If no interface matches the network portion of the destination address, an error message is returned to the application (commonly "Network Unreachable").

SO_ERROR Option

The SO_ERROR socket option permits the application to retrieve the last error that was recorded by the stack for the given socket. In Linux, the errno variable can be used for this purpose (checked after an return from a socket call). Other stacks provide specialized functions to return the last error experienced for the socket. In special cases, for example to determine if a nonblocking **connect** has completed, the SO_ERROR can provide useful error codes.

Retrieving the SO_ERROR value for a given socket is shown in Listing 5.4.

Listing 5.4   Retrieving the SO_ERROR value for a socket.

```
int sock;
int val, len, ret;

sock = socket( AF_INET, SOCK_DGRAM, 0 );

...

len = sizeof( val );
ret = getsockopt( sock, SOL_SOCKET, SO_ERROR,
                  (void *)&val, &len );

printf( "so_error = %d\n", val );
```

Setting of the error variable with **setsockopt** is not permitted. After the error value has been read in the **getsockopt** function, it is cleared. Prior to utilizing this functionality in your application, a review of the source or documentation should be done to ensure that SO_ERROR is indeed supported.

SO_LINGER Option

Before discussing the purpose of the SO_LINGER option, let's review what happens when the application performs a socket write operation (**write**, **send**, **sendto**, and so on). The data is moved from the application into the context of the stack and buffered awaiting transmission to the peer. Based upon a number of factors including the advertised window from the peer (receiver flow control) and the congestion window (sender flow control), the data may not be sent immediately. What happens if this buffered data is still waiting to be sent when the sending application closes its end of the socket? This is where SO_LINGER comes into play. The SO_LINGER option tells the stack how to deal with this data that remains to be sent (the default action is to continue to try to send the data).

The SO_LINGER option includes two elements, an enable and a time value. The enable value (l_onoff) is obvious and enables or disables this option for lingering on close with send data present. The time value (l_linger) specifies the number of seconds to linger before closing the socket and discarding the unsent data. This data is encapsulated into a structure called linger on most systems:

```
struct linger {
    int l_onoff;    // enable(1)/disable(0)
    int l_linger;   // Linger time in seconds
};
```

Three interesting cases are important to understand when using the SO_LINGER option. These are shown in Table 5.2.

TABLE 5.2   Interesting combinations for the SO_LINGER socket option

l_onoff	l_linger	*Description*
0	N/A	Linger is disabled, normal behavior
1	0	Discard data immediately after close is issued
1	> 0	Linger for the number of seconds defined and then close

Let's look at the third case in a simple example. In this example, we'll give the socket ten seconds before discarding any data (see Listing 5.5).

Listing 5.5   Example of the SO_LINGER socket option.

```
int sock;
int ret;
struct linger ling;

sock = socket( AF_INET, SOCK_STREAM, 0 );

...

ling.l_onoff = 1;
ling.l_linger = 10;

ret = setsockopt( sock, SOL_SOCKET, SO_LINGER,
                    (void *)&ling, sizeof(ling) );

...

ret = close( sock );
```

The **close** function commonly blocks until the buffered data has been sent. If the time value (l_linger) specified with SO_LINGER times out, the **close** function will fail with an error.

SO_KEEPALIVE Option

The SO_KEEPALIVE option is used to enable or disable the TCP keep-alive probes. These probes are used to maintain a TCP connection and regularly test the connection to ensure that it's still available. The keep-alive probe packet solicits an Ack from the peer, identifying that the connection (and sometimes the peer) is still available.

The keep-alive probe is sent once every two hours, but only if there is no traffic on the given connection. If traffic exists on the connection, there is no point for the keep-alive probe because the peer stack should be acknowledging data and can, therefore, be ruled alive. This option is enabled by default, but can be disabled using this socket option. In the TCP socket options section, the `TCP_KEEPALIVE` can be used to modify the time between keep-alive probes.

The following example (Listing 5.6) illustrates how to disable the keep-alive probes for a given connection.

Listing 5.6   Disabling the TCP keep-alive probes.

```
int sock;
int ret, on;

sock = socket( AF_INET, SOCK_STREAM, 0 );

...

on = 1;
ret = setsockopt( sock, SOL_SOCKET, SO_KEEPALIVE,
                  (void *)&on, sizeof( on ) );
```

SO_SNDBUF / SO_RCVBUF Options

The `SO_SNDBUF` and `SO_RCVBUF` options permit an application to change the size of the socket buffers used to queue data for transmission and queue data for receipt. These options are a very important mechanism to increase the performance of a connection (explained in more detail in the Chapter 7, Optimizing Sockets Applications).

Because these options change the size of the queue between the Sockets layer and the transport protocol, they must be defined prior to a connection being established. This means that a client must set these options before the **connect** function is called and a server must perform it before the **accept** function is called.

An example of setting the send and receive buffers to 32 KB is shown in Listing 5.7.

Listing 5.7   Modifying the send and receive socket buffer sizes.

```
int sock;
int value, ret;

sock = socket( AF_INET, SOCK_STREAM, 0 );

...

value = 32768;
```

```
ret = setsockopt( sock, SOL_SOCKET, SO_SNDBUF,
                    (void *)&value, sizeof(value) );

value = 32768;
ret = setsockopt( sock, SOL_SOCKET, SO_RCVBUF,
                    (void *)&value, sizeof(value) );
```

An application can also retrieve the default socket buffer sizes using the **getsockopt** function. This value typically differs based upon the stack being used. Therefore, the relevant socket option documentation should be consulted.

SO_RCVLOWAT Option

The SO_RCVLOWAT socket option defines the minimum number of bytes that should be used for input operations with the **select** function. Receive calls block if no data is available to be read. If data is available, the call will return the smaller of the number of bytes requested in the receive call or the SO_RECVLOWAT count.

An example of setting the SO_RCVLOWAT option to a value of 48 (wait for at least 48 bytes before returning) is shown in Listing 5.8.

Listing 5.8 Setting SO_RCVLOWAT to await 48 bytes before read operation return.

```
int sock;
int value, ret;

sock = socket( AF_INET, SOCK_STREAM, 0 );

...

value = 48;
ret = setsockopt( sock, SOL_SOCKET, SO_RCVLOWAT,
                    (void *)&value, sizeof( value ) );
```

For Listing 5.8, the read API function will return when at least 48 bytes have been received for the particular connection. If an error occurs for the given connection, fewer bytes may be returned. The default value for SO_RCVLOWAT is one.

SO_SNDLOWAT Option

The SO_SNDLOWAT option is the opposite of the SO_RCVLOWAT option discussed previously. This option sets the minimum number of bytes necessary for output operations. See Listing 5.9 for an example of setting a minimum of 48 bytes for a select write operation.

Listing 5.9   Setting SO_SNDLOWAT to await 48 bytes before write operation.

```
int sock;
int value, ret;

sock = socket( AF_INET, SOCK_STREAM, 0 );

...

value = 48;
ret = setsockopt( sock, SOL_SOCKET, SO_SNDLOWAT,
                  (void *)&value, sizeof( value) );
```

SO_SNDTIMEO/SO_RCVTIMEO Options

The SO_SNDTIMEO and SO_RCVTIMEO socket options are used to retrieve the timeout values that apply to input and output operations. For SO_SNDTIMEO, the values represent the timeout value that will be observed for blocking send operations. If a send operation blocks for SO_SNDTIMEO or more, a partial send could occur or an EWOULDBLOCK error if no data was sent (on Linux systems).

The SO_RCVTIMEO provides the same timeout functionality for receive operations. If a blocking-receive operation requires more time than is defined by the SO_RCVTIMEO operation, a short count or an EWOULDBLOCK is returned.

Getting the SO_SNDTIMEO and SO_RCVTIMEO sock options is illustrated in Listing 5.10.

Listing 5.10   Retrieving the SO_SNDTIMEO timeout values.

```
#include <sys/socket.h>
#include <sys/time.h>

int sock, ret, len;
struct timeval timeo;

sock = socket( AF_INET, SOCK_STREAM, 0 );

len = sizeof( timeo );
ret = getsockopt( sock, SOL_SOCKET, SO_SNDTIMEO,
                  (void *)&timeo, &len );

printf( "Timeout %d seconds, %d microseconds\n",
        timeo.tv_sec, timeo.tv_usec );
```

These options can only be retrieved, and can never be set by the application. Before attempting to use it, the availability of this socket option should be verified with the stack implementation.

SO_REUSEADDR **Option**

The SO_REUSEADDR socket option is used to permit reuse of local addresses within the **bind** function. By local address, we refer here to the address that was bound to a local socket.

To better understand why this option is important, let's look at a couple of examples that illustrate its purpose. In the first example, we show how the problem most commonly occurs, in which SO_REUSEADDR provides a workaround.

When a server application binds a local address to a socket and then begins accepting connections on it, the local address is bound to the local socket. If we were to halt the server Socket application and restart, the bind would fail. Let's first look at the server code to better understand why (see Listing 5.11).

Listing 5.11 Sample server code for the "address in use" error.

```
int sock, ret;
struct sockaddr_in servaddr;

sock = socket( AF_INET, SOCK_STREAM, 0 );

memset( &servaddr, 0, sizeof(servaddr) );
servaddr.sin_family = AF_INET;
servaddr.sin_addr.s_addr = htonl( INADDR_ANY );
servaddr.sin_port = htons( MY_PORT );

ret = bind( sock, (struct sockaddr_in *)&servaddr,
            sizeof(servaddr) );

printf( "bind returned %d\n", ret );

...
```

This standard server code pattern illustrates creating a socket and then binding a local address to it. If we execute this code (and the subsequent **accept** function that would be present), and then fail the sequence for some reason, a subsequent attempt to bind the local address will fail. This is commonly known as the "address in use" error. The reason that this occurs is that the server socket is in a state known as a "WAIT_STATE." For two minutes, the socket remains in this state, and is then freed, permitting the local address to be reused. To force the ability to reuse the local address before the expiration of the two-minute period, the SO_REUSEADDR socket option can be used. To enable reuse, the option must be enabled prior to the call to bind (see Listing 5.12).

Listing 5.12 Enabling local address reuse with SO_REUSEADDR.

```
int sock, ret;
int on;
```

```
struct sockaddr_in servaddr;

sock = socket( AF_INET, SOCK_STREAM, 0 );

on = 1;
ret = setsockopt( sock, SOL_SOCKET, SO_REUSEADDR,
                    (void *)&on, sizeof( on ) );

memset( &servaddr, 0, sizeof(servaddr) );
servaddr.sin_family = AF_INET;
servaddr.sin_addr.s_addr = htonl( INADDR_ANY );
servaddr.sin_port = htons( MY_PORT );

ret = bind( sock, (struct sockaddr_in *)&servaddr,
               sizeof(servaddr) );

printf( "bind returned %d\n", ret );
```

This particular option is very common and can be observed in the initialization code of almost any socket server.

TCP Layer Options

The TCP layer options are those defined within the context of level IPPROTO_TCP and focus on the TCP layer. The typical options for the TCP layer are defined in Figure 5.2.

Option Name	Description	get/set	value
TCP_KEEPALIVE	Modifies number of seconds between TCP keepalives	g/s	int
TCP_MAXRT	Modifies the maximum TCP retransmit time	g/s	int
TCP_MAXSEG	Modifies the TCP maximum segment size	g/s	int
TCP_NODELAY	Enable/Disable TCP's Nagle algorithm	g/s	int

FIGURE 5.2 TCP layer options.

TCP_KEEPALIVE Option

The TCP_KEEPALIVE option is used to define the number of seconds that the keep-alive probes will be sent when the SO_KEEPALIVE socket option is enabled. Recall from the discussion of SO_KEEPALIVE, that these probes are sent only when the particular connection is inactive.

The TCP_KEEPALIVE option can be set and retrieved. Setting the TCP_KEEPALIVE socket option is illustrated in Listing 5.13.

Listing 5.13   Setting keep-alive probes to 10-second intervals.

```
int sock, ret, interval;
struct sockaddr_in servaddr;

sock = socket( AF_INET, SOCK_STREAM, 0 );

interval = 1;
ret = setsockopt( sock, IPPROTO_TCP, TCP_KEEPALIVE,
                  (void *)&interval, sizeof( interval ) );
```

TCP_MAXRT Option

The TCP_MAXRT option can be used to define how long to retransmit data over a TCP connection. Once retransmission occurs on a TCP connection, the value specified with the TCP_MAXRT option defines the desired behavior (see Table 5.3).

TABLE 5.3   Meaning of values for the TCP_MAXRT socket option

Option Value	Description
-1	Retransmit forever
0	Use the system default behavior
> 0	Number of seconds before the connection is broken

An example of setting the TCP_MAXRT option to three seconds is shown in Listing 5.14.

Listing 5.14   Defining a three-second TCP_MAXRT.

```
int sock, ret, duration;
struct sockaddr_in servaddr;

sock = socket( AF_INET, SOCK_STREAM, 0 );

duration = 1;
ret = setsockopt( sock, IPPROTO_TCP, TCP_KEEPALIVE,
                  (void *)&duration, sizeof( duration ) );
```

TCP_NODELAY Option

The TCP_NODELAY option permits us to enable or disable the Nagle algorithm within the TCP layer of the stack. The Nagle algorithm (created by John Nagle in the early 1980s at Ford Aerospace) was an important optimization in the TCP stack because it minimizes the number of small segments that can be sent by a device.

Consider an application that sends a small amount of data for each send operation. Without the Nagle algorithm, these small amounts of data would be packaged within TCP and IP headers and sent onto the wire. The Nagle algorithm delays the transmission of data, with the hope that within some small amount of time, more data will arrive for the socket that can be accumulated and sent as a larger packet.

This is an important optimization because maximizing network utilization depends upon the transmission of maximum-sized segments onto the wire (the maximum amount of payload data that can be sent within a packet).

Disabling the Nagle algorithm, via the TCP_NODELAY socket option, can be done as is shown in Listing 5.15.

Listing 5.15   Disabling the Nagle algorithm.

```
int sock, ret, off;
struct sockaddr_in servaddr;

sock = socket( AF_INET, SOCK_STREAM, 0 );

off = 1;
ret = setsockopt( sock, IPPROTO_TCP, TCP_NODELAY,
                  (void *)&off, sizeof( off ) );
```

We discuss the Nagle algorithm in Chapter 7, Optimizing Sockets Applications, and identify where and when this socket option should be used.

TCP_MAXSEG Option

The TCP_MAXSEG option permits us to change the size of the Maximum Segment Size, otherwise known as the MSS. Let's first understand the purpose of the MSS and how it relates to other elements.

First, an interface operates with what's known as a Maximum Transmission Unit, or MTU. This is the largest size packet that may be communicated over the particular interface (known as the "Interface MTU"). When communicating over a network, our packet may encounter a device whose Interface MTU is smaller than yours. If so, the device will fragment your packet into two or more packets to ensure that they fit the given MTU. If we extend this out to the endpoint, the smallest MTU that is supported is called the "Path MTU."

Returning to the MSS, the MSS is the MTU minus the packet headers (the payload size of a packet). Figure 5.3 illustrates this concept.

The stack automatically determines the MSS for a given connection (because the Path MTU can be different for each connection in addition to the size of the packet headers). Using the TCP_MAXSEG option, we can statically define this to a size of our liking. This is illustrated in Listing 5.16.

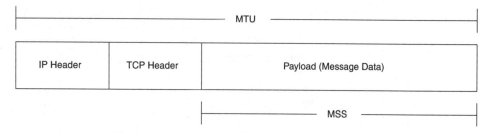

FIGURE 5.3 Relationship of a packet with MTU and MSS.

Listing 5.16 Defining a static MSS for a given socket.

```
int sock, ret, sz;
struct sockaddr_in servaddr;

sock = socket( AF_INET, SOCK_STREAM, 0 );

sz = 128;
ret = setsockopt( sock, IPPROTO_TCP, TCP_MAXSEG,
                    (void *)&sz, sizeof( sz ) );
```

The MSS can be both set and retrieved, but setting it should be done with caution. If the MSS is set to a value that causes the MTU to exceed the Path MTU, fragmentation will occur with a result of performance loss.

IP Layer Options

The IP layer options are those defined within the context of level IPPROTO_IP and focus on the IP layer. The typical options for the IP layer are defined in Figure 5.4.

IP_HDRINCL Option

The IP_HDRINCL option permits an application developer to write raw IP frames onto the wire and provide the IP header that will be attached. Because the point of this option is to provide our IP header given a raw socket (which is prefixed to each outgoing datagram), the argument for this option is whether an IP header is included in the outgoing datagram. In Listing 5.17, the sample source illustrates sending a raw IP datagram with an IP header of our choosing.

Option Name	*Description*	*get/set*	*value*
`IP_HDRINCL`	IP header precedes data in buffer	g/s	`int`
`IP_TOS`	Modifies the IP Type-Of-Service header field	g/s	`int`
`IP_TTL`	Modifies the IP Time-To-Live header field	g/s	`int`
`IP_ADD_MEMBERSHIP`	Join a multicast group	s	`struct mreq`
`IP_DROP_MEMBERSHIP`	Leave a multicast group	s	`struct mreq`
`IP_MULTICAST_IF`	Modify the outgoing multicast interface	g/s	`struct in_addr`
`IP_MULTICAST_TTL`	Modify the outgoing multicast TTL	g/s	`int`
`IP_MULTICAST_LOOP`	Enable/Disable loopback of outgoing datagrams	g/s	`int`

FIGURE 5.4   IP layer options.

Listing 5.17   Providing an IP header for an IP datagram (ipdgram.c).

```
#include <sys/types.h>
#include <sys/socket.h>
#include <netinet/in.h>

/* Standard IP Header */
typedef struct {
  unsigned char verHdrLen;
  unsigned char tos;
  unsigned short len;
  unsigned short ident;
  unsigned short flags;
  unsigned char ttl;
  unsigned char protocol;
  unsigned short checksum;
  struct in_addr sourceIpAdrs;
  struct in_addr destIpAdrs;
} ipHdr_t;

int main()
{
  int sock, on, ret;
  char buffer[255];
  ipHdr_t *ipDatagram;
  struct sockaddr_in addr;
```

```
extern int errno;

/* Checksum function on the CD-ROM... */
unsigned short checksum( unsigned short *, int );

ipDatagram = (ipHdr_t *)buffer;

sock = socket( AF_INET, SOCK_RAW, 255 );

on = 1;
ret = setsockopt( sock, IPPROTO_IP, IP_HDRINCL,
                  &on, sizeof(on) );

ipDatagram->verHdrLen = 0x45;
ipDatagram->tos = 0;
ipDatagram->len = 20;                    /* Just a header */
ipDatagram->ident = htons( 1 );
ipDatagram->flags = htons( 0x4000 ); /* Don't fragment */
ipDatagram->ttl = 64;
ipDatagram->protocol = 255;
ipDatagram->checksum = 0;
ipDatagram->sourceIpAdrs.s_addr = 0;

addr.sin_family = AF_INET;
addr.sin_addr.s_addr = inet_addr("192.168.1.1");
ipDatagram->destIpAdrs.s_addr = addr.sin_addr.s_addr;

ipDatagram->checksum =
  checksum( (unsigned short *)&ipDatagram,
            sizeof(ipHdr_t) );

ret = sendto( sock, buffer, sizeof(ipHdr_t), 0,
              (struct sockaddr *)&addr, sizeof(addr) );

close( sock );

return 0;
}
```

In Listing 5.17, the first step is to create a socket of type SOCK_RAW. This socket permits us to communicate using IP datagrams. Next, we enable the IP_HDRINCL option using the **setsockopt** function. Next, we create our IP header using the previously defined ipHdr_t typedef. We populate the IP header with a set of common values. The source and destination IP addresses are interesting fields to note. For the source address (sourceIpAdrs), we leave this field blank (set to zero), which notifies the stack to fill this in with the local source address for the outgoing interface. The destination address is constructed using the inet_addr function. We must provide a standard sockaddr_in structure with the **sendto** function, so we piggyback the generation of the destination address to also fill in the destination field of the IP header (destIpAdrs).

Note that in some cases, the **htons** function is used (ident and flags). These are required because what we provide will be the IP header for the datagram. Therefore, the fields that are expected to be in network byte order are converted. The checksum of the IP header must also be calculated; here, we use a checksum routine that is provided on the CD-ROM at /software/ch5/ipdgram.c.

Finally, we send our IP datagram using the **sendto** function. We include the destination address in the argument list (even though it also exists in our embedded IP header).

IP_TOS **Option**

The IP_TOS option permits an application to change the Type of Service (TOS) field within the IP header for a given socket. The TOS field is commonly used to specify service precedence within networks that support this feature. The TOS field permits segmenting traffic using quality of service parameters (see Figure 5.5).

7	6	5	4	3	2	1	0
	Precedence		D	T	R	0	0

		Bit = 0	Bit = 1
Bits 7-5	Precedence		
Bit 4	Delay (D)	Normal	Low
Bit 3	Throughput (T)	Normal	High
Bit 2	Reliability (R)	Normal	High
Bit 1-0	Future Use		

FIGURE 5.5 IP TOS field in detail.

The three-bit precedence field defines a category of service (see Figure 5.6). The Delay, Throughput, and Reliability bits indicate a requested quality of service (along three axes).

Setting the IP_TOS value is performed easily through an integer option, as shown in Listing 5.18.

Precedence

0	0	0	Routine
0	0	1	Priority
0	1	0	Immediate
0	1	1	Flash
1	0	0	Flash Override
1	0	1	CRITIC/ECP
1	1	0	Internetwork Control
1	1	1	Network Control

FIGURE 5.6 Precedence values and meanings.

Listing 5.18 Defining an IP Type of Service for a given connection.

```
int sock, ret, tos;

#define LOW_DELAY          0x10
#define HIGH_THROUGPUT     0x08
#define HIGH_RELIABILITY   0x04

sock = socket( AF_INET, SOCK_STREAM, 0 );

tos = (HIGH_THROUGHPUT | HIGH_RELIABILITY);

ret = setsockopt( sock, IPPROTO_IP, IP_TOS,
                  (void *)&tos, sizeof( tos ) );
```

The IP TOS field is commonly ignored on the WAN, but can be used in LAN environments to segment traffic based upon quality of service needs. Some implementations provide quality of service APIs that simplify the manipulation and management of this field.

IP_TTL Option

The IP_TTL option is used to initialize the Time To Live (TTL) field within the IP header. As an IP datagram traverses a network, the TTL field is decremented for each device that it passes through. Once the TTL field reaches zero, the datagram is dropped. The purpose of this field is to prevent a datagram from cycling forever through a network, and specifies the maximum number of hops an IP datagram may take.

Setting the IP_TTL value is performed easily through an integer option, as shown in Listing 5.19. This restricts datagrams to the current subnet. No datagrams will traverse a router or gateway on the current subnet.

Listing 5.19 Defining an IP Time To Live of one for a given connection.

```
int sock, ret, ttl;

sock = socket( AF_INET, SOCK_DGRAM, 0 );

ttl = 1;

ret = setsockopt( sock, IPPROTO_IP, IP_TTL,
                  (void *)&ttl, sizeof( ttl ) );
```

IP_ADD_MEMBERSHIP Option

The IP_ADD_MEMBERSHIP option is used to enable receipt of multicast packets for a given multicast address. The mreq structure is used to configure multicast packet receipt and has the following structure:

```
struct ip_mreq {
    struct in_addr imr_multiaddr;
    struct in_addr imr_interface;
};
```

The imr_multiaddr is a 32-bit network-byte-order IP address that represents the multicast address for which we want to subscribe. The imr_interface represents the interface on which we want to subscribe to that multicast communication (such as is done with the **bind** function). This can be the address of an available interface on the host (such as "192.168.1.1") or INADDR_ANY.

Joining a multicast group is illustrated in Listing 5.20. In this example, the multicast group address "239.255.255.253" is joined over all available interfaces on the current host (defined by INADDR_ANY). This provides the ability to receive packets sent to address "239.255.255.253" through the socket defined by sock.

Listing 5.20 Subscribing to a multicast group for a given connection.

```
int sock, ret;
struct ip_mreq mreq;

sock = socket( AF_INET, SOCK_DGRAM, 0 );

...
```

```
bzero( (void *)&mreq, sizeof(mreq) );
mreq.imr_multiaddr.s_addr = inet_addr("239.255.255.253");
mreq.imr_interface.s_addr = htonl( INADDR_ANY );

ret = setsockopt( sock, IPPROTO_IP, IP_ADD_MEMBERSHIP,
                      (void *)&mreq, sizeof( mreq ) );
```

If we want to restrict multicast receipt to a specific interface, we would specify the address in the imr_interface field of the mreq structure, such as:

```
mreq.imr_interface.s_addr = inet_addr( "192.168.1.1" );
```

The IP_ADD_MEMBERSHIP can be set with setsockopt, but not read with getsockopt.

IP_DROP_MEMBERSHIP Option

The IP_DROP_MEMBERSHIP option is used to drop membership to a multicast group for a given socket, as shown in Listing 5.21. The preparation for this call is identical to the IP_ADD_MEMBERSHIP function. We specify the multicast address in imr_multiaddr and the interface over which we've subscribed as imr_interface.

Listing 5.21   Leaving a multicast group for a given connection.

```
int sock, ret;
struct ip_mreq mreq;

sock = socket( AF_INET, SOCK_DGRAM, 0 );

...

bzero( (void *)&mreq, sizeof(mreq) );
mreq.imr_multiaddr.s_addr = inet_addr("239.255.255.253");
mreq.imr_interface.s_addr = htonl( INADDR_ANY );

ret = setsockopt( sock, IPPROTO_IP, IP_ADD_MEMBERSHIP,
                      (void *)&mreq, sizeof( mreq ) );

...

ret = setsockopt( sock, IPPROTO_IP, IP_DROP_MEMBERSHIP,
                      (void *)&mreq, sizeof( mreq ) );
```

As shown in Listing 5.21, the ip_mreq structure used to join a multicast group should be used to leave it.

The IP_DROP_MEMBERSHIP can be set with setsockopt, but not read with getsockopt.

`IP_MULTICAST_IF` Option

The previous options, `IP_ADD_MEMBERSHIP` and `IP_DROP_MEMBERSHIP`, are used to configure receipt of multicast datagrams. The `IP_MULTICAST_IF` option is used to configure which interface will be used to send multicast datagrams (for hosts that have multiple interfaces). The primary interface is the default interface from which multicast datagrams are transmitted, but this can be changed using the `IP_MULTICAST_IF` option.

The `IP_MULTICAST_IF` option uses an `in_addr` structure to define the interface for outgoing datagrams. The `in_addr` structure must contain the IP address for the interface of choice. Consider the following example in Listing 5.22.

Listing 5.22 Setting the outgoing multicast interface for a given connection.

```
int sock, ret;
struct in_addr intf_addr;

sock = socket( AF_INET, SOCK_DGRAM, 0 );

...

intf_addr.s_addr = inet_addr("192.168.1.2");

ret = setsockopt( sock, IPPROTO_IP, IP_MULTICAST_IF,
                  (void *)&intf_addr, sizeof( intf_addr ) );
```

In Listing 5.22, we initialize our `s_addr` structure (`intf_addr`) with a network byte order IP address (converted by `inet_addr`). The **setsockopt** call is then used with `intf_addr` to configure the interface represented by "192.168.1.2" as our outgoing multicast interface.

`IP_MULTICAST_TTL` Option

The `IP_MULTICAST_TTL` option is used to change the TTL field for outgoing multicast packets. This value defaults to one for multicast sockets, but can be adjusted up to 255 (representing the number of multicast router hops possible through the network before the packet is dropped). Listing 5.23 illustrates the use of the `IP_MULTICAST_TTL` option.

Listing 5.23 Setting the `IP_MULTICAST_TTL`.

```
int sock, ret, mttl;

sock = socket( AF_INET, SOCK_DGRAM, 0 );

mttl = 10; /* 10 hops */

ret = setsockopt( sock, IPPROTO_IP, IP_MULTICAST_TTL,
                  (void *)&mttl, sizeof( mttl ) );
```

In the example from Listing 5.23, a maximum of 10 hops is defined for the given multicast socket.

IP_MULTICAST_LOOP Option

For hosts on which multiple applications subscribe to a given multicast address, all outgoing datagrams are looped back so that other applications can read it (the default is enabled). The IP_MULTICAST_LOOP option permits the disabling of the loopback. This can be done to optimize, because not looping datagrams back increases the performance of the application (and underlying multicast layer).

The following example, Listing 5.24, illustrates disabling the loopback option.

Listing 5.24   Setting the IP_MULTICAST_LOOP.

```
int sock, ret, on;

sock = socket( AF_INET, SOCK_DGRAM, 0 );

on = 0;

ret = setsockopt( sock, IPPROTO_IP, IP_MULTICAST_LOOP,
                  (void *)&on, sizeof( on ) );
```

This option should be used with care, because disabling loopback means that all outgoing traffic will be seen by other subscribing hosts, but not applications on the current host.

SUMMARY

In this chapter, we summarized the major socket options that can be used to change the default behavior of sockets. The **setsockopt** call is used to set socket options, whereas the **getsockopt** call is used to read them. Socket options exist at three levels, socket (SOL_SOCKET), TCP (IPPROTO_TCP), and IP (IPPROTO_IP). The layers correspond to the specific layers within the TCP/IP stack. Although most socket options utilize integer options, some structures are also used for more complex configurations.

RESOURCES

Stevens, W. Richard, *UNIX Network Programming*, 2nd edition, Volume 1, Prentice Hall PTR, 1998.

6 Advanced Sockets Programming Topics

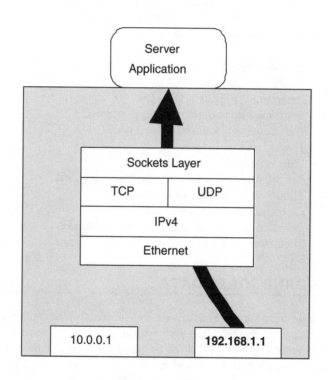

In this chapter, we investigate a variety of topics that lie beyond the basics of Sockets programming. We look at multi-homing, name resolution, error handling, blocking and nonblocking communication, and many other topics of Sockets application development. Although advanced, these topics are important for the development of efficient and effective Sockets applications. All code for this chapter can be found on the companion CD-ROM at /software/ch6.

INTRODUCTION

A number of disparate topics are discussed in this chapter; some are useful in everyday applications, whereas others are useful only in niche applications. The advanced Sockets programming topics covered here include:

- Out-of-band (OOB) data
- Nonblocking sockets
- Determining peer socket closure
- Resolving a domain name to an IP address
- Writing servers for multi-homed nodes
- Timing out a socket connect
- Data framing
- Connectionless and connected datagram sockets
- Timing out a read or write operation
- Determining peer information
- Determining the protocol argument for **socket**
- Identifying service ports and protocols

OUT-OF-BAND (OOB) DATA

Out-of-band (or OOB) data is a stream socket abstraction that creates the concept of a dual stream for communication over a pair of sockets. This secondary stream is a logically independent stream of data that can be used for signaling outside of the transfer of regular data. As the regular data stream and the OOB data stream are independent, the Sockets API must provide a way for the application to specify which of the streams should be manipulated. This specification is provided through the MSG_OOB flag of the **send/recv** and **sendto/recvfrom** Sockets API functions. For example:

```
ret = send( sock, buffer, len, MSG_OOB );

ret = recv( sock, buffer, len, MSG_OOB );
```

For the **send** function, the MSG_OOB flag defines that the data in buffer is out-of-band data. Conversely, with the **recv** function, the MSG_OOB flag specifies that receipt of out-of-band data be requested.

With the TCP protocol, only one byte of data may be transferred as out-of-band at a time. TCP utilizes the urgent mode (described in the TCP headers) as its means to identify and transport out-of-band data between sockets.

Reading or writing out-of-band data looks relatively simple and painless, but how does an application know when out-of-band data is available to read? One way to know is the **sockatmark** function, which tells the application that the next data to read is out-of-band data. Consider the example in Listing 6.1.

Listing 6.1   Receiving out-of-band data using `sockatmark`.

```
int sock, ret, on;
char buffer[100], oobdata;

...

on = 1;
ret = setsockopt( sock, SOL_SOCKET, SO_OOBINLINE,
                    &on, sizeof(on) );

...

if (sockatmark( sock )) {

  /* We're at the OOB byte position */
  ret = read( sock, &oobdata, 1 );

} else {

  /* Normal data coming... */
  ret = read( sock, buffer, sizeof(buffer) );

}
```

The **sockatmark** function simply tells the application that the next byte to be read is out-of-band data. Therefore, if **sockatmark** returns 1, we know that out-of-band data follows and we read it accordingly (in this case, using the **read** function). Otherwise, we read as we typically would, expecting a number of bytes. Note, in this case, that we've specified that out-of-band data will be received in line with our regular data (using the SOL_SOCKET option, SO_OOBINLINE). When normal in-band data is read, the read automatically stops at the out-of-band data. Therefore, a normal **read** reads only up to the point of the out-of-band data, and no further.

Unix systems may also use the SIGURG signal to identify when out-of-band data is available to be read. This method is covered within [Stevens98].

NONBLOCKING SOCKETS

Let's first understand the difference between blocking and nonblocking semantics. In a typical system, when a call is made to a socket read function such as **recv**, if no data is available to be read at that time, the socket blocks. By blocking, we mean that the call pends and does not return until either data is available to read or an error occurs. Therefore, the return of the **recv** call may happen quickly, or it may take a long time; it's very dependent upon data to read. This standard behavior can be altered to the reverse semantics, or nonblocking. In nonblocking semantics, functions return immediately and do not block (or pend) awaiting some action to be performed. This means that a call to **recv** will return an error if no data is available to read. The error message returned is commonly EWOULDBLOCK, which means that the call would normally block, but due to the nonblocking nature of the socket, blocking is disallowed and the resulting error is returned.

Making sockets nonblocking is very implementation dependent. Some stacks provide socket options to enable nonblocking behavior. Unix systems provide an **ioctl** call to change the blocking nature of a socket. Consider the following example in Listing 6.2.

Listing 6.2   Making a socket nonblocking on a Unix system.

```
int sock, mode, ret;

#define BLOCKING        0
#define NONBLOCKING     1

sock = socket( AF_INET, SOCK_STREAM, 0 );

mode = NONBLOCKING;

ret = ioctl( sock, FIONBIO, &mode );
```

Nonblocking sockets are useful for performance and are commonly assisted by the use of the **select** call for I/O multiplexing.

DETERMINING PEER SOCKET CLOSURE

Handling return errors from Sockets functions is a common mistake found in many Sockets applications. A simple way to identify whether a peer socket has closed its connection is to check the return status of Sockets functions.

When the peer closes its socket, either intentionally or unintentionally, the proper action is to detect the closure and permit the application to take the necessary action. This is commonly closing the socket and then attempting to reconnect (if needed).

Detecting a peer socket close is commonly performed with data input/output functions (**send**, **recv**, **sendto**, **recvfrom**, **read**, **write**). The input functions return a zero on peer socket closure, whereas output functions return an error (commonly broken-pipe). The following code snippets (Listing 6.3 and Listing 6.4) illustrate detecting peer closure for read and write.

Listing 6.3 Detecting peer closure using input calls.

```
int sock, ret;
char buffer[MAX_BUFFER_SIZE+1];

sock = socket( AF_INET, SOCK_STREAM, 0 );

/* Client connect... */

ret = recv( sock, buffer, MAX_BUFFER_SIZE, 0 );

if ( ret > 0 ) {

  /* Received data, deal with it accordingly */

} else if ( ret < 0 ) {

  /* An error occurred, check error status */

} else if ( ret == 0 ) {

  /* Peer closed, error leg */

  close( sock );

}
```

In Listing 6.3, after a return of zero is detected from the **recv** call, the socket is unusable. Clients commonly close the socket and attempt a reconnect, whereas servers simply close the connection and await a reconnect.

The peer socket closure can also be detected through a **write** function, such as is shown in Listing 6.4. In this case, an error results from the **write** and the error status must be checked to identify whether the peer socket closed or if some other error occurred. In Unix systems, the error value that results from peer closure is EPIPE (pipe broken). Additionally, Unix systems typically generate a signal.

Listing 6.4 Detecting peer closure using output calls.

```
int servsock, clisock, ret, size;
char buffer[MAX_BUFFER_SIZE+1];
```

```
servsock = socket( AF_INET, SOCK_STREAM, 0 );

/* Server setup... */

clisock = accept( servsock, (struct sockaddr *)NULL, NULL );

/* buffer setup for size... */

ret = write( sock, buffer, size, 0 );

if ( ret == size ) {

  /* Data written */

} else if ( ret < 0 ) {

  /* An error occurred, check error status */

  if ( errno == EPIPE ) {

    /* Peer closed */
    close( clisock );

  }

}
```

Handling this kind of error is necessary in any Sockets application. An application can't guarantee the behavior of the remote application nor can it guarantee the reliability of the network; therefore, detecting peer closure is essential.

RESOLVING A DOMAIN NAME TO AN IP ADDRESS

Although converting a host name to an IP address is not really an advanced topic, code dealing with what could be an FQDN or an IP address in a consistent manner is commonly missing in many Sockets applications. In terms of configuration, an IP address or host name could be present and the application should be able to resolve either of these to a binary network address uniformly.

The code pattern shown in Listing 6.5 illustrates the handling of an unknown address type. Regardless of whether a dotted-notation IP address is provided, or an FQDN, the result is a binary network address (assuming a valid entry or resolvable address).

Listing 6.5   Resolving an unknown type of address.

```
struct sockaddr_in saddr;
char address[]={"192.168.1.1"};
// or
char address[]={"www.address.com"};

saddr.sin_addr.s_addr = inet_addr( address );

if ( saddr.sin_addr.s_addr = 0xffffffff ) {

  struct hostent *hptr =
    (struct hostent *)gethostbyname( address );

  if (hptr == NULL) {

    /* Can't resolve the address... */

  } else {

    struct in_addr **addrs;
    addrs = (struct in_addr **)hptr->h_addr_list;
    memcpy( &saddr.sin_addr, *addrs, sizeof(struct in_addr) );

  }

}
```

The code pattern shown in Listing 6.5 is very useful for automatically generating a 32-bit network byte order IP address from a string representing either a string dotted-notation IP address or FQDN.

WRITING SERVERS FOR MULTI-HOMED NODES

A multi-homed node is simply a device that contains more than one network interface and, therefore, connects to more than one network. Writing simple servers for multi-homed nodes is commonly as simple as single-homed nodes, unless the server treats the network interfaces differently. For example, is the server accessible over each of the networks connected to the network interfaces, or only a subset of them?

Let's now look at a few scenarios and the code patterns that are necessary to provide the needed functionality.

The first case is the simplest; the server accepts connections over any interface (from any connected network). This is the default of most server applications built today. Recall that the **bind** function is used to "bind" a socket to a given address. The application may specify the address of a given network interface, or a wildcard for all interfaces. This is shown in Listing 6.6.

Listing 6.6   Server accepting connections from all interfaces.

```
int serverSock;
struct sockaddr_in saddr;

serverSock = socket( AF_INET, SOCK_STREAM, 0 );

memset(&saddr, 0, sizeof(saddr));
saddr.sin_family = AF_INET;
saddr.sin_addr.s_addr = htonl( INADDR_ANY );
saddr.sin_port = htons( MY_PORT );

bind( serverSock, (struct sockaddr *)&saddr, sizeof(saddr));
```

The INADDR_ANY wildcard address specifies that any available network interface may be used to accept incoming connections. However, what if the server wanted to restrict connections only to those from a specific interface? This is common in devices that consist of a WAN and LAN interface. For example, configuration may only be possible through the LAN interface (internal local network). If the LAN interface had the IP address "192.168.1.1", the following code pattern in Listing 6.7 could be used to restrict connections to only those from that interface.

Listing 6.7   Server accepting connections from a specific interface.

```
int serverSock;
struct sockaddr_in saddr;

serverSock = socket( AF_INET, SOCK_STREAM, 0 );

memset(&saddr, 0, sizeof(saddr));
saddr.sin_family = AF_INET;
saddr.sin_addr.s_addr = inet_aton( "192.168.1.1" );
saddr.sin_port = htons( MY_PORT );

bind( serverSock, (struct sockaddr *)&saddr, sizeof(saddr));
```

The only difference here is that we bind the sin_addr element of our sockaddr_in structure with the network byte order of the address using inet_aton.

The final example is a bit more complicated. Let's assume in this case that we want to present a server on multiple interfaces, but we'll treat connections differently based upon the interface over which they arrived. The first interface is represented by "192.168.1.1" and the second by "10.0.0.1". In the example shown in Listing 6.8, we create two sockets and bind each of them with a different sin_addr address, but both with port 8192. Then, using the **select** function, we identify whether a connect request has arrived by sensing that the server socket is readable. After determining that a socket is readable, we perform the **accept** on the given socket, and then do whatever action is necessary. This allows us to have a single server represented by a number of sockets with the ability to differentiate based upon the interface over which a connection is requested.

Listing 6.8 Server differentiating based upon connect interface (bindtest.c).

```
#include <sys/socket.h>
#include <sys/types.h>
#include <arpa/inet.h>
#include <unistd.h>
#include <stdio.h>

#define MAX(x, y)        ((x > y) ? x : y)

main()
{
  int s1, s2, s3, ret, max, on;
  struct sockaddr_in sa1, sa2;
  fd_set rfds;

  s1 = socket( AF_INET, SOCK_STREAM, 0 );
  s2 = socket( AF_INET, SOCK_STREAM, 0 );

  max = MAX(s1, s2) + 1;

  on = 1;
  ret = setsockopt( s1, SOL_SOCKET, SO_REUSEADDR,
                    &on, sizeof(on) );
  ret = setsockopt( s2, SOL_SOCKET, SO_REUSEADDR,
                    &on, sizeof(on) );

  memset( &sa1, 0, sizeof(sa1) );
  memset( &sa2, 0, sizeof(sa2) );

  /* Set up the two address structures for the same */
  /* port, different interfaces */
  sa1.sin_family = sa2.sin_family = AF_INET;
  sa1.sin_port = sa2.sin_port = htons(8192);
  sa1.sin_addr.s_addr = inet_addr( "192.168.1.1" );
  sa2.sin_addr.s_addr = inet_addr( "10.0.0.1" );
```

```
/* Bind address1 to s1 */
ret = bind( s1, (struct sockaddr *)&sa1, sizeof(sa1) );
ret = listen(s1, 5);

/* Bind address2 to s2 */
ret = bind( s2, (struct sockaddr *)&sa2, sizeof(sa2) );
ret = listen(s2, 5);

while (1) {

  /* Set up the read socket descriptor list */
  FD_ZERO( &rfds );
  FD_SET( s1, &rfds );
  FD_SET( s2, &rfds );

  /* Await an incoming connection */
  ret = select( max, &rfds, NULL, NULL, NULL );

  /* Check that a read (connect) arrived */
  if (ret > 0) {

    /* Was it socket1? */
    if ( FD_ISSET( s1, &rfds ) ) {

      printf("Received connect request over 192.168.1.1\n");

      s3 = accept( s1, (struct sockaddr_in *)NULL, NULL);
      close( s3 );

    /* Was it socket2? */
    } else if ( FD_ISSET( s2, &rfds ) ) {

      printf("Received connect request over 10.0.0.1\n");

      s3 = accept( s2, (struct sockaddr_in *)NULL, NULL);
      close( s3 );

    }

  } else {

    printf("Error!\n");

  }

}
```

```
    close(s1); close(s2);
}
```

Although two sockets were used in the multiple interface example (Listing 6.8), this is transparent to all external peers. The only difference seen at the peers is the externally observable behavior that is coded within the server.

TIMING OUT A SOCKET CONNECT

Timing out a socket **connect** call can sometimes be useful, especially in systems that do not support blocking function semantics. Consider the problem of trying to connect to many nodes (or ports) at a given time. Either a large number of threads must be created for each **connect** request, or the **connect** call can be made nonblocking so that many connects can be performed simultaneously by a single thread.

One useful example of this technique is in the construction of a simple port scanning utility. Port scanning is the name given to applications that search a network for nodes that have applications sitting on ports of interest (a common hacking tool). In this example, we want to know for our given subnet, which nodes provide telnet servers. We know that telnet daemons commonly sit on port 23, so if we connect to port 23 on all available nodes, we should be able to identify the location of all telnet daemons (based upon successful completion of the **connect** call).

Consider the simple port scanner shown in Listing 6.9. We begin by setting up our timeout structure for five seconds (the maximum amount of time we'll await the **connect** requests) and then set up our address structure with everything but the address to which we'll connect. We then perform two loops. The first loop connects to each address and the second loop awaits responses.

In the first loop, we complete the address structure with the address to which we'll connect, and then attempt a **connect** to that address. The socket is made nonblocking so that our **connect** does not block for each attempt, and the socket descriptor is added to our descriptor list that we'll later pass to **select**. The **connect** function begins the TCP three-way handshake, so at the completion of the **connect**, we should have received a SYN-ACK from the peer, which lets us know that an application is actually sitting on the designated port. Otherwise, if no application sits on the port, a RST is returned, indicating that no server with that port designation is present.

All of our connect requests have now been made, and we enter the second loop that awaits the responses (the SYN-ACK). In this **select** example, we use the write descriptors to know when a socket has been connected (if the socket is writable, then the **connect** succeeded and a server sits on the port on the given node). When **select** returns in the loop, we check to see if a positive value was returned indicating that some file descriptors are writable. We then walk through the descriptor list looking for descriptors that were set and indicate to the user which IP addresses housed the telnet daemon. Note that we also clear the descriptor so that it won't be shown as writable in our next iteration. If a nonpositive value was returned from **select**, then either an error occurred or the process has timed out and we exit. As a final cleanup step, we close all of our sockets at the end of the port scanner application.

Listing 6.9 Simple port scanner illustrating timing out the **connect** function (portscan.c).

```
#include <sys/socket.h>
#include <sys/types.h>
#include <sys/time.h>
#include <arpa/inet.h>
#include <stdio.h>
#include <unistd.h>
#include <time.h>
#include <fcntl.h>
#include <unistd.h>

#define SUBNET     "192.168.1.%d"
#define MAX_HOSTS  254

int main ( )
{
  int i, count, ret;
  struct sockaddr_in servaddr;
  int  socks[MAX_HOSTS], maxfd=0;
  char buf[30+1];
  fd_set fds;
  struct timeval timeout;

  /* Set up a timeout of 5 seconds */
  timeout.tv_sec = 5;
  timeout.tv_usec = 0;

  /* Partially set up the address structure */
  memset(&servaddr, 0, sizeof(servaddr));
  servaddr.sin_family = AF_INET;
  servaddr.sin_port = htons(23);

  FD_ZERO( &fds );

  /* Attempt to connect to all nodes on the subnet */
  for (i = 1 ; i <= MAX_HOSTS ; i++) {

    /* Complete the address structure */
    sprintf(buf, SUBNET, i);
    servaddr.sin_addr.s_addr = inet_addr(buf);
```

```
/* Create a new socket for this node */
socks[i] = socket(AF_INET, SOCK_STREAM, 0);

/* Add it to the socket descriptor list */
FD_SET( socks[i], &fds );

/* Make the socket nonblocking */
fcntl(socks[i], F_SETFL, O_NONBLOCK);

/* Attempt a connect (nonblocking) */
connect( socks[i], (struct sockaddr_in *)&servaddr,
         sizeof(servaddr));

if (socks[i] > maxfd) maxfd = socks[i];

}

count = MAX_HOSTS;

while (1) {

  /* Await connect responses (writable sockets) */
  ret = select(maxfd, NULL, &fds, NULL, &timeout);

  if (ret > 0) {

    count -= ret;

    /* Walk through the socket descriptor list */
    for (i = 1 ; i <= MAX_HOSTS ; i++) {
      if ( FD_ISSET( socks[i], &fds ) ) {
        printf("Port open at 192.168.1.%d\n", i);
        FD_CLR( socks[i], &fds );
      }
    }

    /* If we've found everything, exit */
    if (count == 0) break;

  } else {

    /* Timeout, exit */
    break;

  }

}
```

```
        /* Cleanup, close all of the sockets */
        for (i = 1 ; i < MAX_HOSTS ; i++) {
          close(socks[i]);
        }

        return(0);
      }
```

DATA FRAMING (TCP VS. UDP)

Although the topic of data framing is not really an advanced Sockets API topic, it is a behavior that is commonly misunderstood. Framing defines the basic units of information that are transported between two socket endpoints. In the case of TCP, information is transported as a stream, and, therefore, no framing exists. For UDP, information is transported as datagrams, which means that data is framed. UDP provides a message-oriented protocol in which the unit of information sent by the sender is what is received by the receiver. Even if more data was sent after the first segment of information, the receiver still receives the unit sent by the **write** call. TCP, being stream-based, operates in a very different manner. If the sender writes three blocks of information, the receiver may receive all three blocks in a single **read**. This is because TCP is stream-oriented and has no concept of framing.

Let's look at some examples that illustrate what's happening here. In Listing 6.10, we analyze TCP from the perspective of framing.

Listing 6.10 Source illustrating framing (lack of) in TCP stream sockets (tcpframe.c).

```
      #include <sys/types.h>
      #include <sys/socket.h>
      #include <arpa/inet.h>
      #include <unistd.h>
      #include <fcntl.h>

      #define MAX_BUF           100

      main()
      {
        int s1, s2, s3;
        struct sockaddr_in saddr;
        int ret, on=1;
        char outbuf[10];
        char inbuf[MAX_BUF+1];

        /* Create two stream sockets */
        s1 = socket( AF_INET, SOCK_STREAM, 0 );
```

```
s2 = socket( AF_INET, SOCK_STREAM, 0 );

/* Make s1 (the server) reusable from an
 * address perspective.
 */
ret = setsockopt( s1, SOL_SOCKET, SO_REUSEADDR,
                  &on, sizeof(on) );

/* Create the address for the server socket */
saddr.sin_family = AF_INET;
saddr.sin_port = htons( 8192 );
saddr.sin_addr.s_addr = htonl( INADDR_ANY );

/* Bind the address to s1 */
ret = bind( s1, (struct sockaddr_in *)&saddr, sizeof(saddr) );

/* Specify willingness to accept connections */
ret = listen( s1, 1 );

/* Make the client socket nonblocking */
fcntl( s2, F_SETFL, O_NONBLOCK );

/* Connect the two sockets together */
ret = connect( s2, (struct sockaddr_in *)&saddr,
               sizeof(saddr) );

/* Allow the server to accept a new client socket */
s3 = accept( s1, (struct sockaddr_in *)NULL, NULL );

/* Generate some output data */
outbuf[0] = 'a';
outbuf[1] = 0;

/* Make three distinct calls with the data */
ret = write( s3, outbuf, 1 );
ret = write( s3, outbuf, 1 );
ret = write( s3, outbuf, 1 );

/* Sleep for one second */
sleep( 1 );

/* Read the data from the client */
ret = read( s2, inbuf, MAX_BUF );

/* Should have read 3 bytes */
printf( "read %d bytes\n", ret );

/* Close the sockets */
```

```
  close(s1);
  close(s2);
  close(s3);

}
```

From Listing 6.10, we see a single application that provides both a TCP server and client that demonstrates passing data. The server and client setup are performed, and then the client is made nonblocking (so that both can coexist within the same application). The interesting aspect of this application begins near the end, where three writes are made of a single byte of data. We sleep for a second to allow the stack to properly migrate the data and then perform a **read** operation. The numbers of bytes that are read are three, which is the three writes that were performed. From this experiment, we see that no framing occurred because TCP streamed the data without regard to the granularity of the writes.

Let's now look at the UDP case. UDP differs from TCP in a number of ways, but an important distinction is the inclusion of framing. When a write is performed in UDP, the block of data is a datagram and the framing of the data is preserved for read. Listing 6.11 shows the UDP source that mirrors our TCP source from Listing 6.10.

Listing 6.11   Source illustrating framing in UDP datagram sockets (udpframe.c).

```
#include <sys/types.h>
#include <sys/socket.h>
#include <arpa/inet.h>
#include <unistd.h>
#include <fcntl.h>
#include <errno.h>

#define MAX_BUF          100

main()
{
  int s1, s2;
  struct sockaddr_in saddr;
  int ret, on=1;
  char outbuf[10];
  char inbuf[MAX_BUF+1];

  /* Create two stream sockets */
  s1 = socket( AF_INET, SOCK_DGRAM, 0 );
  s2 = socket( AF_INET, SOCK_DGRAM, 0 );

  /* Make s1 (the server) reusable from an
   * address perspective.
   */
  ret = setsockopt( s1, SOL_SOCKET, SO_REUSEADDR,
```

```
                        &on, sizeof(on) );

    /* Create the address for the server socket */
    saddr.sin_family = AF_INET;
    saddr.sin_port = htons( 8192 );
    saddr.sin_addr.s_addr = htonl( INADDR_ANY );

    /* Bind the address to s1 */
    ret = bind( s1,
                (struct sockaddr_in *)&saddr, sizeof(saddr) );

    /* Connect the two sockets together (connected
     * datagram socket).
     */
    ret = connect( s2, (struct sockaddr_in *)&saddr,
                   sizeof(saddr) );

    /* Generate some output data */
    outbuf[0] = 'a';
    outbuf[1] = 0;

    /* Make three distinct calls with the data */
    ret = write( s2, outbuf, 1 );
    ret = write( s2, outbuf, 1 );
    ret = write( s2, outbuf, 1 );

    /* Sleep for one second */
    sleep( 1 );

    /* Read the data from the client */
    ret = read( s1, inbuf, MAX_BUF );

    /* Should have read 1 byte */
    printf( "read %d bytes\n", ret );

    /* Close the sockets */
    close(s1);
    close(s2);

}
```

What is notable about Listing 6.11 is that instead of reading the accumulated three bytes that were written, only one byte is read. This is because each **write** specifies the framing of the data, which is preserved at the peer. To read the data, three **reads** must be performed.

This is a very important distinction and is an important feature of UDP. TCP applications can simulate this behavior by including a layer above the socket to preserve the framing (in a data-dependent fashion), but at the cost of additional processing of the data. This makes UDP a very attractive protocol for message-based communication.

CONNECTIONLESS AND CONNECTED DATAGRAM SOCKETS

By default, datagram sockets are connectionless. This useful feature of UDP means that there is no single peer connected to the datagram socket, and, instead, a packet can be directed to the peer with remote address information provided with the data. This differs from TCP sockets in that there is a single peer associated with the socket. For this reason, when referencing a TCP socket, there is no need to specify remote addressing because it's already been cached in the Sockets layer.

Datagram sockets can be connected like stream sockets, but without the setup required by a stream socket. For example, when a stream socket connects to a peer, a handshake takes place to synchronize the two ends of the connection. Because datagram sockets are message-based, there is no such synchronization. Therefore, connectedness in a datagram socket simply means that there is no need to specify peer address information with every write; it's cached in the stack like stream sockets.

Let's look at some examples to better understand what's happening here. In the first example, Listing 6.12, we look at a typical datagram socket that is connectionless. This is the standard mode for datagrams.

Listing 6.12 Datagram socket in an unconnected state.

```
int sock, ret;
struct sockaddr_in saddr;
char buffer[MAX_BUFFER];

sock = socket( AF_INET, SOCK_DGRAM, 0 );

memset( &saddr, 0, sizeof(saddr) );
saddr.sin_family = AF_INET;
saddr.sin_port = htons( MY_PORT );
saddr.sin_addr.s_addr = inet_aton( "192.168.1.1" );

...

ret = sendto( sock, buffer, strlen(buffer),
                  (struct sockaddr_in *)&saddr, sizeof(saddr) );
```

As shown in Listing 6.12, the **sendto** call is needed for datagram sockets in the unconnected state because the peer address information must be specified.

Now let's look at a datagram socket in the connected state. In this example, Listing 6.13, we see that a **connect** has been used to place our datagram socket in the connected state.

Unlike TCP sockets, this simply caches the peer address information in the stack and does not perform the typical three-way handshake as required by TCP. Also shown is use of the **send** call instead of **sendto**. Because the peer address is already cached in the stack, we don't need to recall this information for **sendto**, and **send** works identically in this case.

Listing 6.13 Datagram socket in a connected state.

```
int sock, ret;
struct sockaddr_in saddr;
char buffer[MAX_BUFFER];

sock = socket( AF_INET, SOCK_DGRAM, 0 );

memset( &saddr, 0, sizeof(saddr) );
saddr.sin_family = AF_INET;
saddr.sin_port = htons( MY_PORT );
saddr.sin_addr.s_addr = inet_aton( "192.168.1.1" );

ret = connect( sock,
               (struct sockaddr_in *)&saddr, sizeof(saddr) );

...

ret = send( sock, buffer, strlen(buffer), 0 );
```

One of the primary advantages of this approach is that it's much cleaner and can be less error-prone. In some cases, it can be a performance advantage because there are fewer arguments passed between the Application layer and the stack.

TIMING OUT A READ OR WRITE OPERATION

In some applications, it's important for an application not to block on an I/O operation, but setting the socket to nonblocking means that CPU cycles are wasted in busy-waiting. To combat this problem, the **select** function can be used to time out the read or write operation. The **select** function can be used to time out after some duration of no data I/O, permitting the application to perform some other activity. The following example in Listing 6.14 illustrates using the **select** function to time out after five seconds of awaiting data for read for the given socket.

Listing 6.14 Timing out a read operation.

```
#include <sys/types.h>
#include <sys/socket.h>
```

```
#include <netinet/in.h>
#include <arpa/inet.h>

...

int sock, ret;
fd_set rfds;
struct timeval tv;

sock = socket( AF_INET, SOCK_STREAM, 0 );

...

FD_ZERO( &rfds );
FD_SET( sock, &rfds );

tv.tv_sec = 5;
tv_tv_usec = 0;

ret = select( sock+1, &rfds, NULL, NULL, &tv );

if (ret > 0) {

  if ( FD_ISSET( sock, &rfds ) ) {

    /* Data is available for read */

    ...

  }

} else {

  /* Timeout */
  ...

}
```

The **select** call, as we've seen before, uses a bit vector (fd_set structure) to identify the socket descriptors we want to monitor and which action we're interested in monitoring for. In this case, we specify the read set (argument 2 of the **select** function). Upon completion of **select**, the return value represents the number of socket descriptors contained in the resulting set. We can then use the **FD_ISSET** function to identify which descriptor the event occurred. Otherwise, a zero return from **select** represents a timeout, which results in whatever application-specific action is necessary.

DETERMINING PEER INFORMATION

There are many cases in which it is important for a server to know who has connected (either for bookkeeping/logging or for simple authentication). After a connection is made by a client, the server interrogates the client socket to identify its address and port information. This is very easily done with the Sockets API through the use of the **getpeername** Sockets function. This function simply gathers the peer address and port information and returns it to the user in a sockaddr_in structure. Listing 6.15 illustrates the retrieval and display of the peer information using the **getpeername** function for a stream socket.

Listing 6.15 Identifying the peer using getpeername.

```
int sock, len, ret;
struct sockaddr_in peeraddr;

sock = socket( AF_INET, SOCK_STREAM, 0 );

...

len = sizeof( peeraddr );
ret = getpeername( sock,
                   (struct sockaddr_in *)&peeraddr, &len );

printf( "Peer address : %s\n",
        inet_ntoa( peeraddr.sin_addr ) );
printf( "Peer port    : %d\n",
        ntohs( peeraddr.sin_port ) );
```

The Sockets application can also use the inverse function, **getsockname**, to retrieve information about the local socket.

DETERMINING THE PROTOCOL ARGUMENT FOR SOCKET

Recall that when we created a simple stream socket, the third argument was defined as zero. This meant that the socket function would pick the default protocol for the given type (such as SOCK_STREAM). In this case, the default protocol is TCP, but what is the actual protocol number for TCP?

This question can be answered using the **getprotobyname** function. Function **getprotobyname** can be used to return the actual protocol number associated with the given protocol string (such as "tcp"). Let's look at a simple example that enumerates some of the standard protocols found in the IP suite (see Listing 6.16).

Listing 6.16   Finding the unique protocol number (proto.c).

```c
#include <netdb.h>

main()
{

  struct protoent *pp;

  pp = getprotobyname( "ip" );
  if (pp) printf(" ip = %d\n", pp->p_proto );

  pp = getprotobyname( "icmp" );
  if (pp) printf(" icmp = %d\n", pp->p_proto );

  pp = getprotobyname( "igmp" );
  if (pp) printf(" igmp = %d\n", pp->p_proto );

  pp = getprotobyname( "tcp" );
  if (pp) printf(" tcp = %d\n", pp->p_proto );

  pp = getprotobyname( "udp" );
  if (pp) printf(" udp = %d\n", pp->p_proto );

  pp = getprotobyname( "rdp" );
  if (pp) printf(" rdp = %d\n", pp->p_proto );

}
```

The function **getprotobyname** returns a protocol entry, of which the p_proto field is the protocol number (which can be used as the protocol argument of the **socket** function). The protocol numbers for the protocols shown in Listing 6.16 are shown in Table 6.1.

TABLE 6.1   Protocol numbers for typical IP protocols

Protocol	Protocol Number
IP	0
ICMP	1
IGMP	2
TCP	6
UDP	17
RDP	27

Use of the **getprotobyname** function is shown in Listing 6.17. This illustrates formal use of the **getprotobyname** function to return the protocol number for use by **socket**.

Listing 6.17   Using getprotobyname for the socket function.

```
#include <netdb.h>

...

int sock;
struct protoent *pp;

/* Get the TCP protocol entry */
pp = getprotobyname( "tcp" );

/* Create a TCP socket */
if (pp) {
  sock = socket( AF_INET, SOCK_STREAM, pp->p_proto );
}
```

The code in Listing 6.17 has the same effect as creating a socket using the default protocol, such as:

```
sock = socket( AF_INET, SOCK_STREAM, 0 );
```

The location of the protocol numbers differs based upon the target system, but on Linux systems, this file can be found at /etc/protocols.

IDENTIFYING SERVICE PORTS AND PROTOCOLS

A number of interesting functions exist to retrieve information about known services in a Unix-based system. Each of the functions returns a servent structure that defines not only the service name and port, but also the protocol by which it should be accessed and any aliases that exist for it. Some of the more common service functions are:

```
struct servent *getservbyname( const char *name,
                               const char *proto );
```

which is used to retrieve service information for a service given its name and protocol (such as "tcp"). Services can also be retrieved by port number using the following function prototype:

```
struct servent *getservbyport( int port,
                               const char *proto );
```

We can also simply enumerate the known services of a system using the **getservent** function. This function requires the initialization function **startservent** and the completion function **endservent**, which all have the following prototypes:

```
void startservent( int stayopen );
void endservent( void );
struct servent *getservent( void );
```

We first call **startservent** to begin our enumeration, specifying by default an argument of zero to allow the services file to be closed. This starts us off at the beginning of the services table. Subsequent calls to **getservent** return the next services entry in the table. After we've enumerated all of the services, a call to **endservent** halts the enumeration (and closes the services file if we've requested that it stay open with **startservent**).

The example shown in Listing 6.18 illustrates enumerating the service entries of the services table. After starting our enumeration with **startservent**, we simply walk through the table using **getservent**, emitting some information about the service for each one found. When completed, we call **endservent**.

The information we emit is the service name (official name), port number used by the service (for the given protocol), and finally the protocol that should be used to access the service (such as "tcp" or "udp").

Listing 6.18 Enumerating the services table (servenum.c).

```
#include <netdb.h>

main()
{
  struct servent *sp;

  /* Start the service enumeration */
  setservent( 0 );

  while (1) {

    /* Get the next service row */
    sp = getservent();

    if (sp != NULL) {

      printf("Service %s is at port %d for %s\n",
              sp->s_name, ntohs(sp->s_port), sp->s_proto);

    } else break;

  }
```

```
  /* End the service enumeration */
  endservent();

}
```

It's important to note that this function simply enumerates the services table and does not necessarily mean that these services are available on the given node (or any other node for that matter). What these functions provide is the ability to specify a service name and then identify the port number and protocol that may be used to access the service.

SUMMARY

In this chapter, we investigated a variety of advanced features not typically encountered in everyday Sockets applications. Some of the techniques discussed include out-of-band data with TCP, using nonblocking socket semantics, how to determine peer socket closure, writing services for multi-homed nodes, timing out connects, reads, and writes using the **select** function, data framing with TCP and UDP, and identifying service ports and protocols.

REFERENCES

[Stevens98] Stevens, W. Richard, *UNIX Network Programming,* 2nd edition, Volume 1, Prentice Hall PTR, 1998.

7 Optimizing Sockets Applications

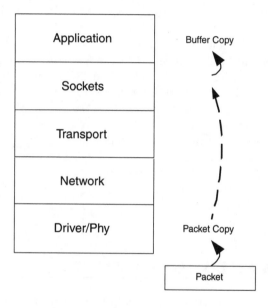

I n this chapter, we investigate a variety of topics that provide mechanisms to increase the efficiency and performance of Sockets applications. Many of these methods take advantage of a basic understanding of the Sockets API, whereas others utilize and understand the networking architecture and communication principles. Finally, other methods take advantage of specialized features of TCP/IP stacks.

INTRODUCTION

A number of disparate topics are discussed in this chapter, each useful whether or not performance is an important aspect of the Sockets application. The optimized Sockets programming topics that are discussed in this chapter include:

- Ensuring full-sized segments
- Optimizing the send and receive buffer sizes
- Minimizing latency with **TCP_NODELAY**
- Reading/writing all available data
- Advantages and disadvantages to `select`
- Using multiple sockets
- Separating control and data connections
- Using stack callbacks
- Using zero-copy buffer functions
- Using UDP instead of TCP
- Stack selection
- Stack configuration

ENSURING FULL-SIZED SEGMENTS

Let's do a quick review to understand the principles at work in this optimization. A datagram on the network is made up of a set of encapsulating packet headers followed by a data payload. The size of the data payload is based upon the particular protocol (we focus here on IPv4), the configuration of the two communicating peers, and the configuration of the intermediate nodes between the two peers.

The IPv4 protocol specifies that a datagram can be of size 64 KB, though this is generally much larger than is normally allowed. The most common MTU (or Maximum Transmission Unit) is 1500 octets. Depending upon the options specified within the IP and/or TCP packet headers, the size of the payload may be around 1448 octets. The largest payload size given the MTU is also known as the MSS (or Maximum Segment Size). This section focuses on ensuring that both the MTU and MSS are the largest supported by the given network and configuration.

The MTU can commonly be configured at the interface. This is known as the Link MTU. The application developer can modify the MSS via a socket option. The most common method to alter the MTU (and MSS) is known as Path MTU Discovery. This method is performed automatically within protocol stacks that support the option. The advantage to Path MTU Discovery is that the method identifies the Path MTU, which should not be confused with the Link MTU. The Path MTU is the largest MTU possible for the path between the two communicating peers on the network. Another important distinction for Path MTUs is that they're only relevant on indirect routes. An indirect route is a path that is not on the same subnetwork. If two peers on are on the same network (can communicate directly with one another without going through a gateway, router, etc.), then they have a direct route to one another. Otherwise, if the path must go through a router, switch, and so on, then the route is indirect. Path MTUs only exist for indirect routes; therefore, Path MTU Discovery isn't possible for peers on the same subnetwork.

Path MTU Discovery works by sending varying sized packets from peer to peer with a special bit set in the IP header called DF, or Don't Fragment. This bit specifies that the datagram should not be fragmented during its trip from peer to peer. If the packet is too large for the Path MTU, then the packet is rejected with the DF bit set (otherwise, the interior node would split the packet up so that it fit within the constraints of the interior node's Link MTU). If the packet is rejected, the originating peer receives a special ICMP message specifying that the packet was too large. The peer then reduces the size of the datagram and tries again until it finds the largest packet possible that makes it from the originator to the recipient. This is the Path MTU and is optimal for the given connection. The Path MTU defines the largest datagram, and the largest segment size that can travel between the two peers, and is, therefore, the optimal MSS for the connection.

The application developer should first check to ensure that their TCP/IP stack supports Path MTU Discovery, and then ensure that it is enabled. Some stacks provide socket option enables on a per-connection basis, but most enable by default if the option is supported.

OPTIMIZING THE SEND AND RECEIVE BUFFER SIZES

One of the most important optimizations that can be made for high-speed Sockets applications is the tuning of the socket send and receive buffer sizes. The **SO_SNDBUF** and **SO_RCVBUF** socket options provide the means to adjust these buffers; the remainder of this section provides some background and discusses how to identify the optimal size of these buffers.

Socket Buffer Sizes and TCP

Although the socket buffer sizes might appear to be detached from the protocol stack, they actually modify the operation of the TCP layer for the receive path. Let's first review flow control for TCP for the receive path.

When TCP initiates a connection, each side advertises a window. The window is the maximum amount of data that it can receive at a given time without the receiver of the data consuming it. Therefore, if a peer advertises a window of 8 KB, and receives 8 KB of data, the sender can transmit no further data until the data is consumed at the receiver. As data is received, the window shrinks to communicate how much new data can be received. This advertisement occurs in each packet sent from the receiver to the sender. The size of this advertised window is the size of the buffer configured at the Sockets layer with the **SO_RCVBUF** socket option.

Conversely, the sending peer also advertises its window for receive data. This is part of flow control for the send path. TCP also maintains another window called the congestion window that operates within the advertised window. Whereas the advertised window provides an upper bound for transmit data, the congestion window provides an optimized window given detected congestion on the link. The size of the send buffer at the Sockets layer is configured with the **SO_SNDBUF** socket option.

Socket Buffers and Performance

What is so fundamentally important about the size of the socket buffer is that it defines how much data can be sent to a peer before expecting an acknowledge to shift the window and send additional data. Consider what happens if we have a socket buffer the same size as our segment (packet payload size). We'd send one packet, and then await a response (acknowledge) from the peer before being able to send another. This lock-step operation isn't very efficient, considering the time that it takes for the peer to receive the packet and then return an acknowledge. Now consider a socket buffer that is the size of two packets. We emit two packets, and then await one (or possibly two) acknowledgments from the peer. In roughly the same amount of time, two packets have been transmitted instead of one. If one considers long-haul networks (geographically distant nodes with very large round-trip times), quite a bit of data could be sent before the peer actually received any.

Another way to think about the socket buffer size is that its size is similar to the amount of data that can be sent before the peer receives it (the bit length of the pipe). If we can tune these to be similar in size, then we have an optimal setting for both transmit and receipt. This concept is known as the "Bandwidth-Delay Product" (BDP), which is the bandwidth of the particular link connecting the two peers times the round-trip time of the connection.

Let's now look at an example of BDP in practice. Consider a 10-Mbps link with a round-trip time of 35 ms. That's ~43KB for an optimal socket buffer size to maintain a full pipe for the given connection. Given the standard of 8 KB in many systems, the default would be suboptimal for this interface and round-trip time.

Configuring the Socket Buffer Sizes

After the optimal socket buffer sizes are known, it's relatively easy to configure the buffer sizes. Note that this must be done prior to a TCP socket going into the connected state, which means prior to an **accept** for servers and prior to **connect** for clients. An example of the socket buffer configuration is shown in Listing 7.1.

Listing 7.1   Configuring the socket buffer sizes.

```
int sock, ret;
int size;

sock = socket( AF_INET, SOCK_STREAM, 0 );

...

size = 44000;
ret = setsockopt( sock, SOL_SOCKET, SO_SNDBUF,
                    &size, sizeof( size ) );

ret = setsockopt( sock, SOL_SOCKET, SO_RCVBUF,
                    &size, sizeof( size ) );
```

If the round-trip time (RTT) can't be determined prior to connection setup, another method is to sample the link to identify what an average RTT will be and then use this statically with the link speed for a socket buffer size using the BDP calculation.

Other Options

Proposals exist to provide an auto-tuning mechanism for socket buffer sizes. Many of the proposals include ICMP-based testing mechanisms to identify the RTT, and they utilize this to set the socket buffer sizes for a given connection. In many cases, setting the socket buffers to known optimistic values works just as well.

MINIMIZING LATENCY WITH TCP_NODELAY

In trying to increase the performance of certain networked applications, reducing latency can go a long way in this endeavor. This particular topic has a number of caveats; it's not a panacea for nonperforming applications.

Consider an Application layer protocol that emits smaller requests to a peer. As the TCP layer receives these requests, it buffers them up for at most 200 ms before emitting them as an aggregated datagram. This means that in the worst case, 200 ms will pass before the request is finally emitted from the stack, from the point when it was introduced by the Application layer. From a performance perspective, the introduction of this latency can have very drastic effects on the throughput of this particular connection.

To resolve the latency problem, the **TCP_NODELAY** option can be used. Recall from Chapter 5, Socket Options, that **TCP_NODELAY** is used to disable the Nagle algorithm (used to minimize the number of small packets that are emitted on the network). Small packets are also bad for performance, but in this particular case, sending a smaller packet is not as bad as holding it for 200 ms (especially when considering higher bandwidth Gigabit connections). Therefore, we'll forego the inefficiency of smaller packets for the sake of getting them out on the wire quickly.

Consider the example shown in Figure 7.1. In this illustration, we see a maximum of 200 ms between the introduction of the data at the Sockets layer until the data is emitted from the TCP layer (to IP).

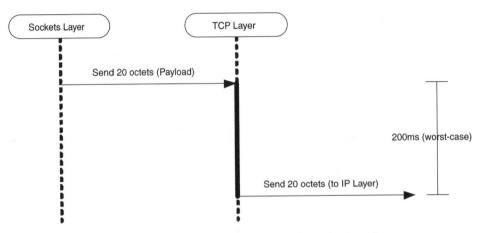

FIGURE 7.1   Example of the potential latency from TCP's Nagle algorithm.

In Figure 7.2, with Nagle disabled (via **TCP_NODELAY**), much less time elapses, as there is no aggregation to minimize small packets. In most cases, the data is pushed through to the IP layer immediately.

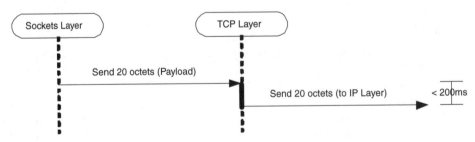

FIGURE 7.2   Example of improved (reduced) latency through disabling Nagle.

The purpose of TCP's Nagle algorithm was to minimize small packets on the network that induced poor performance. For certain applications, in which we know we'll only be emitting a small amount of data before awaiting a response, disabling Nagle can yield significant performance improvement. If the application sends numerous data in a number of unique **write** calls, this option can lead to much degraded performance, and should be avoided.

READING/WRITING ALL AVAILABLE DATA

The basis for this optimization is that the least number of calls should be made to move data between the Application and Sockets layer. This is due to a number of concerns, including user/kernel switches and the cost of buffer copies between the user layer and the kernel layer.

Consider the code in Listing 7.2, which illustrates a number of calls to **write** a series of data through a stream socket.

Listing 7.2   Multiple **writes** for a series of data.

```
int sock, ret;
char cmd1[MAX_CMD_BUFFER+1];
char cmd2[MAX_CMD_BUFFER+1];
char cmd3[MAX_CMD_BUFFER+1];
char resp[MAX_RESP_BUFFER+1];
int cmd1_len, cmd2_len, cmd3_len;

...

sock = socket( AF_INET, SOCK_STREAM, 0 );

...

/* Commands are formatted into the three command arrays */
/* Lengths are stored into cmd?_len */

/* Send the first command */
ret = send( sock, cmd1, cmd1_len, 0 );

/* Send the second command */
ret = send( sock, cmd2, cmd2_len, 0 );

/* Send the third (and final) command */
ret = send( sock, cmd3, cmd3_len, 0 );

/* Await the response */
ret = recv( sock, resp, MAX_RESP_BUFFER, 0 );
```

From an execution standpoint, three **send** calls are made that, depending upon the target operating system can result in transitions between user and kernel space. Each call results in the transition plus a copy of a command between user and kernel space for subsequent transport through the socket. Because TCP is a stream protocol, there is no distinction made to the framing of the data that was sent. Therefore, the receiver could receive **cmd1_len** octets or the total of all three (depending upon how the sender's protocol stack aggregated the data).

Listing 7.2 may be readable and more accurately mirror the intent of the developer, but rather than suffer the performance hit of the three potential kernel calls and three independent buffer copies between user space and kernel space, an alternative is to perform a single **send** call (as shown in Listing 7.3).

Listing 7.3 Single write for higher performance.

```
int sock, ret;
#define MAX_CMDS_BUFFER    (MAX_CMD_BUFFER*3)
char cmd1_2_3[MAX_CMDS_BUFFER+1];
char resp[MAX_RESP_BUFFER+1];
int cmd1_len, cmd2_len, cmd3_len;

...

sock = socket( AF_INET, SOCK_STREAM, 0 );

...

/* Commands are formatted into a single command array */
/* Lengths are stored into cmd?_len */

/* Send the commands */
ret = send( sock, cmd1_2_3,
            (cmd1_len+cmd2_len+cmd3_len), 0 );

/* Await the response */
ret = recv( sock, resp, MAX_RESP_BUFFER, 0 );
```

The advantage to Listing 7.3 over Listing 7.2 is that we perform only one kernel call and rather than three independent buffer copies between user space and kernel space, we perform only one. This has significant performance savings, especially when this activity is performed often.

ADVANTAGES AND DISADVANTAGES TO SELECT

The **select** call in most operating systems is notoriously slow. In most implementations (except for Windows), the **select** call can be used to multiplex I/O not only for socket descriptors, but also for any other type of file descriptor (such as files, serial ports, and so on).

The **select** call is most commonly used to identify the ability to read, write, or identify whether an exception occurs on a set of descriptors. Upon one or more of these events occurring on a set of descriptors, the **select** call returns with this set information to the application so that it can then enumerate through the set and take appropriate action. The advantage to the **select** call is that a single thread can manage a large set of sockets compactly.

Rather than use the **select** call, another approach is to utilize a thread of control for each descriptor. This comes with the cost of context switches for each of the threads, but reduces the latency of I/O operations dramatically.

USING MULTIPLE SOCKETS

A very interesting technique, used effectively in older TCP/IP stacks, is the use of multiple sockets to parallelize data transfer. The technique is also known as "data-striping." In data striping, multiple sockets are opened to a given target and individual segments of data are transferred through each of the open sockets (see Figure 7.3).

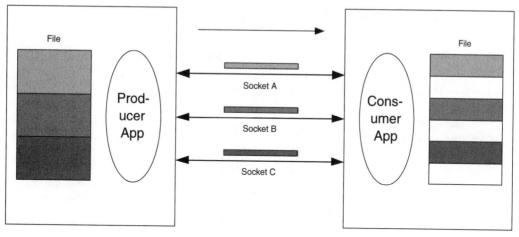

FIGURE 7.3 Data striping over multiple connections.

This technique has also been used successfully in peer-to-peer sharing programs in which a single file is downloaded in pieces from a number of different sources. This permits a user with a high-bandwidth link to download a file from a number of users with slower links, giving the illusion that the file is being downloaded from a single user with a similar link rate.

SEPARATING CONTROL AND DATA CONNECTIONS

A very interesting technique that has been used in production Internet protocols works by separating control and data communication to separate sockets. For example, FTP separates the data transfer from the telnet-like control interface over which the user communicates (see Figure 7.4).

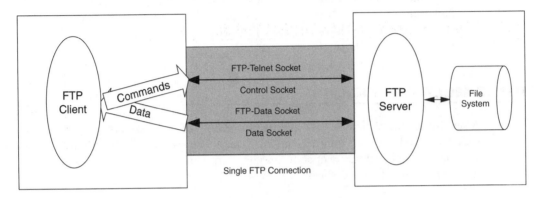

FIGURE 7.4 Illustrating separate control and data connections.

The advantage of independent control and data connections is that the characteristics of the two connections are different, and, therefore, the Sockets application can configure the sockets differently. For example, the control connection will likely emit small requests that will very likely not fill in a packet. Therefore, the **TCP_NODELAY** socket option can be used on the control connection to reduce the latency of small packet transmission (disabling Nagle). As the data connection will very likely transfer large amounts of data, Nagle will remain enabled to make best use of the available bandwidth.

Additionally, the real-time characteristics of the control and data connections will be different (with respect to round-trip times, and so on). Therefore, independent connections provide better bandwidth utilization, as TCP will automatically tune its parameters for the given connection usage.

USING STACK CALLBACKS

Rather than rely on blocking semantics for socket events (such as data available for read), alternatives exist that can provide higher performance. Numerous stacks provide the ability to install user-provided functions for the stack to call when events occur for a given socket. For example, a special user-receive function could be installed for a given socket that would be called when read data was available for the socket.

The greatest advantage to stack callbacks is the reduction of latency in receipt of data to its migration to the Application layer (as the callbacks exist with the application). After data is pushed from the Transport layer (TCP or UDP) to the Sockets layer, the callback is immediately called for the given socket and the data is then available at the Application layer. This can represent a huge performance boost.

Not all Sockets implementations provide callbacks, but when they do, they should be utilized for efficiency. An example stack that provides callback functionality is the Treck stack from Elmic Systems [Elmic].

USING ZERO-COPY BUFFER FUNCTIONS

Buffer copies within protocol stacks are commonly one of the largest performance drains. In a typical TCP/IP stack, a packet is copied from the physical device and then again, between the Sockets layer and the Application layer. Those two copies waste the computing capacity of the processor and should, therefore, be avoided (see Figure 7.5). Newer TCP/IP stacks provide a zero-copy option whereby the initial (potential) buffer copy targets a buffer in the application space. The application then simply passes the pointer of the buffer to the Application layer and no copy is performed.

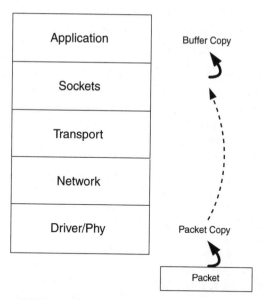

FIGURE 7.5 Illustrating buffer copies in a traditional TCP/IP stack.

Newer protocols such as RDMA (Remote Direct Memory Access) take this concept to the next level by not only providing buffers to the stack, but also providing a means to reserve them from the peer.

USING UDP INSTEAD OF TCP

For developers who are implementing new protocols, the trade-offs of using UDP rather than TCP should be considered. TCP requires connection establishment; therefore, the start-up time for a connection can be considerably longer than that of UDP. TCP also includes a number of features for reliability that weigh down the protocol and its ability to perform. These include congestion avoidance and retransmit. This is not to say that these features are not useful; they provide a very necessary function, but may be too heavy weight for certain applications.

The advantages of UDP over TCP are primarily in performance. Because UDP does not provide retransmit or reliable communication, these elements of the protocol are not provided in UDP and, therefore, a more efficient protocol implementation results. UDP also provides message framing, which can be a great benefit to applications. Rather than the application trying to deal with framing issues, the protocol automatically returns data in the granularity in which it was sent. Therefore, protocols that require message boundaries can yield significant performance benefits over TCP.

One particular arena in which UDP can play a great role is in applications that communicate only over LAN connections. The Internet is a hostile environment for UDP packets and the probability of packet loss increases dramatically compared to communicating over a local LAN. Therefore, if the application communicates solely over LAN connections, then UDP provides a number of advantages over TCP.

Another area in which UDP is important involves applications that can support some loss of data. Streaming audio applications are an interesting example. In TCP, when data is lost, the protocol backs off sending (for congestion avoidance) and then retransmits the old data. For audio, this presents a problem because what is important may not be in the past, but in the present (current audio, rather than audio in the past). With UDP, there is no reliability, and, therefore, lost data is simply lost. Protocols that utilize UDP commonly include sequence numbers within their higher-level protocol to identify when packet loss occurs, but typically do nothing about it from a lower-level protocol perspective.

STACK SELECTION

Although in most cases, selecting a new TCP/IP stack is not an option, for many embedded projects, the selection of a TCP/IP stack is a common effort. When selecting a new TCP/IP stack, there are a number of characteristics that must be satisfied for support of high performance.

One of the most important aspects of a high-performance TCP/IP stack is RFC-1323, or "Support for High Performance Extensions." This RFC defines support for large windows that are critical for high-bandwidth, long-haul networks. Another important RFC is RFC 2018, or "Selective Acknowledgment." This option is also critical for high performance networking in that it permits support for acknowledging out-of-sequence data. Finally, a standard RFC that should be part of any commercial TCP/IP stack is RFC 2001, "TCP Slow

Start, Congestion Avoidance, Fast Retransmit." These options were important for performance, but have undergone recent updates to increase their performance for very high bandwidth links (such as 10-Gigabit links).

STACK CONFIGURATION

In many embedded TCP/IP stacks, the ability to reconfigure them for small footprint or to simply reduce the functionality that's available can result in greater performance. For example, if only one interface is available for the protocol stack, removing the logic required for multiple interfaces can reduce footprint and increase performance. Configuration of upper-layer protocol options such as SACK can also lead to better performance. For example, SACK is less useful for applications that communicate over a LAN, so if your application does not require a WAN, this option may be compiled away for better performance.

Other configurations include support for jumbo frames. If an application can support larger MTU frames, better performance can result. Consider the amount of time required for processing packets at a variety of rates. For line-rate performance at 10 Mb/s, standard MTU packets (1500 octets) require 1.2 ms per packet (833 packets per second). For 9000 MTU packets (standard jumbo frame), only 138 packets per second are processed, or 7.246 ms per packet. This benefit becomes even more clear when operating at very high performance rates. Consider now 10 Gb/s links. Standard MTU packets require 1.2 us of processing per packet (833,333 packets per second). For jumbo frames at 10 Gb/s, 7.2 us are required per packet (at 138,888 packets per second). Clearly, increasing the packet size has a beneficial effect on the time required to process each packet, making it more likely that line rate can be supported.

SUMMARY

In this chapter, we investigated a variety of methods to increase the performance of Sockets API applications. The methods could be split into a few different categories, such as Application layer modifications (adjusting socket buffer sizes for the bandwidth delay product), TCP/IP stack configuration, and simple design choices (such as using UDP over TCP when possible). Luckily, the Sockets API provides a variety of choices for optimizing Sockets applications. Most are very simple configurations (such as socket buffer size changes), whereas others require more thought about the effects of the change.

REFERENCES

[Elmic] Elmic Systems, "Turbo Treck Embedded TCP/IP Stack," available online at http://www.elmic.com.

8 Sockets Programming Pitfalls

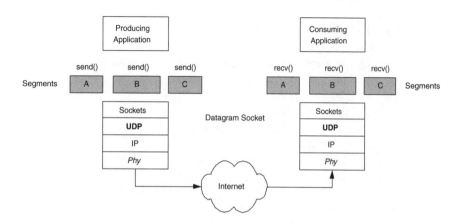

Although Sockets programming is conceptually simple, improperly using the API or not taking into account protocol behaviors can induce errors within network applications. In this chapter, we investigate some common and not-so-common pitfalls that can be found in common network applications, and then explain how to correct them.

INTRODUCTION

A number of programming pitfalls are discussed in this chapter. Some pitfalls concentrate on improper usage of the BSD API, whereas others explore protocol behavior effects on the API. The common pitfalls that are discussed in this chapter include:

- Failing to use function returns
- Ignoring peer socket closure
- Ignoring endianness of parameters in API functions
- Making protocol framing assumptions
- Causing "Address in use" error for **bind**
- Listening on multiple interfaces
- Initial UDP datagrams disappearing
- Defensive programming
    - Using safe functions to avoid buffer overflow
    - Rigorously checking error returns
    - Rigorously checking input and output parameters
    - Declaring string arrays
    - Minimizing protocol feedback
    - Initializing all variables
    - Enabling all compiler warnings

FAILING TO USE FUNCTION RETURN VALUES

This particular pitfall should be understood without question, but in many cases (even in some of the examples in this book), return values of the Sockets API functions are ignored leading to subsequent erroneous behavior by the application. Many of the BSD API functions return error indications, but some return other relevant information.

For example, consider this simple accept code snippet:

```
server = socket( AF_INET, SOCK_STREAM, 0 );

...

listen( server, SO_MAXCONN );

while ( 1 ) {
```

```
    clisock = accept( server, (struct sockaddr_in *)NULL, NULL );

    nread = read( clisock, buffer, MAX_BUFFER);

...
```

This code assumes that the client socket (`clisock`) returned by the **accept** function is a valid socket descriptor. The **accept** call could return an error, which would need to be dealt with very easily as:

```
server = socket( AF_INET, SOCK_STREAM, 0 );

...

listen( server, SO_MAXCONN );

while ( 1 ) {

  clisock = accept( server, (struct sockaddr_in *)NULL, NULL );

  if (clisock == -1) {

    printf("Error on accept (%d)\$$\n", errno);

  } else {

    nread = read( clisock, buffer, MAX_BUFFER);

...
```

A more complex example that relates to the underlying Sockets layer is provided with the **send** API function. The **send** function for stream sockets returns –1 upon error, or the number of octets that were actually sent. What this means is that if insufficient space exists for the socket's send buffer, the function may block or enqueue what it can and then return the count of bytes that were actually enqueued. The following code snippet makes use of the return value to ensure that all data is actually written:

```
/* count is number of bytes to send */
written = 0;

while (written < count) {

    sts = send( socket, &buffer[written], (count-written), 0 );

    if (sts == -1) break;
```

```
        else written += sts;

    }
```

This ensures that all data is written from buffer, regardless of the chunk rate of the underlying Sockets layer.

A final point to note is that error return behavior in blocking sockets must be treated differently than in nonblocking sockets. For example, in a nonblocking socket, an error return of EAGAIN from the **recv** call indicates that no data currently exists to read (would need to block). Similarly, for the **send** call, EAGAIN indicates that the call is unable to queue to data for transmission (likely because the socket's send buffer is full).

IGNORING PEER SOCKET CLOSURE

Ignoring peer socket closure is related to ignoring return statuses, but we cover this separately because it is one of the most common errors. The **read** function returns the number of octets that were read from a socket or a –1 if an error occurred. Additionally, a zero return commonly represents the end-of-file indication, which for a socket represents peer socket closure. The following code snippet illustrates this functionality:

```
bytesRead = read( clisock, buffer, BUF_SIZE );

if (bytesRead > 0) {

    /* Successful return, data in buffer */

} else {

    /* 0 or –1 returned, either way, close the socket */

}
```

The **read** function permits testing the socket for closure and once detected can produce the necessary local socket closure. Failure to deal with peer socket closure can result in subsequent error behavior that may not be easily traceable to the closure condition.

IGNORING ENDIANNESS OF PARAMETERS IN API FUNCTIONS

A common mistake in Sockets applications is not taking into account the parameters that must be presented in network byte order (particularly on hosts that utilize little-endian byte ordering). Recall that non-byte data on the Internet is in what is called network byte order, specifically big endian. On architectures that are little endian, such as Intel processors, bytes must be swapped in order to maintain consistency of control data (such as packet headers).

For this reason, the byte ordering functions **ntohs**, **htons**, **ntohl**, and **htonl** exist to provide the swapping capability. All network programs should use these, even if they're currently running on a big-endian architecture because porting them to run on little-endian architectures will then be much simpler. It should be noted that byte-swapping functions are NULL functions on architectures that are already big-endian.

One of the most common errors that can be found in network applications comes in defining a socket address, such as:

```
struct sockaddr_in saddr;

memset( &saddr, 0, sizeof(saddr) );

saddr.sin_family = AF_INET;
saddr.sin_port = 80;
saddr.sin_addr.s_addr = inet_addr( "192.168.1.1" );
```

At first glance, this socket address appears to specify the interface identified by IP address "192.168.1.1" and port 80. The problem is that the sin_port (and sin_addr) must be in network byte order. On a little-endian architecture, the port bound here is actually 20480 (due to improper byte ordering). This bug is easily fixed by using the **htons** function (host-to-network-short).

```
struct sockaddr_in saddr;

memset( &saddr, 0, sizeof(saddr) );

saddr.sin_family = AF_INET;
saddr.sin_port = htons(80);
saddr.sin_addr.s_addr = inet_addr( "192.168.1.1" );
```

Note that the **inet_*** functions (**inet_addr**, **inet_ntoa**, **inet_aton**) automatically perform the byte swapping internally and, therefore, their result is in the proper order. Although this discussion is true for the C language, it may not actually apply to other languages discussed in this text.

MAKING PROTOCOL FRAMING ASSUMPTIONS

Framing protocols are those that maintain message boundaries for messages sent through a socket connection. TCP, which is a stream-oriented protocol, maintains no framing on the data sent, which means that data may be accumulated en route to the destination (see Figure 8.1). The application can't guarantee that the data sent with the **send/write** function is the same granularity that will be read by the peer. UDP, which is a datagram-oriented protocol, maintains the message boundaries for transmitted data. This means that the unit of data sent in the **send/write** function is the same unit that is read by the peer (see Figure 8.2).

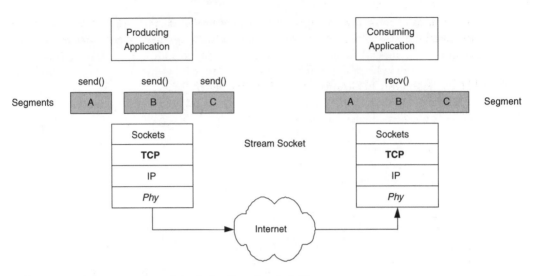

FIGURE 8.1 Demonstration of the lack of framing in TCP.

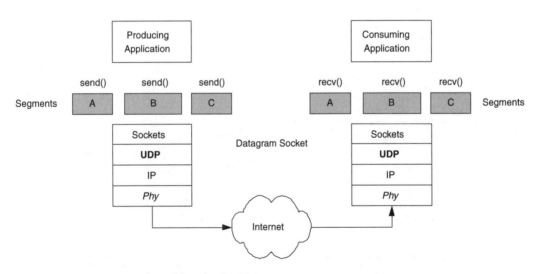

FIGURE 8.2 Demonstration of framing in UDP.

The framing characteristic of UDP makes it an ideal protocol for message-based protocols. TCP could be used, but the Application layer would be required to perform the framing. The lesson to be learned here is that unless the protocol supports framing (such as UDP or SCTP), one cannot assume that the granularity of data sent is the same that will be received. As illustrated in Figure 8.1, the blocks of data sent can be accumulated together (within the given MSS) by the sender or receiver TCP layer.

CAUSING "ADDRESS IN USE" ERROR FOR Bind

When a socket is opened and a local address is bound to it, the act of closing this socket causes it to enter the TIME_WAIT state. The purpose of this state is to ensure that the local address is not immediately used again, with the possibility of confusing the local stack by a remote socket that still believes that it's connected. By waiting TIME_WAIT seconds (typically 120), one can be reasonably assured that remote connections will be closed as well.

The problem that TIME_WAIT creates is that the local address previously bound to the socket cannot be bound again within the TIME_WAIT period. Therefore, one must wait for the TIME_WAIT timeout to occur, or use the SO_REUSEADDR socket option. This option allows a socket to locally bind to the local address in the TIME_WAIT state. To set this option for a given socket, the following code snippet can be used:

```
int socket, on, ret;

socket = socket( AF_INET, SOCK_STREAM, 0 );

on = 1;

ret = setsockopt( socket, SOL_SOCKET, SO_REUSEADDR,
            &on, sizeof(on) );
```

LISTENING ON MULTIPLE INTERFACES

A common problem with the use of the **bind** call is what the address represents in a server socket scenario. Consider the following example:

```
int servsock, ret;
struct sockaddr_in saddr;

servsock = socket( AF_INET, SOCK_STREAM, 0 );

memset( &saddr, 0, sizeof(saddr) );
saddr.sin_family = AF_INET;
saddr.sin_port = htons( 80 );
saddr.sin_addr.s_addr = inet_addr( "192.168.1.1" );

ret = bind( servsock, (struct sockaddr_in *)&saddr,
            sizeof( saddr ) );
```

The developer has identified that the available Ethernet interface is configured with the IP address "192.168.1.1". When the developer directs a client on the external network to the server using this snippet over the interface identified by "192.168.1.1", all is well. However, what happens now if the address of the interface changes, or another interface is added? This lacks both portability and scalability. If the interface changes, all is lost as the **bind** will very likely fail. Additionally, external client sockets may not be able to access the server, even if the primary interface remained the same address ("192.168.1.1"). This is because the **bind** operation ties the server to the specific interface. Therefore, client connections may only connect through the interface having been configured with the address "192.168.1.1". If the developer desires that all available interfaces be capable of routing client connections, the **inet_addr** line may be changed as follows:

```
saddr.sin_addr.s_addr = htonl( INADDR_ANY );
```

The INADDR_ANY symbolic (also known as the wildcard address) permits incoming connections from all available interfaces.

INITIAL UDP DATAGRAMS DISAPPEARING

A commonly seen behavior with UDP sockets (typically in embedded environments) consists of an initial set of UDP datagrams disappearing when communication first begins with a new peer. Typically, a set of UDP messages is emitted quickly upon boot of the TCP/IP stack, but the first few (depending upon the rate of transmission) are lost. Hooking up a trace tool shows that the packets are never actually seen being emitted from the interface (recall that UDP is an unreliable protocol).

What is happening here is a side effect of ARP, the Address Resolution Protocol. The lowest-level protocol on an Ethernet network is the Ethernet protocol. The Ethernet protocol, which provides for local communication between nodes on an Ethernet LAN, utilizes a 48-bit address that is unique across the entire world. ARP takes care of taking an IPv4 address and then identifying the Ethernet address to which datagrams should be directed. ARP is a discovery process, which commonly operates when the address is required. Therefore, when the first UDP datagram arrives for transmission, ARP sends out an ARP request to identify the Ethernet address for the given IPv4 address. This may take some number of milliseconds before the response arrives. Once the ARP reply is received, the datagram is loaded with the Ethernet address and the packet is emitted onto the wire.

During this period that ARP is awaiting a reply, our initial datagram is stalled. Now, what happens when an additional datagram is emitted on the socket while the ARP reply is still outstanding? Commonly, the UDP layer destroys the datagram and replaces it with the current. This means that all UDP datagrams that are written before the ARP response is received are destroyed, leaving only the last datagram that was provided to the stack before the ARP request was satisfied.

DEFENSIVE PROGRAMMING

Defensive programming is the practice of trying to anticipate where errors can occur in programs, and then adding code to identify or work around the issue to avoid program failures or security holes. Not only does this make programs more reliable and secure, but in the early stages, it makes debugging the programs easier. In many cases, an error isn't apparent where it actually occurred, but where the error was propagated in the system.

Because this practice is so important, a variety of techniques are collected and discussed here together.

Using Safe Functions to Avoid Buffer Overflow

Buffer overflows (or overruns) are a huge concern in the security industry because in some cases they can permit external applications to take control of a system. A buffer overrun occurs when more space is used for a given resource than was actually reserved for it by the application. Buffer overruns occur not only in heap variables, but also for stack variables, allowing the stack to be manipulated by a devious application.

Let's look at an example of how a buffer overrun can occur. Consider the following fragment of code that is used to extract a username from a buffer received through a socket:

```
void getUsername( ... ) {

  char buffer[40];

  retBytes = recv( sock, buffer, 256, 0 );

  if (!strncmp(buffer, "Username: ", 10)) {

    strcpy( username, buffer[10] );

    ...
```

What happens in the previous code if the length of the username in the buffer is greater than the previous 40 bytes reserved for it? If this happens, a buffer overrun writes over whatever follows the buffer's allocation in memory (in this case, the stack). We could correct this situation as follows, using both error checking and the safe version of strcpy:

```
void getUsername( ... ) {

  char buffer[MAX_BUFFER];
  retBytes = recv( sock, buffer, 256, 0 );

  if (!strncmp(buffer, "Username: ", 10)) {

    if ( strlen(&buffer[10]) ) {
```

```
      printf("username overflow detected in getUserName\$$\n");

   }

   /* Truncate the username... */
   strncpy( username, &buffer[10], 38 );
   buffer[39] = 0;

   ...
```

In this case, we detect the overflow situation and alert the user via a `printf`. Additionally, we use the safe `strcpy` function called `strncpy` that includes a third argument defining the maximum number of characters that can be copied into buffer.

Other safe functions exist with the standard C library including `snprintf` (for `sprintf`), `strncmp` (for `strcmp`), `strncat` for `strcpy`, and others. In all cases, the safe functions should be used instead.

Rigorously Checking Error Returns

This item was covered previously in this chapter, but from the perspective of defensive programming, it is one of most important from the aspect of internal consistency. Failing to recognize error returns from internal functions makes debugging very difficult because isolating an error that has been propagated through other functions is never easy.

Rigorously Checking Input and Output Parameters

In addition to error returns, a function should validate both what it receives as well as what it provides. From an error propagation perspective, debugging can be made simpler by identifying where an error first occurs rather than trying to debug the later effects of the error. C provides the **assert** function that can be used to detect and abort on erroneous situations. Other less catastrophic mechanisms can be used as well. Consider the following example:

```
int validateUser( char *username ) {

  int ret = -1;

  assert( username );

  // or...
  if ( username == NULL ) {
    printf("validateUser:  NULL received as input\$$\n");
    return( -1 );
  }
```

```
...

if ( ret == -1 ) {
  printf("validateUser:  Returning failure for %s\n", username );
}

return( ret );

}
```

The application could do an **assert** if a NULL username was provided, or simply test the username and indicate the situation to the user, in addition to returning an error. Later, we could identify when a function failure occurs, and identify this before exiting the function (just in case the caller does not check the return status).

Note that some languages, such as Eiffel, provide language features for this type of checking called interface contracts.

One additional point here is the validation of the internal consistency of a function or application. Emitting error messages when erroneous switch default statements are encountered, or entering nonexistent states within an internal state machine should emit errors to immediately localize any errors.

Declaring String Arrays

A very simple technique that is missed in many applications is the usable size of a declared string. If a developer creates a string of size 10 bytes, only 9 bytes are actually usable by the developer as 1 byte will be used for the NULL terminator. If the string fills up, it's very easy then to overrun the buffer.

A simple solution to this problem is shown in the following code:

```
#define MAX_BUFFER_SIZE          10

char myString[MAX_BUFFER_SIZE+1];
```

We declare the size that we want for our buffer as a symbolic constant (MAX_BUFFER_SIZE) and then when we declare our actual string, we simply use this constant and add one to it. We can now use the MAX_BUFFER_SIZE constant in our application to identify the maximum size of the buffer, but because we've added one to the buffer size, the NULL is taken care of and a potential error is removed.

Minimizing Protocol Feedback

Where possible, minimizing feedback from applications that identify version information should be performed. If an external application can identify the version of your networked application, they can then take advantage of any known exploits.

Consider the SSH protocol (secure shell). If an external hacker telnets to the SSH port on a host, it immediately emits the version information, which can then be used to identify if any exploits exist.

```
telnet theirdomain.com 22
Trying 192.168.1.1...
Connected to theirdomain.com.
Escape character is '^]'.
SSH-2.01-OpenSSH_4.1.3p2
```

The purpose of this version information could be to identify the server to the client so that it knows if it's compatible, or how to tailor the conversation for a given scenario. The issue is that the hacker then knows the version and can work to exploit it with known version-specific exploits.

This issue exists not only in SSH, but in NNTP, HTTP, SMTP, and many other Application layer protocols.

Not emitting version information won't stop an external hacker trying to take advantage of an exploit, but it will make it more difficult for them because the information won't be immediately available.

Initializing All Variables

Although this seems to be a simple problem, it's still very widespread. If a variable isn't initialized, the default comes from the current contents of the location. Consider the following contrived example:

```
int validateUser( char *username ) {

  int validated;

  // Test the username with known users...
  // set validated to one if trusted

  return( validated );

}
```

Because the validated variable isn't initialized, the function can erroneously return a validated status regardless of whether the user was actually trusted. This is because the calling application can load the location in the stack that will be validated with a non-zero value, then call it knowing what it will return. This is, of course, a contrived example, but illustrates how an application can manipulate other functions.

Enabling All Compiler Warnings

A final suggestion for developers is to always enable compiler warnings when building the application (or enable warnings for the interpreter, as is done with 'perl -w'). Many problems can be found, such as a failure to initialize variables, in the compilation stage as long as warnings are enabled.

SUMMARY

In this chapter, we explored a number of potential pitfalls that are faced by both beginner and advanced network application developers. The first set of pitfalls focused on improper usage of the BSD Sockets API and behavioral side effects of protocols. The second set focused on basic coding issues that give rise to security exploits and software failures. For each pitfall, a means was discussed to correct the error. Further, defensive programming techniques were discussed to provide the means to avoid security exploits and externally induced software failures.

RESOURCES

Wheeler, David A., "Secure Programming for Linux and Unix HOWTO," available online at *http://www.dwheeler.com/secure-programs/Secure-Programs-HOWTO/index.html*

PART II

Sockets Programming from a Multi-Language Perspective

The second part of this book is devoted to Sockets programming from the perspective of a variety of different languages. Each of the six languages under investigation are provided individual treatment in a unique chapter. The languages to be discussed are C, Java, Python, Perl, Ruby, and TCL.

Each chapter includes an overview of the language itself, including the language's origin and heritage. The compiler, interpreter, and tools used within the chapter are also identified, including their versions and where to download.

The networking API for the language is finally discussed, bringing attention to how sockets are created, identified, and destroyed. Special attention is then given to Internet addresses, because they are treated very differently depending upon the language being investigated. The available socket primitives, as they relate to the BSD API are also covered, to identify their similarities and differences to the BSD standard.

Other topics that are discussed include input and output, available socket options, and other miscellaneous functions such as asynchronous notification.

Finally, specialized networking APIs for the given language is introduced, including standard packages for HTTP and SMTP clients. When applicable, other modules are also discussed where specialization is possible.

9 Network Programming in the C Language

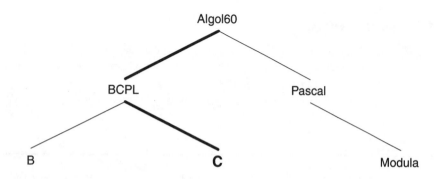

C LANGUAGE OVERVIEW

In this chapter, we investigate network programming in the C language. The C language provided our base language from which we analyzed the Sockets API.

The C language is a general-purpose, imperative language that is commonly used in Unix systems and is the lingua franca of embedded software development.

Origin of the C Language

The C language was originally developed for the Unix operating system by Dennis Ritchie and Brian Kernighan as a systems programming language. Since its inception in the early 1970s, it has evolved to be one of the most popular computer languages of all time.

C Language Heritage

The C language was derived from another language called BCPL. BCPL was the "Basic Combined Programming Language." BCPL, like C, was a block-structured, procedural language used for systems programming.

The C language is now defined as an ANSI standard under ISO/IEC 9899.

TOOLS

In this section, we discuss the compiler used to compile the examples, the networking API, and where to download.

Compiler/Tools Used

ON THE CD

To demonstrate Sockets programming in the C language, we utilize the GNU C compiler suite called GCC. The sample code patterns presented (in Chapter 16, and on the CD-ROM) will compile and run on any standard Linux distribution (Red Hat®, SuSE®, and others). These examples will also compile and run on the Cygwin (port of the popular GNU tools to the Windows environment).

Networking API Used

The current standard BSD4.4 Sockets library is presented here and is present by default in Linux distributions and under the Cygwin Unix environment.

Where to Download

The Sockets library is provided by default under Linux and other Unix operating systems. The Cygwin environment for Windows may be downloaded at *http://www.cygwin.com/*.

NETWORKING API FOR C

As we've seen much of the C language's networking API in Part I of this book, this chapter simply provides a summary of the API information to provide a pattern for future language chapters in this second part of the book.

Sockets API Summary

The networking API for C provides a mixed set of functions for the development of client and server applications. The available functions are shown in Figure 9.1.

Function	Description	Server	Client	Stream	Dgram	Blocking
socket	Create a new socket	•	•	•	•	
bind	Bind a name to a socket	•	•	•	•	
listen	Convert to a listening socket	•		•		
connect	Connect to a peer socket		•	•	•	•
accept	Accept a connection from a listening socket	•		•		•
send	Send data through a socket	•	•	•	•	•
sendto	Send data through a socket to a named destination	•	•		•	•
recv	Send data through a socket	•	•	•	•	•
recvfrom	Send data through a socket to a named destination	•	•		•	•
select	Notify upon socket event	•	•	•	•	•
close	Close a socket	•	•	•	•	
shutdown	Close a socket with user control	•	•	•	•	
getsockopt	Get the value of a specified socket option	•	•	•	•	
setsockopt	Set the value of a specified socket option	•	•	•	•	
getpeername	Get peer information about a socket	•	•	•	•	
getsockname	Get local information about a socket	•	•	•	•	
gethostbyname	Get network host address	•	•			•
gethostbyaddr	Get network host name	•	•			•
inet_addr	Convert a string address to a 32-bit address	•	•	•	•	
inet_aton	Convert a string address to a 32-bit address	•	•	•	•	
inet_ntoa	Convert a 32-bit address to a string address	•	•	•	•	

FIGURE 9.1 The BSD API for the C Language.

For each of the functions shown in Figure 9.1, its classification is shown for a number of different attributes. Some functions are used by only server-side sockets, whereas others are used solely by client-side sockets (most are available to both). Another axis defines whether the function exists for stream sockets (TCP) or datagram sockets (UDP). Finally, some of the functions block until a satisfying event occurs, whereas others will never block.

Sockets API Discussion

Because much of this material has been discussed in Part I of this book, we present a light treatment of the Sockets API here to serve as a model for later chapters that discuss languages other than C.

Creating and Destroying Sockets

A socket is necessary to be created as the first step of any socket-based application. The **socket** function provides the following prototype:

```
int socket( int domain, int type, int protocol );
```

The socket object is represented as a simple integer and is returned by the **socket** function. Three parameters must be passed to define the type of socket to be created. We're interested primarily in stream (TCP) and datagram (UDP) sockets, but many other types of sockets may be created. In addition to stream and datagram, a raw socket is also illustrated by the following code snippets:

```
myStreamSocket = socket( AF_INET, SOCK_STREAM, 0 );

myDgramSocket = socket( AF_INET, SOCK_DGRAM, 0 );

myRawSocket = socket( AF_INET, SOCK_RAW, IPPROTO_RAW );
```

The AF_INET symbolic constant defines that we are using the IPv4 Internet protocol. After this, the second parameter (type) defines the semantics of communication. For stream communication (using TCP), we use the SOCK_STREAM type and for datagram communication (using UDP), we specify SOCK_DGRAM. The third parameter could define a particular protocol to use, but only the types exist for stream and datagram, so it's left as zero.

When we're finished with a socket, we must close it. The close prototype is defined as:

```
int close( sock );
```

After close is called, no further data may be received through the socket. Any data queued for transmission would be given some amount of time to be sent before the connection physically closes.

Socket Addresses

For socket communication over the Internet (domain AF_INET), the sockaddr_in structure is used for naming purposes.

```
struct sockaddr_in {
    int16_t sin_family;
    uint16_t sin_port;
    struct in_addr sin_addr;
    char sin_zero[8];
};

struct in_addr {
    uint32_t s_addr;
};
```

For Internet communication, we'll use AF_INET solely for sin_family. Field sin_port
defines our specified port number in network byte order. Therefore, we must use **htons** to
load the port and **ntohs** to read it from this structure. Field sin_addr is, through s_addr, a
32-bit field that represents an IPv4 Internet address. Recall that IPv4 addresses are four-byte
addresses. We'll see quite often that the sin_addr is set to INADDR_ANY, which is the wildcard. When we're accepting connections (server socket), this wildcard says we accept connections from any available interface on the host. For client sockets, this is commonly left
blank. For a client, if we set sin_addr to the IP address of a local interface, this restricts outgoing connections to that interface.

Let's now look at a quick example of addressing for both a client and a server. First,
we'll create the socket address (later to be bound to our server socket) that permits incoming connections on any interface and port 48000.

```
int servsock;
struct sockaddr_in servaddr;

servsock = socket( AF_INET, SOCK_STREAM, 0);

memset( &servaddr, 0, sizeof(servaddr) );
servaddr.sin_family = AF_INET;
servaddr.sin_port = htons( 48000 );
servaddr.sin_addr.s_addr = inet_addr( INADDR_ANY );
```

Next, we'll create a socket address that permits a client socket to connect to our previously created server socket.

```
int clisock;
struct sockaddr_in servaddr;

clisock = socket( AF_INET, SOCK_STREAM, 0);
```

```
memset( &servaddr, 0, sizeof(servaddr) );
servaddr.sin_family = AF_INET;
servaddr.sin_port = htons( 48000 );
servaddr.sin_addr.s_addr = inet_addr( "192.168.1.1" );
```

Note the similarities between these two code segments. The difference, as we'll see later, is that the server uses the address to bind to itself as an advertisement. The client uses this information to define to whom it wants to connect.

Socket Primitives

In this section, we look at a number of other important socket control primitives.

Bind

The **bind** function provides a local naming capability to a socket. This can be used to name either client or server sockets, but is used most often in the server case. The **bind** function is provided by the following prototype:

```
int bind( int sock, struct sockaddr *addr, int addrLen );
```

The socket to be named is provided by the sock argument and the address structure previously defined is defined by addr. Note that the structure here differs from our address structure discussed previously. The **bind** function may be used with a variety of different protocols, but when using a socket created with AF_INET, the sockaddr_in structure must be used. Therefore, as shown in the following example, we cast our sockaddr_in structure as sockaddr.

```
err = bind( servsock, (struct sockaddr *)&servaddr,
                sizeof(servaddr));
```

Using our address structure created in our server example in the previous address section, we bind the name defined by servaddr to our server socket servsock.

Recall that a client application can also call bind in order to name the client socket. This isn't used often, as the Sockets API will dynamically assign a port to us.

Listen

Before a server socket can accept incoming client connections, it must call the **listen** function to declare this willingness. The **listen** function is provided by the following function prototype:

```
int listen( int sock, int backlog );
```

The sock argument represents the previously created server socket and the backlog argument represents the number of outstanding client connections that may be queued.

Within Linux, the backlog parameter (post 2.2 kernel version) represents the numbers of established connections waiting for by the Application layer protocol. Other operating systems may treat this differently.

Accept

The **accept** call is the final call made by servers to accept incoming client connections. Before **accept** can be called, the server socket must be created, a name must be bound to it, and **listen** must be called. The **accept** function returns a socket descriptor for a client connection, and is provided by the following function prototype:

```
int accept( int sock, struct sockaddr *addr, int *addrLen );
```

In practice, two examples of **accept** are commonly seen. The first represents the case in which we need to know who connected to us. This requires the creation of an address structure that will not be initialized.

```
struct sockaddr_in cliaddr;
int cliLen;

cliLen = sizeof( struct sockaddr_in );

clisock = accept( servsock,
                    (struct sockaddr *)cliaddr, &cliLen );
```

The call to **accept** will block until a client connection is available. Upon return, the clisock return value will contain the value of the new client socket and cliaddr will represent the address for the client peer (host address and port number).

The alternate example is commonly found when the server application isn't interested in the client information. This one typically appears as:

```
cliSock = accept( servsock, (struct sockaddr *)NULL, NULL );
```

In this case, NULL is passed for the address structure and length. The **accept** function will then ignore these parameters.

Connect

The **connect** function is used by client Sockets applications to connect to a server. Clients must have created a socket and then defined an address structure containing the host and port number to which they want to connect. The **connect** function is provided by the following function prototype:

```
int connect( int sock,
                (struct sockaddr *)servaddr, int addrLen );
```

The sock argument represents our client socket, created previously with the Sockets API function. The servaddr structure is the server peer to which we want to connect (as illustrated previously in the Socket Addresses section). Finally, we must pass in the length of our servaddr structure so that **connect** knows we're passing in a sockaddr_in structure. The following code shows a complete example of **connect**:

```
int clisock;
struct sockaddr_in servaddr;

clisock = socket( AF_INET, SOCK_STREAM, 0);

memset( &servaddr, 0, sizeof(servaddr) );
servaddr.sin_family = AF_INET;
servaddr.sin_port = htons( 48000 );
servaddr.sin_addr.s_addr = inet_addr( "192.168.1.1" );

connect( clisock,
         (struct sockaddr_in *)&servaddr, sizeof(servaddr) );
```

The **connect** function blocks until either an error occurs, or the three-way handshake with the server completes. Any error is returned by the **connect** function.

Sockets I/O

A variety of API functions exists to read data from a socket or write data to a socket. Two of the API functions (**recv**, **send**) are used exclusively by sockets that are connected (such as stream sockets), whereas an alternative pair (**recvfrom**, **sendto**) is used exclusively by sockets that are unconnected (such as datagram sockets).

Connected Socket Functions

The **send** and **recv** functions are used to send a message to the peer socket endpoint and to receive a message from the peer socket endpoint. These functions have the following prototypes:

```
int send( int sock, const void *msg, int len,
          unsigned int flags );

int recv( int sock, void *buf, int len,
          unsigned int flags );
```

The **send** function takes as its first argument the socket descriptor from which to send the msg. The msg is defined as a (const void *) because the object referenced by msg will not be altered by the **send** function. The number of bytes to be sent in msg is contained by the len argument. Finally, a flags argument can be used to alter the behavior of the **send** call. An example of sending a string through a previously created stream socket is shown as:

```
strcpy( buf, "Hello\n");
send( sock, (void *)buf, strlen(buf), 0);
```

In this example, our character array is initialized by the `strcpy` function. This buffer is then sent through `sock` to the peer endpoint, with a length defined by the string length function, `strlen`. To illustrate `flags` usage, let's look at one side effect of the **send** call. When **send** is called, it may block until all of the data contained within `buf` has been placed on the socket's send queue. If not enough space is available to do this, the **send** function blocks until space is available. If we want to avoid this blocking behavior, and instead want the **send** call to simply return if sufficient space was available, we could set the `MSG_DONTWAIT` flag, such as:

```
send( sock, (void *)buf, strlen(buf), MSG_DONTWAIT);
```

The return value from **send** represents either an error (less than 0) or the number of bytes that were queued to be sent. Completion of the **send** function does not imply that the data was actually transmitted to the host, only queued on the socket's send queue waiting to be transferred.

The **recv** function mirrors the **send** function in terms of argument list. Instead of sending the data pointed to be `msg`, the **recv** function fills the `buf` argument with the bytes read from the socket. We must define the size of the buffer, so that the network protocol stack doesn't overwrite the buffer, which is defined by the `len` argument. Finally, we can alter the behavior of the **read** call using the `flags` argument. The value returned by the **recv** function is the number of bytes now contained in the `msg` buffer or `-1` on error. An example of the **recv** function is:

```
#define MAX_BUFFER_SIZE        50
char buffer[MAX_BUFFER_SIZE+1];
...
numBytes = recv( sock, buffer, MAX_BUFFER_SIZE, 0);
```

At completion of this example, `numBytes` will contain the number of bytes that are contained within the `buffer` argument.

We could peek at the data that's available to read by using the `MSG_PEEK` flag. This performs a read, but doesn't consume the data at the socket. This requires another **recv** to actually consume the available data. An example of this type of read is illustrated as:

```
numBytes = recv( sock, buffer, MAX_BUFFER_SIZE, MSG_PEEK);
```

This call requires an extra copy (the first to peek at the data, and the second to actually read and consume it). More often than not, this behavior is handled instead at the Application layer by actually reading the data and then determining what action to take.

Unconnected Socket Functions

The **sendto** and **recvfrom** functions are used to send a message to the peer socket endpoint and receive a message from the peer socket endpoint. These functions have the following prototypes:

```
int sendto( int sock, const void *msg, int len,
                unsigned int flags,
                const struct sockaddr *to, int tolen );

int recvfrom( int sock, void *buf, int len,
                unsigned int flags,
                struct sockaddr *from, int *fromlen );
```

The **sendto** function is used by an unconnected socket to send a datagram to a destination defined by an initialized address structure. The **sendto** function is similar to the previously discussed **send** function, except that the recipient is defined by the to structure. An example of the **sendto** function is shown in the following code example:

```
struct sockaddr_in destaddr;
int sock;
char *buf;

...

memset( &destaddr, 0, sizeof(destaddr) );

destaddr.sin_family = AF_INET;
destaddr.sin_port = 581;
destaddr.sin_addr.s_addr = inet_addr("192.168.1.1");

sendto( sock, buf, strlen(buf), 0,
            (struct sockaddr *)&destaddr, sizeof(destaddr) );
```

In this example, our datagram (contained with buf) is sent to an application on host 192.168.1.1, port number 581. The destaddr structure defines the intended recipient for our datagram.

Like the **send** function, the number of characters queued for transmission is returned, or −1 if an error occurred.

The **recvfrom** function provides the ability for an unconnected socket to receive datagrams. The **recvfrom** function is again similar to the **recv** function, but an address structure and length are provided. The address structure is used to return the sender of the datagram to the function caller. This information can be used with the **sendto** function to return a response datagram to the original sender.

An example of the **recvfrom** function is shown in the following code:

```
#define MAX_LEN     100

struct sockaddr_in fromaddr;
int sock, len, fromlen;
char buf[MAX_LEN+1];

...

fromlen = sizeof(fromaddr);

len = recvfrom( sock, buf, MAX_LEN, 0,
                (struct sockaddr *)&fromaddr, &fromlen );
```

This blocking call returns when either an error occurs (represented by a −1 return), or a datagram is received (return value of 0 or greater). The datagram will be contained within buf and have a length of len. The fromaddr will contain the datagram sender, specifically the host address and port number of the originating application.

Socket Options

Socket options permit an application to change some of the modifiable behaviors of sockets and the functions that manipulate them. For example, an application can modify the sizes of the send or receive socket buffers or the size of the maximum segment used by the TCP layer for a given socket.

The functions for setting or retrieving options for a given socket is provided by the following function prototypes:

```
int getsockopt( int sock, int level, int optname,
                void *optval, socklen_t *optlen );

int setsockopt( int sock, int level, int optname,
                const void *optval, socklen_t optlen );
```

First, we define the socket of interest using the sock argument. Next, we must define the level of the socket option that is being applied. The level argument can be SOL_SOCKET for Sockets layer options, IPPROTO_IP for IP layer options, and IPPROTO_TCP for TCP layer options. The specific option within the level is applied using the optname argument. Arguments optval and optlen define the specifics of the value of the option. The optval is used to get or set the option value and optlen defines the length of the option. This slightly complicated structure is used because structures can be used to define options.

Let's now look at an example for both setting and retrieving an option. In the first example, we'll retrieve the size of the send buffer for a socket.

```
int sock, size, len;

...
```

```
getsockopt( sock, SOL_SOCKET, SO_SNDBUF,
                (void *)&size, (socklen_t *)&len );

printf( "Send buffer size is &d\$$\n", size );
```

Now we'll look at a slightly more complicated example. In this case, we're going to set the linger option. Socket linger allows us to change the behavior of a stream socket when the socket is closed and data is remaining to be sent. After **close** is called, any data remaining will attempt to be sent for some amount of time. If after some duration, the data cannot be sent, then the data to be sent is abandoned. The time after the **close** to when the data is removed from the send queue is defined as the linger time. This can be set using a special structure called linger, as shown in the following example:

```
struct linger ling;
int sock;

...

ling.l_onoff = 1;   /* Enable */
ling.l_linger = 10; /* 10 seconds */

setsockopt( sock, SOL_SOCKET, SO_LINGER,
                (void *)&ling, sizeof(struct linger) );
```

After this call is performed, the socket will wait 10 seconds after the socket **close** before aborting the send.

Other Miscellaneous Functions

Let's now look at a few miscellaneous functions from the Sockets API and the capabilities they provide. The three function prototypes we discuss in this section are shown in the following code:

```
struct hostent *gethostbyname( const char *name );

int getsockname( int sock,
                struct sockaddr *name, socklen_t *namelen );

int getpeername( int sock,
                struct sockaddr *name, socklen_t *namelen );
```

Function **gethostbyname** provides the means to resolve a host and domain name (otherwise known as a Fully Qualified Domain Name, or FQDN) to an IP address. For example, the FQDN of *www.microsoft.com* might resolve to the IP address 207.46.249.27. Converting an FQDN to an IP address is important because all of the Sockets API functions work with number IP address (32-bit addresses) rather than FQDNs. An example of the **gethostbyname** function is shown next:

```
struct hostent *hptr;

hptr = gethostbyname( "www.microsoft.com");

if (hptr == NULL) // can't resolve...
  else {

    printf( "Binary address is %x\$$\n", hptr->h_addr_list[0] );

  }
```

Function **gethostbyname** returns a pointer to a structure that represents the numeric IP address for the FQDN (hptr->h_addr_list[0]). Otherwise, **gethostbyname** returns a NULL, which means that the FQDN could not be resolved by the local resolver. This call blocks while the local resolver communicates with the configured DNS servers.

Function **getsockname** permits an application to retrieve information about the local socket endpoint. This function, for example, can identify the dynamically assigned ephemeral port number for the local socket. An example of its use is shown in the following code:

```
int sock;
struct sockaddr localaddr;
int laddrlen;

// socket for sock created and connected.
...

getsockname( sock, (struct sockaddr_in *)&localaddr, &laddrlen );

printf( "local port is %d\n", ntohs(localaddr.sin_port) );
```

The reciprocal function of **getsockname** is **getpeername**. This permits us to gather addressing information about the connected peer socket. An example, similar to the **getsockname** example, is shown in the following code:

```
int sock;
struct sockaddr remaddr;
int raddrlen;

// socket for sock created and connected.
...

getpeername( sock, (struct sockaddr_in *)&remaddr, &raddrlen );
```

```
printf( "remote port is %d\$$\n", ntohs(remaddr.sin_port) );
```

In both examples, the address can also be extracted using the `sin_addr` field of the `sockaddr` structure.

SUMMARY

This quick introduction to Sockets programming in C provided a definition of the tools used for the examples provided. Then, a quick tour of the Sockets API using C was provided, identifying how to create a socket, name it, and perform a variety of other operations on it.

RESOURCES

Dictionary of Programming Languages, available online at *http://cgibin.erols.com/ziring/cgi-bin/cep/cep.pl*.

Feather, Clive D.W., "A Brief(ish) Description of BCPL," Available online at *http://www.lysator.liu.se/c/clive-on-bcpl.html*.

Stevens, W. Richard, *UNIX Network Programming*, 2nd edition, Volume 1, Prentice Hall PTR, 1998.

10 Network Programming in Java

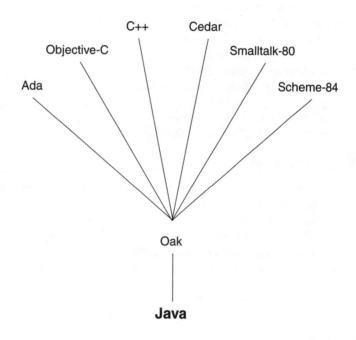

C++   Cedar

Objective-C   Smalltalk-80

Ada   Scheme-84

Oak

Java

JAVA LANGUAGE OVERVIEW

In this chapter, we investigate network programming in the object-oriented language Java™. Java is a very interesting object-oriented language that has a wide multitude of uses, outside of its focus as Web-glue. As we'll see in this chapter, network programming can be very simple in the Java language. All code for this chapter can be found on the companion CD-ROM at /software/ch10.

ON THE CD

Origin of the Java Language

The Java language was created by James Gosling of Sun Microsystems™ in the early 1990s as a new type of language for a new world. Java created new possibilities for an Internet connected world, and much to the chagrin of Microsoft, is still the most popular language for Internet applications. Java is not only object oriented, but also supports the development of distributed and portable applications. What made Java so interesting in the mid 1990s was that a Java application could be written on one platform (such as Windows), but then be run on another platform with no changes. This made it possible for developers to write Web-based applications that could be executed by peer computers on the Internet without having to worry about their target architecture. In some cases, it's not always this easy, but, overall, the goal of Java was reached and the popularity of Java is a testament to this.

Java Language Heritage

Java has a large number of influences that include a variety of languages in different domains. James Gosling cites on his Web site [Gosling] that Simula was one of the most important influences because it introduced the concept of a class (one of the first object-oriented languages). Other influences include Oak (an internal Sun Microsystems language project), Scheme, Smalltalk-80, Objective-C, Cedar (a Pascal-like language), C++, and Ada. These represent an eclectic mix of languages, but the resulting Java language is clean and easy to understand.

TOOLS

Although Java is an interpreted language, it is commonly compiled into bytecodes that are then interpreted by the Java Virtual Machine (or JVM). A number of JVM implementations exist; the examples shown here were tested in the Blackdown JVM for Linux. Sun Microsystems provides freely usable JVMs and JREs (Java Runtime Environments). The version of Java used in this chapter (and Chapter 17, Network Code Patterns in Java) is 1.2 (otherwise known as Java 2 Platform).

Interpreter/Tools Used

The Java Virtual Machine operates on Linux and native-Windows (or Windows using the Cygwin environment), in addition to many other environments. Java includes its own threading model and, therefore, provides multithreading capabilities that are transparent to the underlying operating system.

The sample code patterns presented here (and in Chapter 17) and on the CD-ROM will execute on any of the supported environments.

Networking API Used

Java is bundled with its own networking APIs; therefore, no other downloads are necessary for the Sockets API.

Where to Download

The Java tools may be downloaded at the source, *http://java.sun.com* (under "Download"), or may be downloaded for Linux at *http://www.blackdown.org.*

NETWORKING API FOR JAVA

Java includes a very rich set of networking APIs that operate at three layers. A standard Sockets API, an abstract set of classes for higher-level functionality, and even higher-level Application layer protocols, such as HTTP, FTP, SMTP, and Telnet, are provided.

In this chapter, we focus on the more popular classes within the java.net framework. Each class in this framework is derived from the **Object** class (as shown in Figure 10.1). Class **DatagramSocket** provides the ability to write datagram-based servers and clients (using protocol UDP). Class **DatagramPacket** is used by objects that use the **DatagramSocket** as a means to build packets complete with payload and destination address and port. Class **MulticastSocket** is derived from class **DatagramSocket** to provide multicast capabilities. **InetAddress** is the address class and is used to build Internet addresses. Class **ServerSocket** is a class that abstracts some of the server concepts for building stream socket applications. Finally, **Socket** provides a BSD-like Sockets API for those who desire the standard API.

Sockets API Summary

We saw in Figure 10.1 that Java provides a number of classes for Sockets programming; let's now take a brief look at the classes and methods that are available.

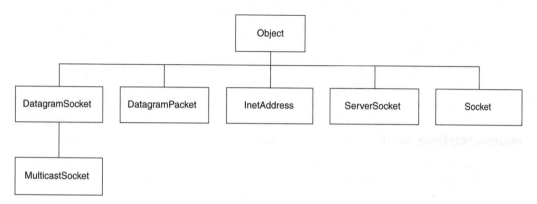

FIGURE 10.1   Java simplified Sockets API class hierarchy.

DatagramSocket Class

The **DatagramSocket** class provides the base for UDP socket classes. This class provides the set of features required to send and receive datagram packets (although not the ability to create them, more to come with **DatagramPacket**). Methods for the **DatagramSocket** class are shown in Figure 10.2.

Method	*Returns*	*Description*
DatagramSocket()	DatagramSocket	Create a new Datagram socket and bind it to an ephemeral port.
DatagramSocket(int port)	DatagramSocket	Create a new Datagram socket and bind it to the defined port.
DatagramSocket(int port, lnetAddress laddr)	DatagramSocket	Create a new Datagram socket and bind it to the defined port and local interface (laddr).
Close()	void	Close the previously "connected" Datagram socket.
connect(lnetAddress raddr, int port	void	Set the remote address association for the Datagram socket.
disconnect()	void	Remove the remote address association for the given Datagram socket.
getlnetAddress()	lnetAddress	Get the remote address association (peer) for the given Datagram socket (set via connect
getLocalAddress()	lnetAddress	Get the local address for the Datagram socket.
getPort()	int	Get the remote port to which this Datagram socket is connected (set in connect).
getLocalPort()	int	Get the local port for this Datagram socket.
getReceiveBufferSize()/ setReceiveBufferSize()	int/void	Get or Set the receive socket buffer size (SO_RCVBUF).
getSendBufferSize()/ setSendBufferSize()	int/void	Get or Set the send socket buffer size (SO_SNDBUF).
getSoTimeout()/setSoTimeout()	int/void	Get or set the Timeout value (in milliseconds) for SO_TIMEOUT.
receive(DatagramPacket packet)	void	Receive a DatagramPacket through the DatagramSocket.
send(DatagramPacket packet)	void	Send a DatagramPacket through the DatagramSocket.

FIGURE 10.2 DatagramSocket class methods. (Adapted from [Java] Java 2 Platform, Standard Edition, v1.2.2 API Specification, available online at *http://java.sun.com/products/jdk/1.2/docs/api/java/net/ DatagramSocket.html*.)

DatagramPacket **Class**

The **DatagramPacket** class provides the ability to create and initialize datagram packets for receipt or transmission by the **DatagramSocket** class. The methods provided by the **DatagramPacket** class are shown in Figure 10.3.

Method	Returns	Description
DatagramPacket(byte[] buf, int buflen)	DatagramPacket	Create a new Datagram socket and bind it to an ephemeral port.
DatagramPacket(byte[] buf, int buflen, lnetAddress peeraddr, int peerport)	DatagramPacket	Create a new Datagram socket and bind it to the defined port.
DatagramPacket(byte[] buf, int offset, int buflen)	DatagramPacket	Create a new Datagram socket and bind it to the defined port and local interface (laddr).
DatagramPacket(byte[] buf, int offset, int buflen, lnetAddress peeraddr, int peerport)	DatagramPacket	Close the previously "connected" Datagram socket.
getAddress()/setAddress	lnetAddress	Get or Set the IP address to which this datagram is being sent or from which it was received.
getPort()/setPort()	void	Get or Set the port to which this datagram is being sent or from which it was received.
getData()	byte[]	Get the data received (or to be sent) in this DatagramPacket.
getLength()/setLength()	int	Get or Set the length of data to be sent or which was received.
getOffset()	int	Get the offset of the data to be sent or from which was received.
setData(byte[] buf)	void	Set the data buffer for this DatagramPacket.
setData(byte[] buf, int offset, int buflen)	int/void	Sets the data buffer for this DatagramPacket.

FIGURE 10.3 **DatagramPacket** class methods. (Adapted from [Java] Java 2 Platform, Standard Edition, v1.2.2 API Specification, available online at *http://java.sun.com/products/jdk/1.2/docs/api/java/net/DatagramPacket.html.*)

MulticastSocket Class

The **MulticastSocket** class provides the an extension of the **DatagramSocket** class for multicast. The methods provided in the **MulticastSocket** class are shown in Figure 10.4.

Method	Returns	Description
DatagramPacket(byte[] buf, int buflen)	DatagramPacket	Create a new Datagram socket and bind it to an ephemeral port.
DatagramPacket(byte[] buf, int buflen, lnetAddress peeraddr, int peerport)	DatagramPacket	Create a new Datagram socket and bind it to the defined port.
DatagramPacket(byte[] buf, int offset, int buflen)	DatagramPacket	Create a new Datagram socket and bind it to the defined port and local interface (laddr).
DatagramPacket(byte[] buf, int offset, int buflen, lnetAddress peeraddr, int peerport)	DatagramPacket socket.	Close the previously "connected" Datagram
getAddress()/setAddress	lnetAddress	Get or Set the IP address to which this datagram is being sent or from which it was received.
getPort()/setPort()	void	Get or Set the port to which this datagram is being sent or from which it was received.
getData()	byte[]	Get the data received (or to be sent) in this DatagramPacket.
getLength()/setLength()	int	Get or Set the length of data to be sent or which was received.
getOffset()	int	Get the offset of the data to be sent or from which was received.
setData(byte[] buf)	void	Set the data buffer for this DatagramPacket.
setData(byte[] buf, int offset, int buflen)	int/void	Sets the data buffer for this DatagramPacket.

FIGURE 10.4 MulticastSocket class methods. (Adapted from [Java] Java 2 Platform, Standard Edition, v1.2.2 API Specification, available online at *http://java.sun.com/products/jdk/1.2/docs/api/java/net/MulticastSocket.html.*)

InetAddress Class

The **InetAddress** class provides a small number of methods to deal with domain names, name resolution, and testing addresses. The methods provided in the **InetAddress** class are shown in Figure 10.5.

Method	Returns	Description
equals(Object obj)	boolean	Compares two lnetAddress objects.
getAddress()	byte[]	Return the raw IP address for the lnetAddress object.
getAllByName(String host)	lnetAddress[]	Return the IP addresses for the specified host.
getByName(String Host)	lnetAddress	Return the IP address for the specified host.
getHostAddress()	String	Return the dotted-notation String.
getHostName()	String	Return the hostname for the current host.
getLocalHost	lnetAddress	Return the local host address.
isMulticastAddress()	boolean	Check if the lnetAddress is a multicast address.

FIGURE 10.5 **InetAddress** class methods. (Adapted from [Java] Java 2 Platform, Standard Edition, v1.2.2 API Specification, available online at *http://java.sun.com/products/jdk/1.2/docs/api/java/net/InetAddress.html.*)

Socket **Class**

The **Socket** class provides a set of methods for the creation of client Sockets applications. These methods are shown in Figure 10.6.

Method	Returns	Description
Socket()	Socket	Creates an unconnected socket.
Socket(lnetAddress raddr, int rport)	Socket	Create a socket and connect it to the lnetAddress/port.
Socket(lnetAddress raddr, int rport, lnetAddress laddr, int lport)	Socket	Create a socket and connect it to raddr/rport, binding it locally to laddr/lport.
Socket(String host, int port)	Socket	Create a socket and connect it to the named host and port.
close()	void	Close the socket
getlnetAddress()	lnetAddress	Return the peer address to which this socket is connected.
getLocalAddress	lnetAddress	Return the local address from which this is connected.
getLocalPort()/getPort()	int/int	Get the Local or Remote port.
getInputStream/ getOutputStream	InputStream/ OutputStream	Convert the IP address to a dotted-notation String.

FIGURE 10.6 **Socket** class methods (*continues*).

getReceiveBufferSize()/ setReceiveBufferSize()	int/void	Gets or Sets the SO_RCVBUF socket option.
getSendBufferSize()/ setSendBufferSize()	int/void	Gets or Sets the SO_SNDBUF socket option.
getSoLinger()/setSoLinger()	int/void	Gets or Sets the SO_LINGER socket option.
getPcpNoDelay()/ setTcpNoDelay()	boolean/void	Gets or Sets the TCP_NODELAY socket option.
getSoTimeout()/ setSoTimeout()	int/void	Gets or Sets the SO_TIMEOUT socket option.

FIGURE 10.6 Socket class methods (*continued*). (Adapted from [Java] Java 2 Platform, Standard Edition, v1.2.2 API Specification, available online at *http://java.sun.com/products/jdk/1.2/docs/api/java/ net/Socket.html.*)

ServerSocket **Class**

Finally, the **ServerSocket** class provides a number of methods specifically for the creation of TCP server sockets. These methods are shown in Figure 10.7.

Method	*Returns*	*Description*
ServerSocket(int port)	Socket	Create a server socket on the specified port.
ServerSocket(int port, int backlog)	Socket	Create a server socket on the specified port with the specified backlog.
ServerSocket(int port, int backlog, lnet Address laddr)	Socket	Create a socket and connect it to raddr/rport, binding locally to laddr/lport.
accept()	Socket	Accept a new client connection.
close()	void	Close the socket.
getlnetAddress()	lnetAddress	Return the local address for this socket.
getLocalPort	int	Return the port on which this port is listening.
getSoTimeout()/setSoTimeout()	int/void	Gets or Sets the SO_TIMEOUT socket option.

FIGURE 10.7 ServerSocket class methods. (Adapted from [Java] Java 2 Platform, Standard Edition, v1.2.2 API Specification, available online at *http://java.sun.com/products/jdk/1.2/docs/api/java/net/ ServerSocket.html.*)

Sockets API Discussion

Let's now look at how the Java networking API is used. We'll exercise many of the methods shown in the previous class method figures, including how using methods from the **ServerSocket** can simplify the development of server Sockets programming.

Creating and Destroying Sockets

As with any sockets-based application, the creation of a socket is the first step before communication may occur. Java provides a number of methods for socket creation, depending upon the type of socket needed by the developer. We look at the primitive example first, and then move on to Java's more expressive methods that can simplify application development.

All of the methods discussed in this section require that the application developer make the networking classes visible. At the beginning of the Java application, a line specifying, "import `java.net.*;` *must be present.*

The **Socket** class provides the primitive standard socket function that is most familiar to C language programmers. The **Socket** class methods can be used to create stream (TCP) client sockets.

```
Socket sock = Socket( "192.168.1.1", 25 );
```

The **Socket** constructor of the **Socket** class not only creates the client TCP socket, but also connects it to the remote peer defined by the arguments. In this case, upon successful completion of the **Socket** constructor, sock will be connected to host "192.168.1.1" and port 25. We could also use a variant of the **Socket** constructor to create a socket, connect it to a remote host, and port and bind it to a local address and port. This is accomplished with:

```
Socket sock = Socket( "192.168.1.1",  // Remote Host
                       25,              // Remote Port
                       "10.0.0.1",      // Local Interface
                       25000 );         // Local Port
```

To create a TCP server socket, the following can be done with **ServerSocket**:

```
ServerSocket sockserv = ServerSocket( 8080 );
```

which creates a TCP server socket and binds it to the local port 8080. To bind not only to a local port, but also to a local interface, we could do the following:

```
InetAddress laddr = InetAddress.getByName( "10.0.0.2" );

ServerSocket sockserv = ServerSocket( 8080, 5, laddr );
```

This first creates an address structure (for the interface represented by IP address "10.0.0.2") and stores it in laddr. Next, the server socket is created with the **Server-Socket** constructor with port 8080, a backlog of 5 connections, and the previously created address object, laddr.

To create a datagram socket, we use the **DatagramSocket** constructor:

```
DatagramSocket dgramsock = DatagramSocket();
```

We create a datagram socket for server connections, binding it to address 23000 with:

```
DatagramSocket dgramsock = DatagramSocket( 23000 );
```

Finally, we could bind not only to a port, but also to a local interface using the following:

```
InetAddress laddr = InetAddress.getByName( "10.0.0.2" );
```

```
ServerSocket sockserv = DatagramSocket( 8080, laddr );
```

Multicast sockets are treated differently in Java than in most other languages, where they are simply derived from datagram sockets. The **MulticastSocket** class is derived from **DatagramSocket**, but must be used if multicast communication is desired for the socket. The following two examples illustrate the creation of a **MulticastSocket** and the creation of a **MulticastSocket** that will be bound to the local port, 45000.

```
MulticastSocket mcsock = MulticastSocket();
```

```
MulticastSocket mcsock = MulticastSocket( 45000 );
```

When we're finished with a socket, we must close it. To close a previously created socket, sock, we use the **close** method. The **close** method is part of all of the previously discussed classes (except for **MulticastSocket**, whose **close** method is derived from the **DatagramSocket**).

```
sock.close();
```

After the **close** method is called, no further communication is possible with this socket.

Any data queued for transmission would be given some amount of time to be sent before the connection physically closes.

Socket Addresses

Unlike C, Java has no sockaddr_in structure, but instead a simpler object that represents only the address and not the port. Dealing with addresses in Java is much simpler than in C, because methods exist in many of the classes to both set and extract addresses and ports for sockets.

To create an address object, we use the **InetAddress** class. To create an **InetAddress** object from a host name or dotted-notation IP address string, the **getByName** method can be used (illustrated as follows):

```
try {
   InetAddress adrs =
      InetAddress.getByName( "www.microsoft.com" );
   System.out.println("Address is " + adrs.toString() );
}
catch (UnknownHostException e )
{
   System.out.println("Caught exception " + e );
}
```

This example illustrates not only the **getByName** method of the **InetAddress** class, but also some additional attributes of the Java language. Rather than return an error, the **getByName** has the potential to throw an exception. Exceptions are not returned, but instead raised at a higher level and caught within the calling method with the catch statement. In the previous example, the **getByName** method can throw the UnknownHostException, which is caught and a message emitted to standard-out (System.out). If the **getByName** method succeeds, we'll continue within the try statement and emit the address to standard-out using the **toString** method.

The **getByName** method is also used with dotted-notation IP address strings:

```
try {
   InetAddress adrs =
      InetAddress.getByName( "10.0.0.1" );
   System.out.println("Address is " + adrs.toString() );
}
catch (UnknownHostException e )
{
   System.out.println("Caught exception " + e );
}
```

Socket Primitives

In this section, we look at a number of other important socket control methods using the standard **Socket** and **SocketServer** classes. We also investigate the variety of ways that the same results can be achieved using different Java classes.

Bind

The **bind** method traditionally provides a local naming capability to a socket. This can be used to name either client or server sockets, but is used most often in the server case. The **bind** method is implicit in a number of Java classes, such as the **ServerSocket** constructor:

```
InetAddress laddr = InetAddress.getByName( "10.0.0.1" );

ServerSocket servsock = ServerSocket( 8080 );
```

```
// or

ServerSocket servsock = ServerSocket( ladrs, 8080 );
```

The **ServerSocket** constructor is used to not only create the server socket, but also to bind it to a local address. In the first case, we bind only to port 8080 and all available interfaces (INADDR_ANY). In the second case, we bind to a specific interface (represented by IP address "10.0.0.1") and port 8080.

The **bind** can also be used to bind a client socket to a local address. The **Socket** constructor of the **Socket** class (for client sockets) provides a special constructor to not only **bind**, but also to **connect**, as is illustrated:

```
InetAddress local_addr =
                InetAddress.getByName( "10.0.0.1" );
InetAddress remote_addr =
                InetAddress.getByName( "192.168.1.1" );

Socket clisock = Socket( remote_addr, 8080,
                         local_addr, 52000 );
```

The **bind** construct can also be done automatically with the **DatagramSocket** constructor of the **DatagramSocket** class. Two variations are provided, just as is done with the **ServerSocket**:

```
InetAddress laddr = InetAddress.getByName( "10.0.0.1" );

DatagramSocket servsock = DatagramSocket( 8080 );

// or

DatagramSocket servsock = DatagramSocket( 8080, ladrs );
```

In the first example, the datagram socket is bound to port 8080 and all interfaces, while the second example illustrates not only binding to port 8080, but also to the local interface identified by the address "10.0.0.1."

Listen

Before a traditional server socket can accept incoming client connections, it must call the **listen** method to declare this willingness. The **listen** method is not provided in the Java networking API, but is instead implicit in the server socket setup. For example,

```
InetAddress laddr = InetAddress.getByName( "10.0.0.1" );
int port = 8080;
int backlog = 5;
```

```
ServerSocket servsock =
                ServerSocket( port, backlog, laddr );
```

provides not only socket creation and interface binding, but also a call to **listen** to start the server socket up for incoming connections. The backlog parameter is traditionally used for the **listen** call, but is passed here to the **ServerSocket** constructor to set up the connection queue backlog limit.

Accept

The **accept** method is the final call made by servers to accept incoming client connections. In Java, this method is provided in the **ServerSocket** class and returns a new client **Socket** object.

```
InetAddress laddr = InetAddress.getByName( "192.168.110.5" );
int port = 8080;
int backlog = 5;

ServerSocket servsock =
                ServerSocket( port, backlog, laddr );

Socket clisock = servsock.accept();
```

The **accept** method in the **ServerSocket** class accepts a new client connection and returns a client **Socket** object. This new client socket can then be used for further communication with the client's peer.

Connect

The **connect** method is used by client Sockets applications to connect to a server. In Java, the **connect** method is implicit in the creation of client sockets using the **Socket** class. Consider the following example of creating a client socket and connecting it to IP address "192.168.1.2", port 13.

```
InetAddress remote_addr =
                InetAddress.getByName( "192.168.1.2" );
int port = 13;

Socket clisock = Socket( remote_addr, port );
```

Upon completion of the **Socket** constructor, the client socket is connected to the defined remote address and port. Note that the exception handling is not shown here (try / catch).

Recall that datagram sockets can also be connected. This means that the peer address is automatically defined and packets may only be sent to that address. Normally, the peer address must be defined for all datagrams sent from the socket. Creating a connected datagram socket is illustrated as follows:

```
InetAddress raddr =
                    InetAddress.getByName( "192.168.1.2" );
int port = 450;

DatagramSocket dgramcli =
                    DatagramSocket( port, raddr );
```

Upon completion of the **DatagramSocket** constructor, a new datagram socket is created with the peer information cached with the socket. No synchronization is performed with the peer, only the peer address and port configured with the socket. To remove the association of the peer address, the **disconnect** call can be performed:

```
dgramcli.disconnect();
```

Upon completion, the association is removed and datagrams can henceforth be sent to any remote peer.

Sockets I/O

A variety of methods exists to read data from a socket or write data to a socket. Some methods are class specific, and others inherit methods from other classes (such as input and output streams). First, we look at reading and writing from connected sockets and then investigate unconnected (datagram-specific) socket communication.

A number of methods are provided to communicate through a socket. Some of the functions include **recv** and **send**, though these are commonly restricted to datagram sockets. We look at examples of each of these functions in the following sections. Other techniques, such as InputStream/OutputStream methods can also be used, and we investigate examples of their use as well.

Stream Sockets

Stream sockets utilize the stream concept of Java IO. This permits a simple layering of stream filters over a socket for read and write purposes. Let's look at a simple example of this concept before moving on to our Sockets examples.

Consider the operation of reading from a file. We'll use a file stream as our inner filter, but then layer a buffered input stream as the outer filter. Buffered streams are typically more efficient because the stream is buffered to reduce the number of physical accesses of the file.

```
FileInputStream file = new
                    FileInputStream( "file.dat" );

BufferedInputStream bis = new
                    BufferedInputStream( file );

int ret = 0;
```

```
//
do {

    ret = buff.read();

} while (ret != -1);

buff.close();
```

In this example, we open a `FileInputStream` on a file called `file.dat`. Next, we lay a `BufferedInputStream` over our `FileInputStream` to change the mechanism for reading the file. In the simple loop, we read the number of characters that are available until the end of file is reached, at which time we close the stream.

Java provides a variety of stream concepts for data input and output. These include Pipes (to channel data from one thread to another), Pushback (to permit peeking at the data, putting it back if necessary), Buffered (to make physical I/O more efficient), LineNumber (to help keep track of line numbers while reading), and many others.

Let's now look at a few examples of stream sockets utilizing Java's stream I/O concepts. In the first example, we'll illustrate the use of the PrintWriter stream, which provides very convenient methods for emitting data through a stream.

```
Socket clisock;
PrintWriter output;

clisock = socket();

...

output = new PrintWriter( clisock.getOutputStream(), true );

...

output.println( "Hello!" );
```

In this simple example, a client socket is created, and after connection, a `PrintWriter` stream is created to provide a means to emit data through the socket. Note that in order to create the `PrintWriter` stream object, we must identify the output stream for the socket. To do this, we use the **getOutputStream** method for the specific socket, which is passed anonymously to the `PrintWriter` method. The true flag defines that we desire autoflush for data written to the stream. The `PrintWriter` stream is identified by the object output, and in the last line, we see an example of a common `println` to emit a string through the socket.

Let's now look at a simple example for reading from a socket. In this example, we use the `BufferedReader` stream class, which can make socket reads more efficient, and we wrap the `BufferedReader` over the socket's input stream.

```
String line;
sock = new Socket();

...

input = new BufferedReader(
                new InputStreamReader(
                      sock.getInputStream() ) );

line = input.readLine();

System.out.println( line );
```

In this example, we create a client socket called sock. To create a BufferedReader stream, we must also create an InputStreamReader, which is done anonymously over the socket's input stream (retrieved using the **getInputStream** method). With our new BufferedReader stream, we can perform very simple line-oriented operations such as readLine to retrieve a single line of text from the socket. This line, in the example, is simply emitted to standard-out.

These two examples should illustrate the power of Java's stream concept. In essence, the mechanism allows the transformation of an I/O channel to extend the types of operations that can be performed on it.

Datagram Sockets

Because datagram sockets are not stream-related, a different set of mechanisms exists to read and write datagram packets from and to the network. In fact, the construction of datagram packets uses a special class called **DatagramPacket**, which identifies not only the payload of the packet, but also the destination address and port information.

Let's look at a datagram server and client that provide the echo functionality illustrated by alternative languages in other chapters. We omit the try/catch elements of the source to increase its readability. Our datagram server takes on a slightly different form:

```
DatagramSocket socket = new DatagramSocket( 45000 );

...

// Construct a packet for receiving datagrams.
byte[] buffer = new byte[100];
DatagramPacket srcpkt = new DatagramPacket( buffer, 100 );

// Receive a datagram packet from the peer.
socket.receive( packet );

// Get the IP address and port from the source packet.
```

```
InetAddress peeradrs = packet.getAddress();
int peerport = packet.getPort();

// Construct a new datagram packet using the source
// packet's payload, and source address and port.
DatagramPacket destpkt = new
    DatagramPacket( srcpkt.getData(),
                    srcpkt.getLength(),
                    peeradrs, peerport );

// Send the new packet back to the peer.
socket.send( destpkt );
```

In this example, we create our **DatagramSocket** and then a **DatagramPacket** (srcpkt) that will be used to receive a datagram through the socket. We'll accept at most a datagram with 100 octets. Next, we call the **receive** method for the socket object to receive the datagram. Upon receiving the datagram, we extract the source IP address from the packet using the **getAddress** method and the source port is extracted with **getPort**. We use the source address and port in the construction of the new datagram packet that we'll send back to the peer.

We create destpkt, which is our response **DatagramPacket**, and provide not only the source address and port, but also the payload using the **getData()** method and the length of the payload with the **getLength()** method. Finally, the new **DatagramPacket** is sent to the peer (source of the original datagram) using the **send** method of the socket object.

As is illustrated in this example, working with datagrams in Java is slightly more involved than most other languages in that the packet structure itself is another object with its own set of methods and constructors.

Socket Options

Socket options permit an application to change some of the modifiable behaviors of sockets and the methods that manipulate them. For example, an application can modify the sizes of the send or receive socket buffers or the size of the maximum segment used by the TCP layer for a given socket.

Socket options within Java are easier than that provided to the C programmer. The socket options that are available utilize class methods to set or read the parameter for the given socket. Therefore, rather than call the **setsockopt** or **getsockopt** with a void pointer element to a structure, methods can be simply called to manipulate the options for a given socket.

As a first example, let's say that we want to identify the size of the receive buffer for a given socket. This can be done with the following code segment:

```
import java.net.*;

// Create a socket and connect it to a server.
InetAddress adrs = InetAddress.getByName( "192.168.1.1" );
```

```
Socket sock = Socket( adrs, 80 );

int receiveBuffer = sock.getReceiveBufferSize();
```

First, we create our address object, which represents the server to which we're going to connect ("192.168.1.1"). Next, we create our client socket using the **Socket** constructor of the **Socket** class. Within the context of this method, we'll be connected to the remote server. Finally, we retrieve the size of the socket receive buffer (traditionally, SO_RCVBUF) and store it into receiveBuffer using the **getReceiveBufferSize** method of the **Socket** class.

Now, let's look at how an option is set for a given socket, specifically, the linger option. Socket linger allows us to change the behavior of a stream socket when the socket is closed and data is remaining to be sent. After **close** is called, any data remaining will attempt to be sent for some amount of time. If after some duration, the data cannot be sent, then the data to be sent is abandoned. The time after the **close** to when the data is removed from the send queue is defined as the *linger* time. In Java, this is yet another method called **setSoLinger**. Rather than the typical linger structure used in C and other languages, the **setSoLinger** method simply takes two options for the enable and linger time, as illustrated:

```
import java.net.*;


// Create a socket and connect it to a server.
InetAddress adrs = InetAddress.getByName( "192.168.1.1" );
Socket sock = Socket( adrs, 80 );

sock.setSoLinger( true, 10 );
```

The first argument to **setSoLinger** is the enable for the linger option (in this case, true represents enabled, false is disabled). The linger time is measured in seconds. In the example, we're enabling linger and setting the linger value to 10 seconds. We could read back what was configured by:

```
int linger = sock.getSoLinger()
```

A return of −1 means that the option is disabled. Any other value means that the linger option is enabled, and the value represents the number of seconds configured for linger.

Other Miscellaneous Functions

Let's now look at a few miscellaneous functions from the various Sockets APIs and the capabilities they provide. The first methods that we discuss provide information about the current host. Method **getHostName** (of the **InetAddress** class) returns the string name of the host:

```
String hoststr;

try {
```

```
    InetAddress myAddr =
        InetAddress.getByName( "192.168.1.1" );

    hoststr = myAddr.getHostName();

    System.out.println( "Hostname is " + hoststr );

} catch ( UnknownHostException e ) {}
```

We first get an **InetAddress** (our raw address format for Java) given a string IP address in dotted notation. Next, we use the **getHostName** method on that InetAddress object to get the host name. This is then emitted to standard-out with System.out.println.

The DNS resolver permits us to resolve an IP address to a host name, or vice versa. Method **getHostAddress** provides address resolution given a name, for example:

```
String adrsstr;

try {

    InetAddress myAddr =
        InetAddress.getByName( "www.gnu.org" );

    adrsstr = myAddr.getHostAddress();

    System.out.println( "Host Address is " + adrsstr );

} catch ( UnknownHostException e ) {}
```

where the response is a string of the IP address in dotted notation.

We can also identify the address of the current host, and its host name with the following Java snippet:

```
String hoststr
String adrsstr;

try {

    // First, create our InetAddress object
    InetAddress myAddr = InetAddress.getLocalHost();

        // Now use this object to retrieve the address and name
        adrsstr = myAddr.getHostAddress();
        hoststr = myAddr.getHostName();

        System.out.println( "Hostname is " + hoststr );
        System.out.println( "Host Address is " + adrsstr );
```

```
} catch ( UnknownHostException e ) {}
```

Additionally, given an **InetAddress** object, we can always convert this to a string for display using the **toString** method of the **InetAddress** class:

```
InetAddress myAddr = InetAddress.getLocalHost();

System.out.println( "Address is " + myAddr.toString() );
```

which provides both the host name and dotted-notation IP address as follows:

```
Address is www.microsoft.com/207.46.134.190
```

Finally, Java also provides a method to retrieve all of the known IP addresses for a given host. The following snippet illustrates this:

```
try {

    // First, create our InetAddress object
    InetAddress[] addrs =
InetAddress.getAllByName( "www.microsoft.com" );

        int i;
        for (i = 0 ; i < addrs.length ; i++ ) {

            System.out.println( "Address : " +
                addrs[i].toString();

        }

    } catch ( UnknownHostException e ) {}
```

We first create our **InetAddress[]** array object and use the **getAllByName** method to load it with all available addresses for the domain name. We then loop through the number of addresses (identified with the length attribute of the addrs variable) emitting the addresses with the **toString** method.

SPECIALIZED NETWORKING APIS FOR JAVA

Java includes a number of specialized APIs for specialized networking. These include HTTP, URL processing, RMI (remote method invocation), and others. In this section, we look at examples of the HTTP and URL classes.

Java URL Class for HTTP

Java provides a high-level mechanism for dealing with URLs, or Uniform Resource Locators. The URL specifies not only an address, but also a protocol for which the address should be accessed. In Listing 10.1, we use the **URL** class to connect to a host and with the specification of the HTTP protocol, read the default (index) file, and display it. The entire Java application is shown for completeness.

Listing 10.1   Sample URL example (host.java).

```java
import java.net.*;
import java.io.*;

public class host {

  public static void main( String[] args ) {

    URL theUrl;
    String line;

    try {

      theUrl = new URL( "http://www.microsoft.com" );

      try {

        BufferedReader input =
          new BufferedReader(
            new InputStreamReader( theUrl.openStream() ));

        try {

          while ((line = input.readLine()) != null ) {
            System.out.println( line );
          }

        } catch ( Exception e ) { System.out.println( e ); }

      } catch ( Exception e ) { System.out.println( e ); }

    } catch( MalformedURLException e ) { System.out.println( e ); }

  }

}
```

From the example, we create a new **URL** object, and specify our URL as the Microsoft Web site. Note that the address is in URL format; it contains a protocol specification ("http://") and an address ("www.microsoft.com"). We then create a **BufferedReader** stream over our **URL** object. First, we open the **URL** using the **openStream** method, which returns an InputStream object. This is passed anonymously into the **InputStreamReader** class, which is used by the **BufferedReader** class to create a BufferedReader object. We can now perform a **readLine** method on our new BufferedReader object to extract the response to our request (which in the **openStream** call was an HTTP 'Get' request). For each line that is received, we emit it to standard-out.

SUMMARY

This very quick tour of Sockets programming in Java provided only a sampling of the available methods and classes that can simplify the development of networking applications. Java continues to be developed, and third-party classes provide even more networking functionality. This chapter sampled Java applications that are run standalone, but other techniques exist within Java, such as Applets and Servlets, which provide powerful abstractions for the development of both client and server applications. Although Java is interpreted, it is interpreted at a bytecode level (architecture independent instructions that execute within a virtual machine), and is, therefore, more efficient than most interpreted scripting languages. Java continues to be an important tool and is worth the effort to understand and apply when the opportunity arises.

REFERENCES

[Gosling] Gosling, James, James Gosling's Web site, available online at *http://java.sun .com/people/jag/,*
[Java] Java 2 Platform, Standard Edition, v1.2.2 API Specification, available online at *http://java.sun.com/products/jdk/1.2/docs/.*

RESOURCES

Java for Linux, available online at *http://www.blackdown.org.*
Sun Microsystems's Java Web site, available online at *http://java.sun.com.*

11 Network Programming in Python

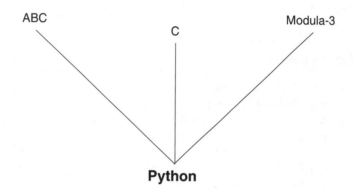

PYTHON LANGUAGE OVERVIEW

In this chapter, we investigate network programming in the popular scripting language Python. The Python language is a very interesting object-oriented, interpreted scripting language that's used in a number of different types of applications. Python continues to evolve and includes a number of classes for network programming. Additionally, Python provides the ability to extend the interpreter with new modules that look just like the native modules provided by the base interpreter. The Python interpreter can also be embedded within an existing application, thus providing internal Python scripting capabilities. All code for this chapter can be found on the companion CD-ROM at /software/ch11.

ON THE CD

One of Python's most interesting features is that it's interactive. By starting the Python shell, applications can be built through experimentation, simplifying the development of both simple and complex applications.

Origin of the Python Language

The Python language began as a Christmas holiday project in 1989 by Guido Van Rossum. Van Rossum had been working in the Amoeba distributed operating system group at CWI (Centrum voor Wiskunde en Informatica). He wanted a language to perform system administration in Amoeba, and using C or Bourne Shell scripts didn't seem all that appealing. By early 1991, Van Rossum had an initial implementation and posted the language to alt.sources on USENET. Today, Python has become one of the most popular of the object-oriented scripting languages.

Python Language Heritage

Python is an interesting language in the number of characteristics that it portrays. For example, Python is an interpreted, object-oriented scripting language that can be used interactively. Python is also general-purpose and can be used to write applications for a variety of purposes.

Python's influences were the ABC language (the first scripting language for the Amoeba distributed operating system), Modula-3, and the C language. The ABC language, C, and Modula-3 influenced Python's syntax, whereas Modula-3 provided the model for exceptions. Some compare Python with BASIC, as it's very portable and easy to learn. The origin of the Python name was due to Guido Van Rossum being a fan of the 1970s BBC comedy "Monty Python" [Python-FAQ].

TOOLS

Although Python was developed by a single person, it's now being developed under the authority of the PSF (Python Software Foundation). The PSF is a group of Python advocates who are responsible for moving Python forward [Python-PSF]. Python is under active development and the PSF seeks proposals for enhancements by the development public. A

Python Enhancement Proposal (PEP) can be submitted at the Python Web site [Python-PEP]. The version of Python used in this chapter (and in Chapter 18, Network Code Patterns in Python) is version 2.2.2.

Interpreter/Tools Used

The Python interpreter operates on Linux, native Windows, and Windows using the Cygwin Unix environment. All examples shown here are self-contained within the standard Python distribution. Python can be extended with dynamically linked libraries, but none are required in this discussion.

ON THE CD

The sample code patterns presented here (and in Chapter 18) and on the CD-ROM will execute on any of the supported environments.

Networking API Used

Python is bundled with its own networking APIs; therefore, no other downloads are necessary for the Sockets API or advanced modules to be discussed in this chapter.

Where to Download

The Python interpreter and tools may be downloaded at the source, *http://www.python.org* (under "Download"). The Python site includes both source and binary versions of the interpreter. Python editors can also be downloaded. The IDLE package provides not only an editor, but also an integrated development environment for Python. If the Cygwin environment for Windows is to be used, it may be downloaded at *http://www.cygwin.com/*.

NETWORKING API FOR PYTHON

Like many of the other object-oriented scripting languages, Python includes a standard Sockets layer (built into classes) and high-level APIs, providing methods to perform the standard Application layer protocols (HTTP, FTP, SMTP, and so on).

We'll divide Python's **socket** library into two distinct parts. The first are the class methods that require no instance of a socket class. For example, when creating a new socket, we'll have no instance to go with. Therefore, we'll use the **socket** class directly, specifically the **socket** method. The second are instance methods that require an instance of a socket. For example, to close a socket, we must have first created one. Therefore, the **close** method operates on an existing instance of the **socket** class. The class methods are shown in Figure 11.1 and the instance methods are listed in Figure 11.2.

Sockets API Discussion

Let's now walk through the Python **socket** API and illustrate how the class and instance methods are used.

Method	Class	Description
socket([family[, type[, proto]]])	socket	Create a new socket of the specified type.
gethostbyaddr(host)	socket	Return the hostname for a given IP address.
gethostbyname(host)	socket	Return the string IP address for a hostname.
gethostname()	socket	Retrun the string hostname for the current host.
getservbyname(servicename, protocolname)	socket	Return the port number for a given string service.
htonl(integer)	socket	Convert from host to network 32-bit.
ntohl(integer)	socket	Convert from network to host 32-bit.
htons(short)	socket	Convert from host to network 16-bit.
ntohs(short)	socket	Convert from network to host 16-bit.
inet_aton(string)	socket	Convert a string IP address to binary format.
inet_ntoa(integer)	socket	Convert a binary format string address to a string.

FIGURE 11.1 Python Sockets class methods.

Method	Type	Description
accept()	Instance	Wait for incoming connections on a server socket instance.
bind(address)	Instance	Bind the socket instance to a local address.
close()	Instance	Close the socket instance.
connect(address)	Instance	Connect the socket instance to the remove address.
fileno()	Instance	Return the integer file descriptor for a socket instance.
getpeername() format.	Instance	Return the address of the remove endpoint in (host, port).
getsockname()	Instance	Return the address of the local socket in (host, port) format.
getsockopt(level, option[, size]))	Instance	Return the value of the socket option for the socket instance.
setsockopt(level, option, value)	Instance	Set the value of a socket option for the socket instance.
listen(backlog)	Instance	Enable the server socket to listen for incoming connections.
recv(size[, flags])	Instance	Receive up to size bytes from the peer socket.
recvfrom(size[, flags])	Instance	Receive up to size bytes. Response is (data, address).
send(data[, flags])	Instance	Send the data bytes to the connected peer.
sendto(data[, flags], address)	Instance	Send the data bytes in datagram format to the peer defined by address.
setblocking(flag)	Instance	Set the socket to blocking or non-blocking (true/false).
shutdown(flag)	Instance	Shutdown the socket (0=read, 1=write, 2=both).

FIGURE 11.2 Python Sockets instance methods.

Creating and Destroying Sockets

We'll begin by investigating how Python creates and destroys local sockets. Python provides a very similar API to the standard BSD4.4 API, as is illustrated in Figures 11.1 and 11.2.

All of the methods discussed in this section require that the application developer make the **socket** class visible. At the beginning of the Python networking application, a line specifying "import socket" must be present. The import statement dynamically loads the module so that the contents of the module are visible. The developer still needs to specify the class and method, such as **socket.bind**.

The **socket** class provides the primitive standard **socket** method that is most familiar to C language programmers. This class method can be used to create stream (TCP), datagram (UDP), and even raw sockets.

```
sock = socket.socket( socket.AF_INET, socket.SOCK_STREAM )
```

Note the similarities of this function to the C language primitive. The primary differences are the Python naming conventions ("**socket.**" to bring the symbol in scope). In this example, we've created a stream socket. To create a datagram socket, we could do the following:

```
sock = socket.socket( socket.AF_INET, socket.SOCK_DGRAM, 0 )
```

Finally, a raw socket could be created with:

```
sock = socket.socket( socket.AF_INET, socket.SOCK_RAW, 0 )
```

Recall that the socket.AF_INET symbol defines that we're creating an Internet protocol socket. The second parameter (type) defines the semantics of the communication (socket.SOCK_STREAM for TCP, socket.SOCK_DGRAM for UDP, and socket.SOCK_RAW for raw IP). The third parameter defines the particular protocol to use, which is useful here only in the socket.SOCK_RAW case. This can be of type IPPROTO_ICMP, IPPROTO_IP, IPPROTO_RAW, IPPROTO_TCP, or IPPROTO_UDP.

When we're finished with a socket, we must close it. To close our previously created socket, **sock**, we use the **close** method.

```
sock.close()
```

Once the **close** method is called, no further communication is possible with this socket. Any data queued for transmission would be given some amount of time to be sent before the connection physically closes.

It appears that in version 2.2.2 of Python, the **close** method aborts the socket connection and, with it, any pending data that has yet to be transmitted to the peer. In cases in which larger amounts of data are being transferred, the **shutdown** method is a better option. The **shutdown** method allows the caller to shut down the receive path, the transmit path, or both. For example:

```
sock.shutdown( 0 )        # Shut down reads

sock.shutdown( 1 )        # Shut down writes

sock.shutdown( 2 )        # Shut down both reads and writes
```

Socket Addresses

Python has no sockaddr_in structure, but instead a tuple of the form '(hostname port)'. Python simplifies the construction of addresses (as compared to other languages, such as Ruby). For example, if we were creating an address representing a host to which we were going to connect, we could do this as:

```
( 'www.mtjones.com', 5930 )
```

if our plan was to connect our TCP client socket to the host identified as www.mtjones.com and the port 5930. Consider the example of binding our server to the loopback address, at the same port number. We'd use an address constructed as:

```
( 'localhost', 5930 )
```

Most times, we'll bind ourselves (from the C language) to the INADDR_ANY wildcard address. This permits a server to accept connections from any interface on the host (assuming a multi-homed host). In this case, we'd provide an empty string for host, as:

```
( '', 5930 )
```

Another example represents the INADDR_BROADCAST host name. In this case, our address is represented as:

```
( '<broadcast>', 5930 )
```

In the examples shown, we've used host names to represents the hosts, but IP addresses can also be used (shown in the following example for localhost):

```
( '127.0.0.1', 5930 )
```

Therefore, as we see from the prior address examples, dealing with addresses in Python is simple and flexible.

Socket Primitives

In this section, we look at a number of other important socket control methods.

Bind

The **bind** method provides a local naming capability to a socket. This can be used to name either client or server sockets, but is used most often in the server case. The **bind** method for the **socket** class is an instance method and is provided by the following prototype:

```
sock.bind( addr )
```

where **addr** is a socket address tuple. Sample usage of this form of **bind** is:

```
sock = socket.socket( socket.AF_INET, socket.SOCK_STREAM )
sock.bind( ('192.168.1.2', 13) )
```

In this example, we're creating a stream server on port 13 that will accept client connections only on the interface defined as 192.168.1.2.

Finally, let's look at a UDP example. The use of bind with a UDP socket uses the **bind** method provided by the **socket** class, the same as provided for TCP sockets. The following example creates a UDP socket and binds it with host INADDR_ANY (accept datagrams from any interface) with port 13:

```
sock = socket.socket( socket.AF_INET )
sock.bind( ('', 13) )
```

Note again the simplicity of this pattern, as provided by Python's address tuples.

Listen

Before a server socket can accept incoming client connections, it must call the **listen** method to declare this willingness. The **listen** method is provided by the following prototype:

```
sock.listen( backlog )
```

The sock argument represents an instance of the server socket and the backlog argument represents the number of outstanding client connections that may be queued. Here's a complete example from socket creation to the **listen** method.

```
sock = socket.socket( socket.AF_INET, socket.SOCK_STREAM )
sock.bind( ('192.168.1.2', 13) )
sock.listen( 5 )
```

The **listen** method is always paired ultimately with an **accept** method, because **accept** is the method used to accept new client connections enabled by the **listen** method.

Accept

The **accept** method is the final call made by servers to accept incoming client connections. Before **accept** can be invoked, the server socket must be created, a name must be bound to it, and the **listen** method must be invoked. The **accept** method accepts a new client connection and returns a tuple in the format (conn, address). The conn element is a new socket object that may be used for client communication and the address is the standard address tuple defined previously. Let's look at a complete simple server example:

```
serv = socket.socket( socket.AF_INET, socket.SOCK_STREAM )
serv.bind( ('192.168.1.2', 13) )
serv.listen( 1 )

cliconn, (remotehost, remoteport) = serv.accept()
cliconn.send("Hello!")
cliconn.close()

serv.close()
```

As illustrated in this example, the **accept** method returns two values. The first, cliconn, is the client socket that can now be used to communicate with the peer (the client that initiated the connection). The second value is the address tuple that identifies the host and port number for the peer socket of the connection. In this example, we accept an incoming client connection, send the string "Hello!" to the peer, and then close both the client and server sockets.

Connect

The **connect** method is used by client Sockets applications to connect to a server. Clients must have created a socket and then defined an address structure containing the host and port number to which they want to connect. This is illustrated by the following code segment (which can be used to connect to the previously illustrated **accept** method example):

```
sock = socket.socket( socket.AF_INET, socket.SOCK_STREAM )

sa = ( '192.168.1.2', 13 )

sock.connect( sa )

print sock.recv( 100 )

sock.close
```

We create our address structure (as with previous examples) and then pass this to our **connect** method using our client socket instance. This short script can be further simplified as follows, by passing the anonymous address tuple to the **connect** method:

```
sock = socket.socket( socket.AF_INET, socket.SOCK_STREAM )

sock.connect( ('192.168.1.2', 13) )

print sock.recv( 100 )

sock.close
```

The **connect** method (for TCP) blocks until either an error occurs, or the three-way handshake with the server completes. For datagram sockets (UDP), the method binds the peer locally so that the peer address isn't required to be specified for every **send** method. In the following example, we use the **connect** method to logically connect our client to an external socket:

```
sock = socket.socket( socket.AF_INET, socket.SOCK_DGRAM )

sock.connect( ('192.168.1.2', 13) )

sock.send( "Hi!" )

sock.close
```

In this example, the **sendto** method is not required because we predefined our destination using the **connect** method. We investigate this concept more in the Sockets I/O section.

Sockets I/O

A variety of API methods exists to read data from a socket or write data to a socket. These methods are socket class specific. First, we look at reading and writing from connected sockets and then investigate unconnected (datagram-specific) socket communication.

A number of methods are provided to communicate through a socket. Some of the functions include **send**, **recv**, **sendto**, **recvfrom**, and **sendall**. We look at examples of each of these functions in the following sections.

Stream Sockets

Stream sockets utilize the **send**, **recv**, and **sendall** socket methods. Let's now look at some sample code that illustrates the stream-specific methods. The first example illustrates a simple echo server built using stream sockets (stream.py):

```
import socket

serv = socket.socket( socket.AF_INET, socket.SOCK_STREAM )
serv.bind( ('', 45000) )
serv.listen( 1 )
```

```
while 1:

    cli, (remhost, remport) = serv.accept()
    cli.send( cli.recv( 100 ) )
    cli.close()
```

In this example, we open a server socket and then await a client connection, storing the newly created client socket in cli. The remote address tuple is ignored in this example. We then simply send what we receive through the client socket back to the source and close the socket. The process then repeats, awaiting a new client connection.

Now, let's look at the TCP client socket that will connect to the previously defined server (client.py):

```
import socket

sock = socket.socket( socket.AF_INET, socket.SOCK_STREAM)
sock.connect( ('localhost', 45000) )

sock.send( "Hello\n", 0 )
mystring = sock.recv( 100 )
print mystring

sock.close()
```

After we create our TCP client socket using the **socket** method, we connect to the server using the **connect** method. For **connect**, we specify the host name and port to which we want to connect. The echo portion of the client comes next, in which we send our message to the server using the **send** method and then await the echo response with the **recv** method. We emit this using the print statement and finally close the socket using the **close** method.

Datagram Sockets

The **sendto** and **recvfrom** methods are used exclusively by datagram sockets. What differentiates these calls from our previously discussed stream calls (**send/recv**) is that these calls include addressing information. Because datagrams are not connected, we must define the destination address explicitly. Conversely, when receiving a datagram, the address information is also provided so that the source of the datagram can be identified.

Let's look at a datagram server and client that provide the echo functionality illustrated previously by our stream socket examples. Our datagram server takes on a slightly different form (serverd.py):

```
import socket

serv = socket.socket( socket.AF_INET, socket.SOCK_DGRAM )
serv.bind( ('', 45000) )
```

```
while 1:
    reply, (remhost, remport) = serv.recvfrom( 100 )
    serv.sendto( reply, (remhost, remport) )

serv.close()
```

After creating our datagram socket using the **socket** method with
`socket.SOCK_DGRAM`, we bind the instance of this socket using the **bind** method. We provide the interface from which we want to accept client connections (all interfaces, or IN-
ADDR_ANY) and the port (45000) to **bind**. Note that we do not use the **listen** method, as
this method is used only for stream sockets.

In our loop, we then receive datagrams using the **recvfrom** method. Note the semantics of this call—the method returns not one parameter, but two. The two parameters returned represent the response data (`reply`) and the source address tuple of the datagram
(`(remhost, remport)`). We use the same source address when returning the datagram
back to the source using the **sendto** method. Returning to the **recvfrom** call, we must also
specify the maximum length of data that we want to receive. The script then continues,
awaiting another datagram from a client.

The datagram client utilizes the datagram **sendto** method with the standard **recv**
method. We can use the standard **recv** here because we're not interested in the source of the
datagram. If we needed to know the source of the datagram, then the **recvfrom** method
would need to be used (clientd.py).

```
import socket

cli = socket.socket( socket.AF_INET, socket.SOCK_DGRAM )

cli.sendto( "Hello\n", ('localhost', 45000) )
print cli.recv( 100 )

cli.close()
```

In the datagram client, after we've created our datagram socket, we send our string to
the echo server using the **sendto** method. The echo server is defined in the **sendto** method
as an address tuple (`'localhost'`, 45000). We then await the response echo datagram using
the **recv** method, and emit it to the terminal using the `print` statement. Finally, the **close**
method is used to destroy the client socket.

Socket Options

Socket options permit an application to change some of the modifiable behaviors of a socket
and change the behavior of the methods that manipulate them. For example, an application
can modify the sizes of the send or receive socket buffers or the size of the maximum segment used by the TCP layer for a given socket.

Socket options can be slightly more complicated than dealing with options in the C environment. This is because we're operating in a scripting environment that must interface with the host environment and its corresponding structures. When we're dealing with scalar types (such as the following example), manipulating socket options can actually be easier in Python.

Let's look at a simple scalar example first. Let's say that we want to identify the size of the receive buffer for a given socket. This can be done with the following code segment (opt.py):

```
import socket

sock = socket.socket( socket.AF_INET, socket.SOCK_STREAM )
size = sock.getsockopt( socket.SOL_SOCKET, socket.SO_RCVBUF )
print size

sock.close()
```

First, we create a new stream socket using the **socket** method. To get the value of a given socket option, we use the **getsockopt** method. We specify the socket.SOL_SOCKET argument (called level) because we're interested in a Sockets layer option (as compared to a TCP or IP layer option). For the receiver buffer size, we specify the option name socket.SO_RCVBUF. In this simple case, a scalar value is returned from the **getsockopt** method. This value is stored into our local variable called size. We print this value using the print statement and then close the socket.

Now let's look at how a more complex option is set for a given socket, specifically, the linger option (opt2.py). This option is more complex because it entails the creation of a structure that maps to an operating system type.

Socket linger allows us to change the behavior of a stream socket when the socket is closed and data is remaining to be sent. After **close** is called, any data remaining will attempt to be sent for some amount of time. If after some duration, the data cannot be sent, then the data to be sent is abandoned. The time after the **close** to when the data is removed from the send queue is defined as the linger time. In Python, we must construct the structure that is expected by the host environment.

```
import socket
import struct

sock = socket.socket( socket.AF_INET, socket.SOCK_STREAM )

ling = struct.pack('ii', 1, 10)
sock.setsockopt( socket.SOL_SOCKET, socket.SO_LINGER, ling )

sock.close()
```

The linger structure (shown here as `ling`) contains first an enable (1 for enable, 0 for disable) and then the linger time in seconds. In our example, we're enabling linger and setting the linger value to 10 seconds (packing the structure into two 32-bit words, the structure that is expected by the host environment). We could read back what was configured by:

```
newlinger = sock.getsockopt( socket.SOL_SOCKET,
                                socket.SO_LINGER, 8 )

(linger_onoff, linger_sec) = struct.unpack( 'ii', newlinger )

print "linger on/off is ", linger_onoff
print "linger sec is ", linger_sec
```

Upon reading the linger array using the **getsockopt** method, we unpack the array using the "ii" template and the unpack method of the **struct** class. This specifies that the two 32-bit integers are packed into the array. Note that the return of unpack is a tuple that contains the enable and the linger time. We emit the tuple elements using the `print` statement.

Other socket options are shown in Figure 11.3 (for Sockets layer options) and Figure 11.4 (for IP layer options). Note that the option names are identical to their C language counterparts.

Option Name	Level	Value	get/set	Description
SO_KEEPALIVE	SOL_SOCKET	[0,1]	g/s	Tries to keep the socket alive using keepalives.
SO_RCVBUF / SO_SNDBUF	SOL_SOCKET	int	g/s	Size of the socket-layer receive or send buffers.
SO_RCVLOWAT / SO_SNDLOWAT	SOL_SOCKET	int	g/s	Number of bytes requested before select notification.
SO_RCVTIMEO / SO_SNDTIMEO	SOL_SOCKET	timeval	g/s	Timeout on recv/send in seconds.
SO_REUSEADDR	SOL_SOCKET	[0,1]	g/s	Permits the local address to be reused.
SO_LINGER	SOL_SOCKET	linger	g/s	Linger on close if data to be transmitted to peer.
SO_OOBINLINE	SOL_SOCKET	[0,1]	g/s	Inlines the out-of-band data.
SO_DONTROUTE	SOL_SOCKET	[0,1]	g/s	Bypass the routing tables.
SO_USELOOPBACK	SOL_SOCKET	[0,1]	g/s	Loop back data sent to the receive path for a routing socket.
SO_BROADCAST	SOL_SOCKET	[0,1]	g/s	Permits sending of broadcast datagrams.
SO_TYPE	SOL_SOCKET	int	g	Return the socket type.
SO_ERROR	SOL_SOCKET	int	g	Return error status.

FIGURE 11.3 SOL_SOCKET socket options provided in Python.

Option Name	Level	Value	get/set	Description
IP_ADD_MEMBERSHIP	IPPROTO_IP	mreg	s	Join a multicast group.
IP_DROP_MEMBERSHIP	IPPROTO_IP	mreg	s	Remove from a multicast group.
IP_MULTICAST_IF	IPPROTO_IP	inaddr	g/s	Specify the outgoing interface for this multicast socket.
IP_MULTICAST_LOOP	IPPROTO_IP	char	g/s	Specify loopback for this multicast socket.
IP_MULTICAST_TTL	IPPROTO_IP	[1..255]	g/s	Configure the IP TTL field for this multicast socket.
IP_HDRINCL	IPPROTO_IP	int	g/s	Notifies that an IP header is part of the outgoing data.
IP_OPTIONS	IPPROTO_IP	String	g/s	Configure the IP options for this socket (a String of up to 44 octets).
IP_TOS	IPPROTO_IP	[0..255]	g/s	Configure the IP TOS field for this socket.
IP_TTL	IPPROTO_IP	[1..255]	g/s	Configure the IP TTL field for this socket.

FIGURE 11.4 IPPROTO_IP socket options provided in Python.

Other Miscellaneous Functions

Let's now look at a few miscellaneous functions from the Sockets API and the capabilities they provide. The first method that we discuss provides information about the current host. Method **gethostname** (of the **socket** class) returns the string name of the host:

```
str = socket.gethostname()
print str
```

The DNS resolver permits us to resolve a host name to an IP address, or vice versa. Method **gethostbyname** provides IP address resolution given an FQDN, for example:

```
str = socket.gethostbyname("www.microsoft.com")
print str
```

where the return string represents the IP address of the FQDN. What if we wanted to go in the opposite direction, providing an IP address and receiving an FQDN? To achieve this, we use the **gethostbyaddr**:

```
str = socket.gethostbyaddr( "207.46.134.155" )
puts str[0]
```

The IP address string is contained within the first element of the first array (in String format).

Now, let's consider the problem of identifying the port number for a given service. To specifically retrieve the port number associated with a service, we use the **getservbyname** method. This method takes two arguments, the first is the string name of the service that is desired (such as "http") and the second is the particular protocol over which the service is run (such as "tcp"). To identify the port number for the SMTP procotol, we could do the following:

```
portnum = socket.getservbyname( "smtp", "tcp" )
```

which would return, in this case, 25.

Notification

Let's now look at the concept of event notification for sockets. This capability is commonly provided by the **select** primitive. In Python, the **select** method is provided by the 'select' module.

As we saw with C, the descriptors representing our sockets are provided to the **select** method for which we desire notification when an event occurs. We can configure the **select** method to tell us when one or more channels are readable, writable, or if an error occurs on them. Further, we can also tell the **select** method to return after a configurable number of seconds if no event occurs. Consider the following Python TCP server in Listing 11.1.

Listing 11.1   Python TCP server illustrating the **select** method.

```
import socket
import select

serv = socket.socket( socket.AF_INET, socket.SOCK_STREAM )
serv.setsockopt( socket.SOL_SOCKET, socket.SO_REUSEADDR, 1 )

sa = ("localhost", 45000)

serv.bind( sa )
serv.listen( 1 )

# Server loop
while 1:

  cli, (remhost, remport) = serv.accept()

  # Client loop
  while 1:

    socki, socko, socke = select.select( [cli], [], [], 5 )
```

```
    if [socki, socko, socke] == [ [], [], [] ]:

      cli.send("Timeout!\n")

    else:

      for sock in socki:

        str = sock.recv( 100 )
        sock.send( str )
```

This very simple server awaits a client connection and then upon receiving one will echo whatever it receives from the subsequently created client socket. To determine when the client has sent something to be echoed, we use the **select** method. The **select** method works with descriptors, which can be either file or socket descriptors. In Windows, the **select** method works only with socket descriptors, but on Unix and on Macintosh, descriptors can be socket or file descriptors.

In the **select** method, we specify three lists representing our request for event notification. The first is a list of descriptors for which we want to be notified if a read event is generated (data is available on the descriptor for read). The second is a list of descriptors for write events, and the third is for error events (or exceptions). The final integer argument represents how many seconds to await an event before timing out. Note that we construct the lists as arguments to **select**. These lists could also be constructed externally using the list module methods.

The return of the **select** method are three lists that represent the descriptors for which events were generated. The three lists are in the same order as were provided to the **select** method (read, write, and exception). Our first check is to see whether a timeout occurred. This can be determined by checking the three lists to see if each is empty. We perform this check first, and if the timeout occurred, we emit a timeout message to the client through the client socket.

If the three lists are not empty, then based upon our configuration, we can assume that something changed in the read list (because this is the only list in which we identified a descriptor to the **select** method). We use an iterator here to step through the returned list, though this is overkill because we know that if a read event was generated, it was based upon our only defined descriptor (cli).

With a read event known, we use the **recv** method to read the available data from the client socket and then write it back out using the **send** method (the echo). We then loop back to the while loop awaiting a new read event for our client socket.

SPECIALIZED NETWORKING APIS FOR PYTHON

Python includes a number of specialized APIs for a number of Application layer protocols. These include SMTP, HTTP, POP3, FTP, Telnet, and others. In this section, we look at examples of the SMTP and HTTP client-side classes.

Python `smtplib` **Module**

SMTP is the Simple Mail Transfer Protocol and is used to transfer e-mail to a mail server that delivers it to the final destination. Python provides a very simple SMTP interface, as is illustrated in Listing 11.2. Three methods are shown, which is all that's necessary in this simple example to send an e-mail.

Listing 11.2   Sample `smtplib` client example.

```
import smtplib

recip = "you@yourdomain.com"
sender = "me@mydomain.com"
server = "yourdomain.com"
message = "From: me\nSubject: Hi\n\nHello\n\n"

smtpSession = smtplib.SMTP( server )

smtpSession.sendmail( sender, recip, message )
smtpSession.quit()
```

The very simple source in Listing 11.2 illustrates sending a short e-mail. We make the **smtplib** module visible using import and then set up our recipient (recip) and source e-mail (sender) addresses. Next, we define the server (where the SMTP client will connect to send the e-mail). Then, we define the e-mail message to be sent in the message string.

The complete SMTP process is then shown in the final three lines. We create a new instance of an SMTP client using **smtplib.SMTP**, defining the server to which we'll connect to send our e-mail. Next, we start the SMTP dialog with the server using the **sendmail** method, to which we provide the sender, recipient, and e-mail body. Finally, to end the session, we use the **finish** method that permits the remote SMTP server to deliver the message.

Python `httplib` **Module**

HTTP is the classic Web transport protocol used to transfer Web pages (and other content) across the Internet. The Python HTTP module provides a very simple and powerful client-side interface that can be used for a variety of purposes. A sample usage of the HTTP client is shown in Listing 11.3.

Listing 11.3   Sample `httplib` client example.

```
import httplib

hcli = httplib.HTTPConnection( "www.mtjones.com" )
hcli.request( 'GET', '/index.html' )
resp = hcli.getresponse()
```

```
if resp.status == 200:

  data = resp.read()
  print data

else:

  print "File not found."

hcli.close()
```

The result of the example shown in Listing 11.3 is the HTML source retrieved, the index.html file, from host www.mtjones.com. We first make the **httplib** module visible using the import statement and then create a new HTTP client instance using the **httplib.HTTPConnection** method, specifying the host to which we want to connect. Next, we request the file using the method **request**, specifying the particular HTTP method that we want to issue (in this case 'GET'). We can identify the status of the request using the **ge-tresponse** method that returns an **HTTPResponse** instance. Using this instance, we can check the success using the status attribute of the **HTTPResponse** object. If this is 200 (successful GET), then we issue a **read** on this **HTTPResponse** object to read the file. We then emit this using the print statement. If anything but a 200 status response was received, then we emit an error message. Finally, we close our HTTP client instance using the **close** method.

Python SocketServer Module

The **SocketServer** module helps simplify the development of TCP and UDP socket servers. It does this by providing a framework for which an application can define server information and then plug in handlers for serving content. Let's look at a simple example that reproduces our echo server discussed earlier in this chapter (see Listing 11.4).

Listing 11.4 Sample **SocketServer** example.

```
import SocketServer
import socket

class EchoHandler( SocketServer.StreamRequestHandler ):

  def handle( self ):

    while 1:

      line = self.rfile.readline()
      self.wfile.write( line )
```

```
# Create a new SocketServer and begin serving
serv = SocketServer.TCPServer( ('', 45000), EchoHandler )
serv.serve_forever()
```

We first create our EchoHandler class that creates only one method, **handle**. This method is called once a client connects to the particular server and performs whatever actions are necessary to service the client. The **handle** method simply operates in an infinite loop reading a line from the socket and writing it back to the client. Note that the methods used here are file-descriptor methods rather than socket methods. The attribute self.rfile is used to read from the connection and self.wfile is used to write to it.

To create our new server, we use the **SocketServer.TCPServer** method, specifying the address to which we'll locally bind and the class that is to be used to handle requests. Finally, we call the method **serve_forever** to begin serving clients.

SUMMARY

This short tour of Sockets programming in Python provided a discussion of the Python language in addition to many of the useful Sockets methods for the network application developer. Although Python has been around for some time, it continues to be actively developed, and has proven to be a popular and powerful object-oriented scripting language for the future.

REFERENCES

[Python-PSF] "Python Software Foundation," available online at *http://www.python.org/psf/.*

[Python-PEP] Warsaw, Barry, Jeremy Hylton, and David Goodger, "PEP Purpose and Guidelines," available online at *http://www.python.org/peps/pep-0001.html.*

[Python-FAQ] "The Whole Python FAQ," available online at *http://www.python.org/cgi-bin/faqw.py?req=all.*

RESOURCES

Python Home Page, available online at *http://www.python.org.*

Python Mailing Lists, available online at *http://www.python.org/psa/MailingLists.html.*

12 Network Programming in Perl

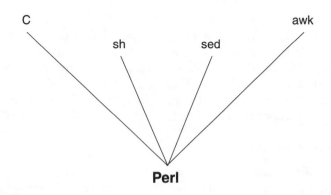

PERL LANGUAGE OVERVIEW

ON THE CD

In this chapter, we investigate network programming in the Perl scripting language. Perl is an acronym for "Practical Extraction and Report Language," (or its creator's alternative "Pathologically Eclectic Rubbish Lister"). Perl can be used as a quick prototyping language or a fully featured, object-oriented language suitable for developing large applications. As we'll see in this chapter, Perl provides a good foundation of services for the network application developer. All code for this chapter can be found on the companion CD-ROM at /software/ch12.

Origin of the Perl Language

Perl began as a quick language to replace awk (a simple scripting language) for the purpose of constructing reports for a bug-tracking system. Larry Wall was the creator and remains the primary maintainer of the language. The earliest records of Perl implementations date back to the late 1980s, but the language has gone through many changes since that time to become the efficient and compact language that it is today. Perl has found uses in a number of different application areas, including not only those reserved by system shells (such as text processing), but also for fast prototyping and Web development (CGI).

Perl Language Heritage

The Perl language, being interpreted, is ideal for quick prototyping and for production software development. It can be used both as a procedural language and as an object-oriented language. Perl has a number of influences ranging from C, to Unix shells, to Unix text-processing utilities. Perl's syntax was inspired primarily by C, but Unix shell (sh) influences are also apparent. Perl also finds some heritage in the older Unix text-processing utilities, sed and awk. Finally, Perl is an interpreted scripting language like Python and Ruby.

TOOLS

The current version of Perl at the time of this writing is Perl 5.8.0 (for which all examples and patterns in this book were written). A new version of Perl is currently under development, called Perl6. This version is the first that includes Perl community input into the design and development process.

Interpreter/Tools Used

Perl began as a tool for Unix platforms, but now runs on almost any operating system imaginable. Perl can be executed on Linux, native-Windows, Mac OS, and many, many others [Perl-Platforms].

ON THE CD

The sample code patterns presented here (and in Chapter 19, Network Code Patterns in Perl) and on the CD-ROM will execute on any of the supported environments.

Networking API Used

Perl is bundled with its own networking APIs, but lacks support for multicast sockets. The **IO-Socket-Multicast** module (0.25) by Lincoln D. Stein was used for the multicast patterns and should be downloaded through CPAN. The LWP (libwww-perl) package is also available on CPAN, which provides a number of HTTP-related methods.

Where to Download

Perl can be downloaded at [CPAN], which is the Comprehensive Perl Archive Network (*http://www.cpan.org*). This site includes both source and binary versions of the interpreter.

NETWORKING API FOR PERL

Perl includes multiple levels for Sockets access. At the bottom is the **Socket** module, which provides direct access to the native Sockets libraries on the host operating system. A higher-level module, **IO::Socket**, provides a mechanism for simplifying the creation of sockets. Finally, optional APIs are available for higher-level Application layer protocols, such as HTTP, FTP, SMTP, Telnet, and even CGI support.

The native socket module provides an API identical to that of the BSD standard API (see Figure 12.1). The **IO::Socket** and **IO::Socket::INET** (see Figure 12.2) provide a single constructor, **new**, as well as a number of methods (in the object-oriented sense). We discuss each of these shortly.

Function	*Purpose*
accept clientsocket, serversocket	Accept a new client socket from a server socket.
bind socket, localname	Bind the localname to the socket.
connect socket, name	Connect the socket to name.
getpeername socket	Get the socket address for the remote peer.
getsockname socket	Get the socket address for the local socket.
getsockopt socket, level, opname	Get the value of a socket option.
setsockopt socket, level, optname, optval	Set the value for a socket option.
listen socket, backlog	Enable accepting incoming client connections.
recv socket, scalar, length, flags	Receive a message from a socket.
send socket, msg, flags [,to]	Send a message to a socket.
shutdown socket, how	Shut down a socket.
socket socket, domain, type, protocol	Create a new socket of the defined type.
close socket	Close the socket.

FIGURE 12.1   Perl Sockets functions.

Function	Type	Purpose
new	Class	Creates a new IO::Socket::INET object.
sockaddr	Method	Return the socket local socket address.
sockport	Method	Return the socket local port.
sockhost	Method	Return the socket local socket address string.
peeraddr	Method	Return the peer socket address.
peerport	Method	Return the peer socket port.
peerhost	Method	Return the peer socket address string.

FIGURE 12.2   Perl IO::Socket::INET capabilities.

Sockets API Discussion

Let's now look at how the Perl Sockets API is used. We'll exercise many of the functions shown in Figure 12.1, including how using functions **IO::Socket** and **IO::Socket::INET** can simplify the creation of sockets.

Creating and Destroying Sockets

As with any socket-based application, the creation of a socket is the first step before communication may occur. Perl provides a number of ways in which a socket can be created. We look at the primitive example first, and then move on to Perl's more expressive methods that can simplify socket creation.

All of the methods discussed in this section require that the application developer make the sockets module visible. At the beginning of the perl script, a line specifying, "use Socket*" must be present. For object interfaces (*IO::Socket::INET*), "use* IO::Socket::INET*" must be present.*

The Socket module provides the primitive standard socket function that is most familiar to C language programmers. This class method can be used to create stream (TCP), datagram (UDP), and raw sockets.

```
socket( myStreamSock, Socket::AF_INET, Socket::SOCK_STREAM, 0 );
```

Note the similarities of this function to the C language primitive. The only real differences are the Perl naming conventions (SOCKET:: to bring the symbol in scope). In this example, we've created a stream socket. To create a datagram socket, we could do the following:

```
socket( myDgramSock, Socket::AF_INET, Socket::SOCK_DGRAM, 0 );
```

Finally, a raw socket could be created with:

```
socket( myRawSock, Socket::AF_INET, Socket::SOCK_RAW, 0 )
```

Recall that the Socket::AF_INET symbol defines that we're creating an Internet protocol socket. The second argument (type) defines the semantics of the communication (Socket::SOCK_STREAM for TCP, Socket::SOCK_DGRAM for UDP, and Socket::SOCK_RAW for raw IP). The third argument defines the particular protocol to use, which is useful here only in the Socket::SOCK_RAW case.

Sockets can also be created using the IO::Socket::INET module. This module contains a constructor, **new**, that is used to create sockets with very specific parameters. For example, rather than calling **socket** and then **bind** to bind a local address, the IO::Socket::INET->**new** function can create the socket and perform the bind in one operation. This simplifies the socket creation process. Additionally, socket options can also be applied (such as the ReuseAddr parameter, which provides the SO_REUSEADDR socket option). The available parameters for the IO::Socket::INET **new** constructor are shown in Figure 12.3.

Key	*Purpose*
PeerAddr	Remote address.
PeerPort	Remote port.
PeerHost	Remote host address.
LocalAddr	Local address.
LocalPort	Local port.
LocalHost	Local host address.
Proto	Protocol (name or number).
Type	Socket type.
Listen	backlog queue length.
ReuseAddr	Allow local reuse of the address.
Timeout	Timeout value for certain operations.
MultiHorned	Attempt to use all local interfaces.
Blocking	Enable blocking mode.

FIGURE 12.3   Options for IO::Socket::INET->**new** constructor.

Let's now look at a few examples of IO::Socket::INET to see how it can be used. In the first example, we simply create a stream socket and automatically connect to a peer:

```
use IO::Socket::INET;
$sock = IO::Socket::INET->new( PeerAddr => '192.168.1.1',
```

```
                                        PeerPort => 25,
                                        Proto    => 'tcp' );
```

In this case, the $sock represents a new client socket that has been connected to the host identified by IP address '192.168.1.1', port 25 using the TCP protocol. Let's now look at a server socket creation example using the same constructor:

```
use IO::Socket::INET;
$sock = IO::Socket::INET->new( LocalAddr => '192.168.1.1',
                               LocalPort => 25,
                               Listen    => 5,
                               Proto     => 'tcp' );
```

In this example, the local address and port keys specify to what the new socket will be locally bound. The Listen key defines the backlog queue length (as is typical with the **listen** call), and, finally, Proto defines the protocol that we want to use, in this case TCP.

When we're finished with a socket, we must close it. To close our previously created socket, $sock, we use the **close** operator.

```
close $sock
```

Once the **close** method is called, no further communication is possible with this socket.

Any data queued for transmission would be given some amount of time to be sent before the connection physically closes. The shutdown operator is used as follows:

```
shutdown( $sock, 0 );     # No further reading possible
shutdown( $sock, 1 );     # No further writing possible
shutdown( $sock, 2 );     # No further use of this socket
```

The IO::Socket::INET object can also use **shutdown** in the following manner:

```
$sock->shutdown( 2 );     # No further use of this socket
```

Socket Addresses

Like C, Perl has the concept of a sockaddr_in structure and a set of associated functions. Recall that the sockaddr_in structure contains not only an address (32-bit IPv4 address), but also a 16-bit port value.

Consider the following C language fragment:

```
struct sockaddr_in sa;

sa.sin_family = AF_INET;
sa.sin_port = htons( 13 );
sa.sin_addr.s_addr = inet_addr("192.168.1.2");
```

This creates a `sockaddr_in` structure and fills it in with IP address "192.168.1.2" and port 13 (using the `AF_INET` family). Consider this synonymous code fragment from Perl:

```
use Socket;

...

my $addr = inet_aton( "192.168.1.2" );

my $paddr = sockaddr_in( 13, $addr );
```

We first convert the dotted-notation string IP address using **inet_aton** (as was done in C with **inet_addr**) and then use this to create the packed address using **sockaddr_in**. The port number is provided (13) resulting in the packed address $paddr.

Given a packed address (such as $paddr in the previous example), we can extract the address and port elements as shown in the following code snippet:

```
( $myport, $myaddr ) = sockaddr_in( $paddr );

print "The port is ", $myport, "\n";

print "The IP address is ", inet_ntoa( $myaddr ), "\n";
```

Using C-style structures as `sockaddr_in` makes it easy for C programmers to build Perl scripts, but because Perl also provides simpler methods, understanding the C address paradigm is not necessary.

Socket Primitives

In this section, we look at a number of other important socket control methods. We also investigate the variety of ways that the same results can be achieved using different Perl methods.

Bind

The **bind** method provides a local naming capability to a socket. This can be used to name either client or server sockets, but is used most often in the server case. The **bind** function is provided as follows:

```
use Socket;

socket( mySock, Socket::AF_INET, Socket::SOCK_STREAM, 0 );

my $port = 25;
my $addr = inet_aton( "192.168.1.1" );
my $paddr = sockaddr_in( $port, $addr );

bind( mySock, $paddr );
```

where $paddr is a packed address of "192.168.1.1", port 25. The same thing could be accomplished using the IO::Socket::INET, as:

```
$mySock = IO::Socket::INET->new( LocalAddr => '192.168.1.1',
                                 LocalPort => 25,
                                 Proto     => 'tcp',
                                 Listen    => 5 );
```

In this example, we're creating a stream server on port 25 that will accept client connections only on the interface defined as '192.168.1.1'. Note also in this example that the **listen** function is also called, which was not performed in the prior example.

If we want to accept client connections from any available interface, we could bind to INADDR_ANY. This behavior is synonymous to the INADDR_ANY symbolic within our C language examples, for instance:

```
use Socket;

socket( mySock, Socket::AF_INET, Socket::SOCK_STREAM, 0 );

my $port = 25;
my $paddr = sockaddr_in( $port, INADDR_ANY );

bind( mySock, $paddr );
```

Note that we could also pass a 0 as the port to allow the networking layer to dynamically assign us a port (ephemeral port).

Finally, let's look at a UDP example. The following example creates a UDP socket and binds it with host INADDR_ANY (accept datagrams from any interface) with port 13:

```
use Socket;

socket( mySock, Socket::AF_INET, Socket::SOCK_DGRAM, 0 );

my $port = 13;
my $paddr = sockaddr_in( $port, INADDR_ANY );

bind( mySock, $paddr );
```

Listen

Before a server socket can accept incoming client connections, it must call the **listen** method to declare this willingness. The **listen** function is illustrated using the mySock TCP server socket:

```
listen( mySock, backlog );
```

The `mySock` variable represents an instance of the server socket and the `backlog` argument represents the number of outstanding client connections that may be queued. This method may only be used on stream sockets. Here's a complete example from socket creation to the **listen** method.

```
use Socket;

socket( mySock, Socket::AF_INET, Socket::SOCK_STREAM, 0 );

my $port = 25;
my $addr = inet_aton( "192.168.1.1" );
my $paddr = sockaddr_in( $port, $addr );

bind( mySock, $paddr );

listen( mySock, 5 );
```

This is identical to the following code fragment as created with IO::Socket::INET:

```
$mySock = IO::Socket::INET->new( LocalAddr => '192.168.1.1',
                                 LocalPort => 25,
                                 Proto     => 'tcp',
                                 Listen    => Socket::SOMAXCONN );
```

It should be clear from this example that the IO::Socket::INET constructor method provides not only server socket creation, but also local address binding and the **listen** call, thus simplifying the code considerably. Rather than specify an arbitrary backlog size, the maximum connections constant can be used instead, Socket::SOMAXCONN.

Accept

The **accept** method is the final call made by servers to accept incoming client connections. Before **accept** can be invoked, the server socket must be created, a name must be bound to it, and **listen** must be invoked. The **accept** function blocks and then returns a new client socket upon the client receiving a new incoming client connection. The following example illustrates a simple socket that accepts a new client socket and then writes a short message to it (acctest.pl).

The **accept** method in the **Sockets** class accepts a new client connection and returns an array containing the new client socket and the client address information (packed in a sockaddr_in string):

```
use Socket;

socket( mySock, Socket::AF_INET, Socket::SOCK_STREAM, 0 );
```

```
my $paddr = sockaddr_in( 5300, INADDR_ANY );

bind( mySock, $paddr );

listen( mySock, Socket::SO_MAXCONN );

accept( cliSock, mySock );

print cliSock "Hello!\n";

close cliSock;
close mySock;
```

In this example, the **accept** function returns a new client socket (cliSock), which is then used for further communication with the client. We could identify the source of the connection as follows:

```
$peer = getpeername cliSock;

($port, $ipaddr) = unpack_sockaddr_in( $peer );

print $port, "\n";
print inet_ntoa( $ipaddr ), "\n";
```

Connect

The **connect** function is used by client Sockets applications to connect to a server. Clients must have created a socket and then defined an address structure containing the host and port number to which they want to connect. This is illustrated (conntest.pl) by the following code segment (which can be used to connect to the previously illustrated **accept** method example):

```
use Socket;

socket( mySock, Socket::AF_INET, Socket::SOCK_STREAM, 0 );

$addr = inet_aton( "127.0.0.1" );
$paddr = sockaddr_in( 5300, $addr );

connect( mySock, $paddr );

$line = <mySock>;

print $line;

close mySock;
```

We create our address structure $paddr (as with previous examples) and then pass this to our **connect** method using our client socket instance. We could also write this using IO::Socket::INET as (conntst2.pl):

```
use IO::Socket::INET;

$mySock = IO::Socket::INET->new(
          PeerAddr => '127.0.0.1',
          PeerPort => 5300,
          Type     => Socket::SOCK_STREAM );

$line = <$mySock>;

print $line;

close $mySock;
```

The **connect** method (for TCP) blocks until either an error occurs, or the three-way handshake with the server completes. For datagram sockets (UDP), the method binds the peer locally so that the peer address isn't required to be specified for every send.

Sockets I/O

A variety of API methods exists to read data from a socket or write data to a socket. Some functions are Socket module specific, and others are inherited from other modules. First, we look at reading and writing from connected sockets and then investigate unconnected (datagram-specific) socket communication.

A number of methods are provided to communicate through a socket. Some of the functions include **recv**, **send**, **print**, **sysread**, and **syswrite**. We will look at examples of each of these functions in the following sections. Other mechanisms can be used for input and output, but for the purposes of brevity, they won't be covered here.

Stream Sockets

Stream sockets can utilize all of the previously mentioned I/O methods. Datagram sockets must utilize methods that allow the specification of the recipient (for example, the **send** function). Let's now look at some sample code that illustrates the stream-specific methods. The first example illustrates a simple echo server:

```
use Socket;

socket( srvSock, Socket::AF_INET, Socket::SOCK_STREAM, 0 );

$addr = inet_aton( '192.168.1.2' );
$paddr = sockaddr_in( 45000, $addr );
```

```
bind( srvSock, $paddr );
listen( srvSock, Socket::SO_MAXCONN );

while( true ) {

    accept( cliSock, srvSock );

    recv( cliSock, $line, 128, 0 );

    print cliSock $line;

    close cliSock;

}

close srvSock;
```

In this example, we create a new server socket using the **socket** function. We create a sockaddr_in structure representing IP address '192.168.1.2' and port 45000 and then bind ourselves to it. After **listen**, we accept a new client socket, read a line from it using **recv**, and then write the line back to the client using the print function. Finally, the client socket is closed with the **close** function, and then the loop repeats.

Now, let's look at the TCP client socket that will connect to the previously defined server:

```
use IO::Socket::INET;

$sock = IO::Socket::INET->new(
        PeerAddr => '192.168.1.2',
        PeerPort => 45000,
        Proto    => 'tcp' );

print $sock "Hello\n";

print <$sock>;

close $sock;
```

After we create our TCP client socket using IO::Socket::INET, we immediately send our test string to the server using the print function. We use the diamond operator (<>) as the means to read from $sock and then anonymously pass this to the print function. Finally, the client socket is closed using the **close** function.

That's simple so far. Now, let's try the **sysread** and **syswrite** functions. The TCP server now looks like (strmtst2.pl):

```
use Socket;

socket( srvSock, Socket::AF_INET, Socket::SOCK_STREAM, 0 );

$addr = inet_aton( '192.168.1.2' );
$paddr = sockaddr_in( 45000, $addr );

bind( srvSock, $paddr );
listen( srvSock, Socket::SO_MAXCONN );

while( true ) {

    accept( cliSock, srvSock );

    sysread( cliSock, $line, 128, 0 );

    syswrite( cliSock, $line, length( $line ) );

    close cliSock;

}
```

After setting up the server socket, we enter the client accept loop. In this example, we use the **sysread** and **syswrite** functions to read from and write to the socket. The prototypes for the **sysread** and **syswrite** functions are:

```
sysread handle, scalar, length, offset;

syswrite handle, scalar, length;
```

Both functions take the socket to use for I/O and the scalar (string buffer, here $line). For **sysread**, we define the total length that we want to receive in the scalar argument and the offset into scalar we want the input data to be placed. For **syswrite**, we provide the scalar buffer and the length that we want to write.

Our client changes similarly (strmcli2.pl):

```
use IO::Socket::INET;

$sock = IO::Socket::INET->new(
        PeerAddr => '192.168.1.2',
        PeerPort => 45000,
        Proto    => 'tcp' );

$line = "Hello!\n";

syswrite( $sock, $line, length($line) );
```

```
sysread( $sock, $nline, 128, 0 );

print $nline, "\n";

close $sock;
```

After sending our single line with the **syswrite** function, we collect the response from the server with the **sysread** function, placing the read buffer into $nline.

Datagram Sockets

A variation of **send** is used exclusively by datagram sockets, while the **recv** function is used also (through the return of **recv**, which is the source address of the datagram). What differentiates these calls from our previously discussed stream calls is that these calls include addressing information. Because datagrams are not connected, we must define the destination address explicitly. Conversely, when receiving a datagram, the address information is also provided so that the source can be identified.

Let's look at a datagram server and client that provide the echo functionality illustrated previously by our stream socket examples. Our datagram server takes on a slightly different form. Additionally, we'll use the IO::Socket::INET module solely for both the server and the client (dgrmserv.pl).

```
# Server example

use IO::Socket::INET;

$sock = IO::Socket::INET->new(
            LocalAddr => 'localhost',
            LocalPort => 45000,
            Proto     => 'udp',
            ReuseAddr => 1 );

while (true) {

    $from = recv( $sock, $line, 128, 0 );

    send( $sock, $line, 0, $from );

}

close $sock;
```

Using the object interface (IO::Socket::INET), we create a new SOCK_DGRAM socket (identified by protocol 'udp') and automatically bind it to localhost and port 45000. Additionally, we specify the ReuseAddr socket option (SO_REUSEADDR) to enable immediate reuse of the local address. We then enter our infinite loop where we await a datagram and then bounce the datagram back to the source. Using the **recv** function (which can also be used on object interface sockets), we receive the datagram and store the payload into $line. We also store the result of **recv** into $from, which represents the source of the datagram. Next, we immediately send the datagram back to the source using the $from variable to direct the datagram in the **send** function.

In the client, we'll explore some of the object-oriented methods that are possible. Rather than call a function with the socket as the argument, we'll use the socket argument and invoke a method attached to it (dgramcli.pl).

```perl
# Client example

use IO::Socket::INET;

$sock = IO::Socket::INET->new(
        PeerAddr => 'localhost',
        PeerPort => 45000,
        Proto    => 'udp' );

$line = "Hello!\n";

$sock->send( $line );

recv( $sock, $nline, 128, 0 );

print $line;

close $sock;
```

We create the UDP socket with the **new** constructor, and this time use the PeerAddr and PeerPort to identify where this socket should be connected. Note that SOCK_DGRAM sockets (UDP) are disconnected by default; the only purpose for connection-based datagram sockets is so that destination address information need not be specified for each outgoing datagram. We construct a string that we'll send ($line) and then send this to the peer using the **send** function. Note the object-oriented usage of this call (object->method) and the fact that no destination was specified. This is because we specified the destination address when we created the socket. We expect a response from the server, so we use the standard **recv** call here to get the response and then emit it to standard-out using the print function. Finally, we close the socket using the **close** function.

Socket Options

Socket options permit an application to change some of the modifiable behaviors of sockets and the methods that manipulate them. For example, an application can modify the sizes of the send or receive socket buffers or the size of the maximum segment used by the TCP layer for a given socket.

Socket options are very simple in Perl. Options can be specified with the IO::Socket::INET with the **new** constructor, or with the standard **setsockopt/getsockopt** API. We'll look at examples of both.

Let's look at a simple example first. Let's say that we want to identify the size of the receive buffer for a given socket (option SO_RCVBUF). This can be done with the following code segment (sockopt1.pl):

```perl
use IO::Socket::INET;

sock = IO::Socket::INET->new(
            LocalAddr => 'localhost',
            LocalPort => 5000,
            Proto     => 'tcp' );

print $sock->getsockopt( Socket::SOL_SOCKET,
                         Socket::SO_RCVBUF );

close $sock;
```

First, we create a new stream server socket (allow incoming client connections on any interface to port 23). Using the **getsockopt** call, we extract the receive buffer size for the socket. We specify the Socket::SOL_SOCKET argument because we're interested in a Sockets layer option (as compared to a TCP layer option). For the receiver buffer size, we specify Socket::SO_RCVBUF. The value returned from **getsockopt** is then emitted to standard-out using the print function.

Now, let's look at how an option is retrieved for a given socket, specifically, the linger option. Socket linger allows us to change the behavior of a stream socket when the socket is closed and data is remaining to be sent. After **close** is called, any data remaining will attempt to be sent for some amount of time. If after some duration, the data cannot be sent, then the data to be sent is abandoned. The time after the **close** to when the data is removed from the send queue is defined as the linger time. In Perl, we must deconstruct the structure that is provided by the host environment (sockopt2.pl).

```perl
use IO::Socket::INET;

$sock = IO::Socket::INET->new( Proto => 'tcp' );

$ling = $sock->getsockopt( Socket::SOL_SOCKET, Socket::SO_LINGER );

($ena, $val) = unpack( 'ii', $ling );

print $ena, "\n";
print $val, "\n";
close $sock;
```

The SO_LINGER option returns a binary packed structure that contains first an enable (1 for enable, 0 for disable) and then the time for linger in seconds. Using the unpack function, we unpack the binary structure into two independent 32-bit values. These are returned to $ena and $val, respectively. The default is 0 for both values, representing that the linger option is disabled. We could configure the linger option for enabled and 10 seconds as:

```
use IO::Socket::INET;

$sock = IO::Socket::INET->new( Proto => 'tcp' );

my $linger = pack('ii', 1, 120);

$sock->setsockopt( Socket::SOL_SOCKET, Socket::SO_LINGER, $linger );
close $sock;
```

The pack function takes the two arguments and packs them into two 32-bit binary values (as specified by the 'ii' template).

Other Miscellaneous Functions

Let's now look at a few miscellaneous functions from the various Sockets APIs and the capabilities they provide. The DNS resolver permits us to resolve a host name to an IP address, or vice versa. Method **gethostbyname** provides name resolution given an FQDN; for example:

```
($name, $aliases, $addrtype, $length, @addrs) =
    gethostbyname( "www.microsoft.com" );
```

where @addrs is a list of IP addresses in binary notation and $length refers to the number of addresses that were found. To emit the first IP address for the given domain in dotted-string notation, we could do:

```
print inet_ntoa( $addrs[0] ), "\n";
```

The other return variables are the official host name ($name), the official alias ($aliases), the address type (2 for Socket::AF_INET), the number of addresses found ($length), and finally the list of addresses (@addrs).

What if we want to go in the opposite direction, providing an IP address and get in return an FQDN? To achieve this, we use the **gethostbyaddr**:

```
$myaddr = inet_aton( "207.46.134.190" );

($name, $aliases, $addrtype, $length, @addrs) =
    gethostbyaddr( $myaddr, Socket::AF_INET );

print $name, "\n";
```

Now, let's consider the problem of identifying the port number for a given service. To retrieve the service information given a service name (such as 'www'), the following can be done:

```
($name, $aliases, $port, $proto) = getservbyname( 'www', 'tcp' );
```

The values for this example include "www" for $name, "http" for $aliases, 80 for $port, and 'tcp' for $proto. We can also retrieve the service by port number, such as:

```
($name, $aliases, $port, $proto) = getservbyport( 25, 'tcp' );
```

which would result in values "smtp" for $name, "mail" for $aliases, 25 for $port, and 'tcp' for $proto.

Notification

As a final topic for Perl, let's look at the concept of event notification for sockets. This capability is commonly provided by the **select** function.

As we saw with C, the descriptors representing our IO channels are provided to the **select** method to identify when some event occurs. We can configure the **select** method to tell us when a channel is readable, writable, or if an error occurs on it. Further, we can also tell the **select** method to return after a configurable number of seconds if no event occurs. Consider the following Perl TCP server in Listing 12.1

Listing 12.1   Perl TCP server illustrating the `select` method.

```
use Socket;

socket( srvSock, Socket::AF_INET, Socket::SOCK_STREAM, 0 );

$addr = inet_aton( "192.168.1.2" );
$paddr = sockaddr_in( 45000, $addr );

setsockopt( srvSock, Socket::SOL_SOCKET, Socket::SO_REUSEADDR, 1 );

bind( srvSock, $paddr );
listen( srvSock, 1 );

while( 1 ) {

  print "In the big loop\n";

  accept( cliSock, srvSock );

  $timeout = 5;

  # Client loop
  while( 1 ) {
```

```
$rin = '';
vec( $rin, fileno(cliSock), 1 ) = 1;

$nfound = select( $rout = $rin, undef, undef, $timeout );

if ($nfound) {

  if ( vec($rout, fileno(cliSock), 1) == 1 ) {

      # Don't mix select recv/send -- use sysread/syswrite instead
      sysread( cliSock, $line, 128, 0 );

      syswrite( cliSock, $line, length($line), 0 );

  }

} else {

  $to = "Timeout!\n";
  syswrite( cliSock, $line, length($line), 0 );

}

  }

}
```

This very simple server awaits a client connection and then upon receiving one will echo whatever it receives from the client. To determine when the client has sent something to be echoed, we use the **select** function. The **select** method works with file handles, so the fileno function must be used to convert the socket descriptor into a file number. We use the vec function to set the file handle in the read-set vector ($rin). In the **select** function, we can specify three vectors representing our request. The first vector indicates the file handles for which we want to be notified if a read event is generated (data is available on a given file handle for read). The second is a vector indicating write events, and the third is for error events. The last argument represents how much time to await an event before timing out. Note in this example, that we only specify read events and the timeout. For vectors that are not used, we pass in undef, which acts like a NULL vector (unspecified).

The return of **select** is the number of file handles that were set within vectors (in this case, the $rin vector). In this usage, the $rout is the test vector that we'll use to check to see what happened. As long as $nfound (number of bits set in the vectors) is non-zero, we'll check to see if our file-handles bit was set in $rout using the vec function. If it was set, we then use **sysread** to read from the socket and then loop the data back to the client using **syswrite**. If the timeout arrives before any event ($nfound is zero), we notify the peer of this event by writing "Timeout!" to the client.

SPECIALIZED NETWORKING APIS FOR PERL

Perl includes a number of specialized APIs for a number of Application layer protocols (some are included, others must be installed via CPAN). These include SMTP, HTTP, CGI, POP3, FTP, Telnet, and others. In this section, we look examples of the SMTP and HTTP client-side modules.

Perl `Net::SMTP` Module

SMTP is the Simple Mail Transfer Protocol and is used to transfer e-mail to a mail server that delivers it to the final destination. Perl's `Net::SMTP` module provides a very flexible SMTP interface, as is illustrated in Listing 12.2.

Listing 12.2 Sample `Net::SMTP` client example (smtpc.pl).

```perl
use Net::SMTP;

my $server = "mail.mtjones.com";

$smtp = Net::SMTP->new( $server ) or
  die "Couldn't connect to the server";

$smtp->mail( "tim\@mtjones.com" );
$smtp->to( "mtj\@mtjones.com" );

$smtp->data();
$smtp->datasend("Subject: Test email\n");
$smtp->datasend("Content-Type: text/html\n");
$smtp->datasend("\n");

$smtp->datasend(
  "<HTML><BODY><H1>This is the test email</H1></BODY></HTML>" );

$smtp->dataend();

$smtp->quit();
```

The very simple source in Listing 12.2 illustrates sending a short HTML encoded e-mail. We make the **Net::SMTP** module visible using use and then create a local string variable to hold our server address ($server). Using the **new** constructor of Net::SMTP, we create a new SMTP object named $smtp. We pass in the server name to which we'll ultimately connect to send our mail (the outgoing mail transfer agent).

We then begin using the $smtp object to load the e-mail to send. Using the **mail** function, we specify from whom the e-mail is coming and using **to**, we specify to whom the e-mail will go. The **data()** function begins the e-mail transaction, which is then followed by a number of **datasend** functions. The **datasend** function is simply a way of providing the $smtp object with a set of data to send to the server. In this case, we send the subject of the e-mail and the content type (to allow the client to understand how to render the final e-mail). We end the **datasend** with a blank line, to separate the optional SMTP headers from the body of the e-mail. The next **datasend** contains the body of the e-mail followed by a **dataend** that specifies that the e-mail transaction is complete (terminates the SMTP session with the server). Finally, we call the **quit** method to close the session with the server.

Perl LWP::Simple **Module**

HTTP is the classic Web transport protocol used to transfer Web pages (and other content) across the Internet. Numerous methods exist for Perl in HTTP, but the simplest is likely the LWP package (libwww-perl). A sample usage of LWP::Simple for HTTP client functionality is shown in Listing 12.3.

Listing 12.3   Sample LWP::Simple client example (httpc.pl).

```
use LWP::Simple;

$url = "http://www.microsoft.com";
getprint( $url ) or die "getprint failed.";
```

In the example shown here (Listing 12.3), the LWP::Simple module is used to grab and print content from a given URL ($url). The getprint function retrieves and then prints the HTML page. LWP provides a number of modules for accessing Web content. The simplest is the Simple module; for more complex operations, the LWP::UserAgent can be used.

That's extremely simple, compared to what actually goes on underneath the covers of HTTP. Therefore, Perl and its accompanying modules provide some very useful and simplifying abstractions for higher-level protocols.

SUMMARY

This very quick tour of Sockets programming in Perl provided a short discussion of the Perl language and the Sockets methods available within it. Perl provides a nice set of features for Sockets programming, including those that mimic the standard BSD API and additional modules that help simplify network application development. A large variety of abstractions for Application layer protocols can also be imported (from CPAN) that further simplifies complex network applications. Perl continues to be developed and advanced. Perl6 is a rewrite of the interpreter and includes much input from the Perl development community. A major feature of this new release will be simplified maintenance of Perl moving forward.

REFERENCES

[CPAN] Comprehensive Perl Archive Network, available online at *http://www.cpan.org*.
[Perl-Platforms] Perldoc.com/Perlport, available online at *http://www.perldoc.com/perl5.8.0/pod/perlport.html#Supported-Platforms*.

RESOURCES

O'Reilly's Perl Web site, available online at *http://www.perl.com*.
Perl Mongers, available online at *http://www.perl.org*.

13 Network Programming in Ruby

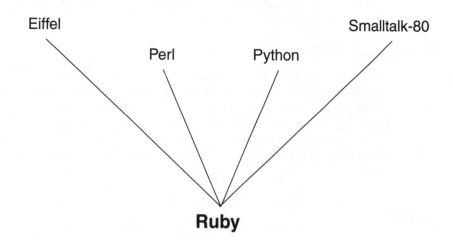

Eiffel   Perl   Python   Smalltalk-80

Ruby

RUBY LANGUAGE OVERVIEW

In this chapter, we investigate network programming in the Ruby scripting language. Ruby is a very interesting pure object-oriented scripting language that has a wide multitude of uses. As we'll see in this chapter, network programming is very simple in the Ruby language. All code for this chapter can be found on the companion CD-ROM at /software/ch13.

ON THE CD

Origin of the Ruby Language

The Ruby language is the brainchild of Yukihiro Matsumoto of Japan. Ruby's first release dates back to 1995, but it has continued to evolve today into a useful production language for a variety of purposes, including text processing, application glue, systems programming, and network application development.

Ruby Language Heritage

The Ruby language is both imperative and object-oriented in nature. (The interpreter turns imperative programs into object-oriented programs.) Ruby has a number of influences, depending upon the particular language feature. Ruby's syntax was inspired by Eiffel and Perl (and to some extend, Ada). Ruby includes useful exception-handling features like Python and Java. Ruby is a pure object-oriented language like Smalltalk. Finally, Ruby is an interpreted scripting language like Perl and Python and can extend other languages such as C [Ruby-FAQ].

TOOLS

There currently exists a single implementation of Ruby, from its creator Yukihiro Matsumoto. Ruby is still under active development, but is stable enough for production software development. The version of Ruby used in this chapter (and Chapter 20, Network Code Patterns in Ruby) is 1.6.8.

Interpreter/Tools Used

The Ruby interpreter operates on Linux, native Windows, and Windows using the Cygwin environment. Ruby includes its own threading model and, therefore, provides multi-threading capabilities that are transparent to the underlying operating system.

The sample code patterns presented here (and in Chapter 20) and on the CD-ROM will execute on any of the supported environments.

ON THE CD

Networking API Used

Ruby is bundled with its own networking APIs; therefore, no other downloads are necessary for the Sockets API.

Where to Download

The Ruby interpreter and tools may be downloaded at its source, *http://www.ruby-lang.org* (under "Download"). The Ruby site includes both source and binary versions of the interpreter. If the Cygwin environment for Windows is to be used, it may be downloaded at *http://www.cygwin.com/*.

NETWORKING API FOR RUBY

Ruby includes a rich set of networking APIs that operate at two layers. The standard Sockets API is provided and higher-level Application layer protocols such as HTTP, FTP, SMTP, Telnet, and even CGI are supported.

The Sockets API is defined as a class hierarchy that provides a range of capabilities (see Figure 13.1). All networking classes are derived from the **IO** base class. Class **BasicSocket** is the abstract base class from which all other sockets classes are derived. The **Socket** class is the low-level Sockets API class that mimics the BSD4.4 API. The **IPSocket** class is the base class for both the **TCPSocket** class and the **UDPSocket** class. Finally, the **TCPServer** class provides specific behaviors for server-side TCP sockets.

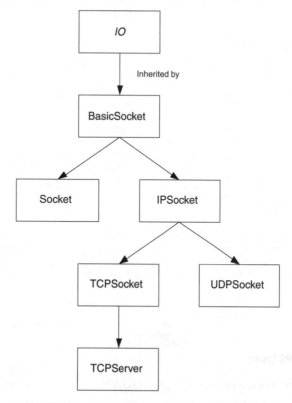

FIGURE 13.1 Ruby Sockets API class hierarchy.

Class vs. Instance Methods

In the following sections, we discuss two distinct types of methods, class methods and instance methods. A class method is a method that operates on the class, whereas an instance method operates on an instance of the class. Let's look at an example using the **Socket** class. We can use the class method **new** to create a new socket, which is performed as:

```
sock = Socket::new( Socket::AF_INET, Socket::SOCK_STREAM, 0 )
```

which creates a new socket (a stream socket), stored in the instance variable sock. An example of an instance method uses the previously defined sock instance. For example, after we've bound this socket with a local address, we could accept new incoming connections with:

```
clisock = sock.accept
```

Just remember that a class method operates directly on the class, whereas an instance method operates on an instance of the class.

Sockets API Summary

We saw in Figure 13.1 that Ruby provides a number of classes for Sockets programming; let's now look at the classes and methods that are available.

BasicSocket Class

The **BasicSocket** class provides the base for all other socket classes. It provides a base set of features that are inherited and used through the other socket classes. The **BasicSocket** class is shown in Figure 13.2.

Method	*Type*	*Description*
BasicSocket::do_no_reverse_ lookup	Class	Returns 'true' if a DNS query returns a numeric IP address instead of a hostname.
BasicSocket::do_not_reverse_	Class	Sets the behavior for reverse lookup.
getpeername lookup=bool	Instance	Returns the sockaddr_in information for the socket peer.
getsockname	Instance	Returns the sockaddr_in information for the local socket.
Getsockopt(level, optname)	Instance	Returns the option value for the specified socket option.
setsockopt(level, optname, value	Instance	Sets the option value for the specified socket option.
shutdown([howl])	Instance	Performs a shutdown of the socket instance.
recv(len, [flags])	Instance	Receives data from the socket and returns a string.
send(buffer, flags [, to])	Instance	Sends data through a socket. The optional to represents a sockaddr_in binary string.

FIGURE 13.2 BasicSocket class methods. (Adapted from [Matsumoto01] Matsumoto, Yukihiro, *Ruby in a Nutshell*, O'Reilly & Associates, 2001.)

All methods provided by the **BasicSocket** class are accessible by instances in all other networking classes (because all are derived from the **BasicSocket** class).

Socket **Class**

The **Socket** class provides the closest abstraction to the BSD4.4 API as is possible with Ruby. This class provides the basic set of methods, most of which are synonymous with the standard Sockets API functions. The methods provided in the **Sockets** class are shown in Figure 13.3.

Method	*Type*	*Description*
`Socket::getaddrinfo( host, port [, family[, type[, proto[, flags]]]] )`	Class	Create a `sockaddr_in` structure using the `port [, family[, type[, proto[, flags]]]] )` provided data.
`Socket::gethostbyaddr( addr [, type] )`	Class	Returns a `sockaddr_in` structure for the address.
`Socket::gethostbyname( name)`	Class	Returns an array containing host information for name.
`Socket::gethostname`	Class	Returns a string of the local hostname.
`Socket::getnameinfo( addr[, flags] )`	Class	Returns an array containing name information for the defined address.
`Socket::getservbyname( service[, proto] )`	Class	Return the port number for the defined service.
`Socket::new( domain, type, proto )Socket::open( domain, type, proto )`	Class	Create a new socket.
`accept`	Instance	Accept a client connection through a server socket.
`addr`	Instance	Returns a `sockaddr_in` structure.
`bind( addr )`	Instance	Binds the socket to the address information contained in addr (`sockaddr_in`).
`connect( addr )`	Instance	Connects the client socket to the server defined in addr.
`listen( backlog )`	Instance	Put the server socket into the listening state with an established queue of size backlog.
`recvfrom( len[, flags] )`	Instance	Returns the data and source address information in an array.
`peeraddr`	Instance	Returns a `sockaddr_in` structure for the peer socket.

FIGURE 13.3 Socket class methods. (Adapted from [Matsumoto01] Matsumoto, Yukihiro, *Ruby in a Nutshell*, O'Reilly & Associates, 2001.)

IPSocket **Class**

The **IPSocket** class provides the base class for all Internet protocol sockets. For this reason, both **TCPSocket** and **UDPSocket** classes are derived from the **IPSocket** class. The methods provided in the **IPSocket** class are shown in Figure 13.4.

Method	*Type*	*Description*
IPSocket::getaddress(host) address.	Class	Resolves the fully qualified domain name to an IP.
addr	Instance	Returns the sockaddr_in structure for the local socket.
peeraddr	Instance	Returns the sockaddr_in structure for the peer socket.
recvfrom(len [, flags])	Instance	Returns the received data from the socket in an array with the source address information.

FIGURE 13.4 **IPSocket** class methods. (Adapted from [Matsumoto01] Matsumoto, Yukihiro, *Ruby in a Nutshell*, O'Reilly & Associates, 2001.)

Although this book focuses on IPv4, Ruby supports IPv6 if the underlying platform supports it.

UDPSocket **Class**

The **UDPSocket** class provides a small set of methods that focus on the User Datagram Protocol (UDP). Although the **Sockets** class provides everything that is needed for both TCP and UDP Sockets programming, the **UDPSocket** class provides two simple constructor methods to create UDP sockets simply. The methods provided in the **UDPSocket** class are shown in Figure 13.5.

Method	*Type*	*Description*
UDPSocket::new([socktype]) UDPSocket::open([socktype])	Class	Create a new UDP Socket.
bind(host, port)	Instance	Binds the host and port information to the local socket.
connect(host, port)	Instance	Connect the socket to the named port on host.
send(buffer, flags [, to]) send(buffer, flags [, host, port])	Instance	Send the buffer using a sockaddr_in binary string (to), or a host and port number. Returns sent length.

FIGURE 13.5 **UDPSocket** class methods. (Adapted from [Matsumoto01] Matsumoto, Yukihiro, *Ruby in a Nutshell*, O'Reilly & Associates, 2001.)

TCPSocket **Class**

The **TCPSocket** class provides two methods for simple construction of TCP sockets. These methods are shown in Figure 13.6.

Method	Type	Description
TCPSocket::new(host, service) TCPSocket::new(host, service)	Class	Creates a new TCP socket and connects it to service on host. Service may be a string or port number.

FIGURE 13.6 **TCPSocket** class methods. (Adapted from [Matsumoto01] Matsumoto, Yukihiro, *Ruby in a Nutshell*, O'Reilly & Associates, 2001.)

Like **UDPSocket**, the **TCPSocket** methods could be done using the **Sockets** class, but with a greater number of method calls to achieve the same result.

TCPServer **Class**

Finally, the **TCPServer** class provides a small number of methods specifically for the creation of TCP server sockets. These methods are shown in Figure 13.7.

Method	Type	Description
TCPServer::new([host,] service) TCPServer::new([host,] service)	Class	Creates a new server socket (synonymous with socket/bind/listen).
accept	Instance	Accept a new client connection (blocking call).

FIGURE 13.7 **TCPServer** class methods. (Adapted from [Matsumoto01] Matsumoto, Yukihiro, *Ruby in a Nutshell*, O'Reilly & Associates, 2001.)

Sockets API Discussion

Let's now look at how the Ruby Sockets API is used. We exercise many of the methods shown in the previous class method figures, including how using methods from the **UDP-Socket** and **TCPServer** can greatly simplify Sockets programming.

Creating and Destroying Sockets

As with any socket-based application, the creation of a socket is the first step before communication may occur. Ruby provides a number of methods for socket creation, depending upon the type of socket needed by the developer. We look at the primitive example first, and then move on to Ruby's more expressive methods that can simplify application development.

All of the methods discussed in this section require that the application developer make the sockets classes visible. At the beginning of the Ruby application, a line specifying, "require 'socket'" must be present.

The Socket class provides the primitive standard socket function that is most familiar to C language programmers. This class method can be used to create stream (TCP), datagram (UDP), and even raw sockets. Note that the **Socket::new** and **Socket::open** methods achieve the same result.

```
sock = Socket::new( Socket::AF_INET, Socket::SOCK_STREAM, 0 )
```

Note the similarities of this function to the C language primitive. The only real differences are the Ruby-specific naming conventions (SOCKET:: to bring the symbol in scope). In this example, we've created a stream socket. To create a datagram socket, we could do the following:

```
sock = Socket::open( Socket::AF_INET, Socket::SOCK_DGRAM, 0 )
```

Finally, a raw socket could be created with:

```
sock = Socket::open( Socket::AF_INET, Socket::SOCK_RAW, 0 )
```

Recall that the Socket::AF_INET symbol defines that we're creating an Internet protocol socket. The second parameter (type) defines the semantics of the communication (Socket::SOCK_STREAM for TCP, Socket::SOCK_DGRAM for UDP, and Socket::SOCK_RAW for raw IP). The third parameter defines the particular protocol to use, which is useful here only in the Socket::SOCK_RAW case.

Depending upon the protocol being used for the socket, Ruby provides three other mechanisms to not only create the socket, but also to connect or bind/listen depending upon the method. We can create a client TCP socket using:

```
sock = TCPSocket::open( "192.168.1.2", 13 )
```

This example not only creates a socket, **sock**, but also connects the socket to the named host and port number. For a TCP server socket, the **TCPServer** class can be used:

```
sock = TCPServer::open( "192.168.1.1", 13 )
```

This example creates the TCP server socket and then binds the address and port number to the newly created socket. This has the effect of restricting the interface from which connections may be accepted to "192.168.1.1" (with port 13).

To create a UDP socket, we can use the **UDPSocket** class:

```
sock = UDPSocket::new( Socket::AF_INET )
sock = UDPSocket::new
```

Each of these examples creates a new UDP socket, as the Socket::AF_INET is the default and is not required to be specified.

When we're finished with a socket, we must close it. To close our previously created socket, **sock**, we use the **close** method. The **close** method is actually part of the **IO** class, but we inherit it and can, therefore, use it on the socket.

```
sock::close
```

After the **close** method is called, no further communication is possible with this socket. Any data queued for transmission would be given some amount of time to be sent before the connection physically closes.

Socket Addresses

Unlike C, Ruby has no sockaddr_in structure, but instead has a binary string that represents this structure as an array. This is a little more complicated than dealing with C socket addresses, but luckily, they're not needed very often.

Because Ruby is interpreted and its networking classes sit on top of the host system's networking interface, we must map our socket address onto what is expected by the host system. In other words, we have to represent the socket address in the binary form of the sockaddr_in structure. Recall from Chapter 9, Network Programming in the C Language, that this is represented in C as:

```
struct sockaddr_in {
    int16_t sin_family;
    uint16_t sin_port;
    struct in_addr sin_addr;
    char sin_zero[8];
};

struct in_addr {
    uint32_t s_addr;
};
```

To map our array to Ruby, the simplest way is to use the pack method of an **Array** class. Consider the following:

```
sa = [Socket::AF_INET, 13, 192, 168, 1, 2, 0, 0].pack("snCCCCNN")
```

Let's break this down piece by piece. First, we create an array (contained within the square brackets) and assign values representing our socket address. In this case, the **sin_family** is AF_INET, the **sin_port** is 13, the **sin_addr** represents our IP address 192.168.1.2, and the **sin_zero** is two 32-bit values (ignored). We use the **pack** method to pack the elements of the array into a string according to the provided template (the argument to the **pack** method). The template is "snCCCCNN", where each letter represents a different directive. The 's' represents an unsigned short, so **Socket::AF_INET** is represented by a 16-bit value. The 'n' represents a big-endian short so the value 13 (the port number) is byte-swapped before being packed into the string. The 'C' represents an unsigned char. Therefore, our IP address (192, 168, 1, 2) is packed as four bytes (note the four 'C's) into the binary string. Finally, the 'N' directive denotes a big-endian long. The final two zeros in our array satisfy the unused **sin_zero** character array.

This is synonymous with the following C language fragment:

```
struct sockaddr_in sa;

sa.sin_family = AF_INET;
sa.sin_port = htons( 13 );
sa.sin_addr.s_addr = inet_addr("192.168.1.2");
```

Although C appears a little more straightforward, we could emulate this structure in Ruby. It would be a little less efficient, albeit more readable, but it's a pure matter of preference.

Socket Primitives

In this section, we look at a number of other important socket control methods. We also investigate the variety of ways that the same results can be achieved using different Ruby methods.

Bind

The **bind** method provides a local naming capability to a socket. This can be used to name either client or server sockets, but is used most often in the server case. The **bind** method for the **Sockets** class is an instance method and is provided by the following prototype:

```
sock.bind( addr )
```

where addr is a sockaddr_in structure packed into a binary string and sock is an instance of a socket. Sample usage of this form of **bind** is:

```
s = Socket::new( Socket::AF_INET, Socket::SOCK_STREAM, 0 )
sa = [Socket::AF_INET, 13, 192, 168, 1, 2, 0, 0].pack("snCCCCNN")
s.bind(sa)
```

In this example, we're creating a stream server on port 13 that will accept client connections only on the interface defined as 192.168.1.2.

A simpler example utilizes the **TCPServer** class. Let's use the **TCPServer** class to reproduce the previous **bind** example.

```
sock = TCPServer::new( "192.168.1.2", 13 )
```

Knowing that port 13 is the "daytime" service port, we could also utilize the service string to make it even more readable:

```
sock = TCPServer::new( "192.168.1.2", "daytime" )
```

If we want to accept client connections from any available interface, we could provide any empty string. This behavior is synonymous to the INADDR_ANY symbolic within our C language examples, for example:

```
sock = TCPServer::new( "", "daytime" )
```

Note that we could also pass a 0 as service to allow the networking layer to dynamically assign us a port (ephemeral port). It's also important to note that along with creating the server socket using **TCPServer**, we've also automatically bound the address and performed the **listen** method. The developer need only call the **accept** method to accept a client connection.

Finally, let's look at a UDP example. The use of bind with a UDP socket uses the **bind** method provided by the **UDPSocket** class. The following example creates a UDP socket and binds it with host INADDR_ANY (accept datagrams from any interface) with port 13:

```
sock = UDPSocket::new( Socket::AF_INET )
sock.bind( "", 13 )
```

Using the **TCPSocket** class permits us to create a client socket. The **TCPSocket** class not only allows us to create a socket, but also allows us to automatically connect to a client socket. For example:

```
sock = TCPSocket::new( "192.168.1.2", 13 )
```

Upon return, our socket named **sock** will have been connected to the server at 192.168.1.2 port 13.

Listen

Before a traditional server socket (one created using the **Sockets** class) can accept incoming client connections, it must call the **listen** method to declare this willingness. The **listen** method is provided by the following prototype:

```
sock.listen( backlog )
```

The sock argument represents an instance of the server socket and the backlog argument represents the number of outstanding client connections that may be queued. This method may only be used on sockets created with the **Sockets** class. Here's a complete example from socket creation to the **listen** method.

```
s = Socket::new( Socket::AF_INET, Socket::SOCK_STREAM, 0 )
sa = [Socket::AF_INET, 13, 192, 168, 1, 2, 0, 0].pack("snCCCCNN")
s.bind(sa)
s.listen( 5 );
```

Recall that the four lines of Ruby code can be performed in a single line using the **TCPServer** class, as:

```
sock = TCPServer::new( "192.168.1.2", 13 )
```

Therefore, Ruby's helper classes should be used wherever possible to minimize source-code length to make the overall script more readable.

Accept

The **accept** method is the final call made by servers to accept incoming client connections. Before **accept** can be invoked, the server socket must be created, a name must be bound to it, and **listen** must be invoked. There are two **accept** methods, one in the **TCPServer** class and one in the **Sockets** class. Because the method differs depending upon the class upon which it's invoked, let's look at both cases.

The **accept** method in the **Sockets** class accepts a new client connection and returns an array containing the new client socket and the client address information (packed in a sockaddr_in string):

```
s = Socket::new( Socket::AF_INET, Socket::SOCK_STREAM, 0 )
sa = [Socket::AF_INET, 13, 192, 168, 1, 2, 0, 0].pack("snCCCCNN")
s.bind(sa)
s.listen( 5 );

cli = s.accept
cli[0].write("Hello!")
cli[0].close

s.close
```

The first element of the array returned from **accept** is the client socket. The second element is the packed sockaddr_in string. If we wanted to know the client information, we could unpack the string as follows:

```
sockinfo = cli[1].unpack("snCCCCNN")
```

Our sockinfo array now contains the unpacked binary string. Element 1 (sockinfo[1]) contains the port number of the client connection and elements 2 through 5 contain the client's IP address (sockinfo[2] through sockinfo[5]).

The **TCPServer** method is much simpler. With **TCPServer**, we simply invoke the **accept** method on our **TCPServer** instance, which returns a client socket that we can immediately use. Recall that the **Sockets** class returned an array; **TCPServer** returns only the socket so that it can be promptly used. For example, the following code performs the same task as our prior **accept** illustration example:

```
s = TCPServer::new( "192.168.1.2", "daytime" )

cli = s.accept
cli.write("Hello!")
cli.close

s.close
```

Connect

The **connect** method is used by client Sockets applications to connect to a server. Clients must have created a socket and then defined an address structure containing the host and port number to which they want to connect. This is illustrated by the following code segment (which can be used to connect to the previously illustrated **accept** method example):

```
s = Socket::new( Socket::AF_INET, Socket::SOCK_STREAM, 0 )

sa = [Socket::AF_INET, 13, 192, 168, 1, 2, 0, 0].pack("snCCCCNN")

s.connect( sa )

print s.recv( 100 )

s.close
```

We create our address structure (as with previous examples) and then pass this to our **connect** method using our client socket instance.

Note the symmetry between the socket address arrays in this example and the server example shown previously with the **accept** method. There is perfect symmetry because the address structure used to name the server is the same address structure that is needed to connect to the server from the client.

The **connect** method (for TCP) blocks until either an error occurs, or the three-way handshake with the server completes. For datagram sockets (UDP), the method binds the peer locally so that the peer address isn't required to be specified for every send.

Sockets I/O

A variety of API methods exists to read data from a socket or write data to a socket. Some methods are class specific, and others are inherited from superclasses such as **IO**. First, we look at reading and writing from connected sockets and then investigate unconnected (datagram-specific) socket communication.

A number of methods are provided to communicate through a socket. Some of the functions include **recv**, **send**, **recvfrom**, **gets**, and **puts**. We look at examples of each of these functions in the following sections. Numerous other IO methods can be used, but for the purposes of brevity, they won't be covered here.

Stream Sockets

Stream sockets can utilize all of the previously mentioned I/O methods except for the datagram-specific methods, **recvfrom** and certain variations of **send**. Let's now look at some sample code that illustrates the stream-specific methods. The first example illustrates a simple echo server built from the **TCPServer** class (acctest.rb):

```ruby
require 'socket'

serv = TCPServer::new( "192.168.1.1", 45000 )
while true
    cli = serv.accept
    cli.send( cli.recv( 100 ), 0 )
    cli.close
end
serv.close
```

In this example, we open a server socket and then await a client connection, storing the newly created client socket in `cli`. We then simply send what we receive through the client socket back to the source and close the socket. The process then repeats.

Now, let's look at the TCP client socket that will connect to the previously defined server (clitest.rb):

```ruby
require 'socket'

sock = TCPSocket::open( "192.168.1.1", 45000 )
sock.send( "Hello\n", 0 )
mystring = sock.recv( 100 )
puts mystring
sock.close
```

After we create our TCP client socket, we immediately send our test string to the server using the **send** method. Knowing that the server will echo our transmission, we await the response with the **recv** method. We then simply put the string to standard-out using **puts** and close the socket using the **close** method.

That's simple so far. Now, let's try the **gets** and **puts** methods. These methods get and put a single line from the IO (in this case, a socket). If we were trying to transfer more than one line around, this would become more complicated. Nevertheless, for our example of sending a single line, this works fine. The TCP server now looks like (acctest2.rb):

```ruby
require 'socket'

serv = TCPServer::new( "192.168.1.1", 45000 )
while true
    cli = serv.accept
    cli.puts( cli.gets )
```

```
        cli.close
    end
    serv.close
```

Our client changes similarly (clitest2.rb):

```
    require 'socket'

    sock = TCPSocket::open( "192.168.1.1", 45000 )
    sock.puts( "Hello\n" )
    mystring = sock.gets
    puts mystring
    sock.close
```

After sending our single line with the **puts** method, we collect the response from the server with the **gets** method.

Datagram Sockets

The **recvfrom** method is used exclusively by datagram sockets, in addition to the **send** variation that includes source information. What differentiates these calls from our previously discussed stream calls is that these calls include addressing information. Because datagrams are not connected, we must define the destination address explicitly. Conversely, when receiving a datagram, the address information is also provided so that the source can be identified.

Let's look at a datagram server and client that provide the echo functionality illustrated previously by our stream socket examples. Our datagram server takes on a slightly different form (dgramsrv.rb):

```
    require 'socket'

    serv = UDPSocket::new( Socket::AF_INET )
    serv.bind( "192.168.1.2", 45000 )

    while true
        reply, from = serv.recvfrom( 100, 0 )
        serv.send( reply, 0, from[2], from[1] )
    end

    serv.close
```

After creating a new **UDPSocket**, we bind the instance of this socket with the interface from which we want to accept client connections ("192.168.1.2") and the port 45000. In our loop, we then receive datagrams using the **recvfrom** call. Note the semantics of this call; the method returns not one parameter, but two. The two parameters returned represent the array of data (reply) and the source address of the datagram (from, an array). Element two of the array contains the source host and element one contains the port number. Note that we use these elements when returning the datagram back to the source using the **send** method. Returning to the **recvfrom** call, we must also specify the maximum length of data that we want to receive and any flags that must be specified. Similarly, the **send** method includes flags as the second parameter (though none are defined in either case here).

The datagram client utilizes the datagram **send** variant with the standard **recv** method (dgramcli.rb). We can use the standard **recv** here because we're not interested in the source of the datagram. If we needed to know the source of the datagram, then the **recvfrom** method would need to be used.

```
require 'socket'
cli = UDPSocket::new( Socket::AF_INET )

cli.send( "Hello\n", 0, "192.168.1.2", 45000 )
puts cli.recv( 100 )

cli.close
```

In the datagram client, after we've created our datagram socket, we send our string to the echo server. The echo server is defined in the **send** method as the IP address and port number combination ("192.168.1.2", 45000). We then await the response echo datagram using the **recv** method, and emit it to the terminal using the **puts** method.

Socket Options

Socket options permit an application to change some of the modifiable behaviors of sockets and the methods that manipulate them. For example, an application can modify the sizes of the send or receive socket buffers or the size of the maximum segment used by the TCP layer for a given socket.

Socket options are a bit more complicated than dealing with options in the C environment. This is because we're operating in a scripting environment that must interface with the host environment and its corresponding structures.

Let's look at a simple example first. Let's say that we want to identify the size of the receive buffer for a given socket. This can be done with the following code segment (sockopt1.rb):

```
require 'socket'

sock = TCPServer::new(0)
sz = sock.getsockopt( Socket::SOL_SOCKET, Socket::SO_RCVBUF )
p sz.unpack("i")
```

First, we create a new stream socket. The 0 argument to the new method forces the stack to define our interface/port pair (which ends up being INADDR_ANY with a random port assignment).

We specify the `Socket::SOL_SOCKET` argument because we're interested in a Sockets layer option (as compared to a TCP layer option). For the receiver buffer size, we specify `Socket::SO_RCVBUF`.

To identify the actual value returned by **getsockopt**, we unpack the returned array using the unpack method. For this socket option, it will contain an integer, so we specify "i". The p method permits us to emit the value that's returned and is an invaluable debugging tool.

Now, let's look at how an option is set for a given socket, specifically, the linger option (sockopt2.rb). Socket linger allows us to change the behavior of a stream socket when the socket is closed and data is remaining to be sent. After **close** is called, any data remaining will attempt to be sent for some amount of time. If after some duration, the data cannot be sent, then the data to be sent is abandoned. The time after the **close** to when the data is removed from the send queue is defined as the linger time. In Ruby, we must construct the structure that is expected by the host environment.

```
require 'socket'

sock = TCPServer::new(0)

ling = [1, 10].pack("i2")
sock.setsockopt( Socket::SOL_SOCKET, Socket::SO_LINGER, ling )
```

The linger structure contains first an enable (1 for enable, 0 for disable) and then the time for linger in seconds. In our example, we're enabling linger and setting the linger value to 10 seconds (packing the structure into two 32-bit words, the structure that is expected by the host environment). We could read back what was configured by:

```
outval = sock.getsockopt( Socket::SOL_SOCKET, Socket::SO_LINGER )
p outval.unpack( "i2" )
```

Upon reading the linger array using the **getsockopt** method, we unpack the array using the "i2" template. This specifies that the two 32-bit integers are packed into the array.

Other Miscellaneous Functions

Let's now look at a few miscellaneous functions from the various Sockets APIs and the capabilities they provide. The first methods that we discuss provide information about the current host. Method **gethostname** (of the **Socket** class) returns the string name of the host:

```
str = Socket::gethostname
puts str
```

The DNS resolver permits us to resolve a host name to an IP address, or vice versa. Method **gethostbyname** provides name resolution given an IP address, for example:

```
str = Socket::gethostbyname( "192.168.1.1" )
puts str[0]
```

where the first element of the return string represents the FQDN of the IP address. What if we wanted to go in the opposite direction, providing a FQDN and returning an IP address? To achieve this, we use the **gethostbyaddr**:

```
str = Socket::gethostbyaddr( "www.microsoft.com", "http" )
puts str[0][3]
```

The IP address string is contained within the third element of the first array (in `String` format). To see the format of the returned array, simply use:

```
p str
```

The 'p' method will emit the object in debugging format.

Now, let's consider the problem of identifying the port number for a given service. Most of the Ruby methods permit us to specify not only a service string, such as "http", but also the raw port number, such as 80. To specifically retrieve the port number associated with a service, we use the **getservbyname** method.

```
portnum = Socket::getservbyname( "http" )
```

which would return, in this case, 80.

Notification

As a final topic for Ruby, let's look at the concept of event notification for sockets. This capability is commonly provided by the **select** primitive, which is an instance method in Ruby's IO class (and, therefore, inherited by **Sockets** classes).

As we saw with C, the descriptors representing our IO channels are provided to the **select** method to identify when some event occurs. We can configure the **select** method to tell us when a channel is readable, writable, or if an error occurs on it. Further, we can also tell the **select** method to return after a configurable number of seconds if no event occurs. Consider the following Ruby TCP server in Listing 13.1 (seltest.rb).

Listing 13.1   Ruby TCP server illustrating the `select` method.

```
require 'socket'

serv = Socket::new( Socket::AF_INET, Socket::SOCK_STREAM, 0 )
serv.setsockopt( Socket::SOL_SOCKET, Socket::SO_REUSEADDR, 1 )

sa = [Socket::AF_INET, 45000,
```

```
        192, 168, 1, 2, 0, 0].pack("snCCCCNN")
serv.bind( sa )
serv.listen( 1 )

# Server loop
while true

  cli = serv.accept
  fd = cli[0].to_io

  # Client loop
  while true

    res = select( [fd], nil, nil, 5 )
    if res != nil

      res = res[0]
      if res.include?(fd)

        str = fd.gets
        fd.puts str

      end

    else

      fd.puts "Timeout!"

    end

  end

end
```

This very simple server awaits a client connection and then upon receiving one will echo whatever it receives from the client. To determine when the client has sent something to be echoed, we use the **select** method. The **select** method works only with IO objects, so we must first convert our socket descriptor (cli) to an IO object using the to_io method. The result is the IO object fd, which we'll use exclusively for communicating with the client. In the **select** method, we specify three arrays representing our request. The first is an array of IO objects for which we want to be notified if a read event is generated (data is available on the IO object for read). The second is an array of IO objects for write events, and the third is for error events. The last number represents how many seconds to await an event before timing out.

The return of **select** is an array of file descriptors for which an event was generated. Note that in this case we use res[0] for the read events. For the write events, we use res[1], and for error events, we use `res[2]`. These arrays contain the file descriptors for which events fired. We use the `include?` method to test the array to see if the file descriptor is a member. If the file descriptor is a member of the array, then the event was generated for the socket.

Knowing that the event was generated, we simply read a line from the socket (represented by the new file descriptor, `fd`) and then write this back out to the client. If the timeout arrives before any event, we notify the peer of this event by writing "Timeout!" to the client.

SPECIALIZED NETWORKING APIS FOR RUBY

Ruby also includes a number of specialized APIs for a number of Application layer protocols. These include SMTP, HTTP, POP3, FTP, Telnet, and others. In this section, we look at examples of the SMTP and HTTP client-side classes.

Ruby `Net::SMTP` Class

SMTP is the Simple Mail Transfer Protocol and is used to transfer e-mail to a mail server that delivers it to the final destination. Ruby provides a very simple SMTP interface, as is illustrated in Listing 13.2 (smtpc.rb).

Listing 13.2 Sample `Net::SMTP` client example.

```
require 'net/smtp'

recip = "you@yourdomain.com"
from = "me@mydomain.com"
server = "yourdomain.com"
message = "From: me\n\nSubject: Hi\n\nHello\n\n"

smtpSession = Net::SMTP::new( server )

smtpSession.start

smtpSession.sendmail( message, from, recip )
smtpSession.finish
```

The very simple source in Listing 13.2 illustrates sending a short e-mail. We make the `Net::SMTP` class visible using `require` and then set up our recipient (`recip`) and source e-mail (`from`) addresses. Next, we define the `server` (where the SMTP client will connect to send the e-mail). Then, we define the e-mail `message` to be sent in the message string.

The complete SMTP process is then shown in the final four lines. We create a new instance of an SMTP client using `Net::SMTP::new`, defining the server to which we'll connect to send our e-mail. We use the `start` method to start the actual SMTP session and then send our e-mail using the **sendmail** method. Finally, to end the session, we use the **finish** method that permits the remote SMTP server to deliver the message.

Ruby `Net::HTTP` Class

HTTP is the classic Web transport protocol used to transfer Web pages (and other content) across the Internet. The Ruby HTTP class provides a very simple and powerful client-side interface that can be used for a variety of purposes. A sample usage of the HTTP client is shown in Listing 13.3.

Listing 13.3 Sample `Net::HTTP` client example (httpc.rb).

```
require 'net/http'

hcli = Net::HTTP::new( "www.mtjones.com" )

resp, data = hcli.get( "/index.html" )

print data
```

The result of the example shown in Listing 13.3 is the HTML source retrieved (the `index.html` file from host `www.mtjones.com`. We first make the HTTP network library visible using the `require` method and then create a new HTTP client using **Net::HTTP:new** method, specifying the host to which we want to connect. Next, we request the file using the get method. Finally, we emit the page (in HTML format) using `print`.

That's simple, compared to what actually goes on underneath the covers of HTTP. Therefore, Ruby provides some very useful and simplifying abstractions for higher-level protocols.

SUMMARY

This very quick tour of Sockets programming in Ruby provided a discussion of the Ruby language in addition to many of the useful Sockets methods for the network application developer. Although Ruby continues to be actively developed, it represents a very clean, object-oriented scripting language that can be a useful part of any programmer's toolbox.

REFERENCES

[Matsumoto01] Matsumoto, Yukihiro, *Ruby in a Nutshell*, O'Reilly & Associates, 2001.
[Ruby-FAQ] "The Ruby FAQ," available online at *http://www.rubygarden.org/iowa/ faqtotum.*

RESOURCES

Ruby Central, available online at *http://www.rubycentral.com/*.
Ruby Home Page, available online at *http://www.ruby-lang.org/en/*.
Ruby-Talk, Ruby Mailing-List Archive, available online at *http://www.ruby-talk.org/*.

14 Network Programming in Tcl

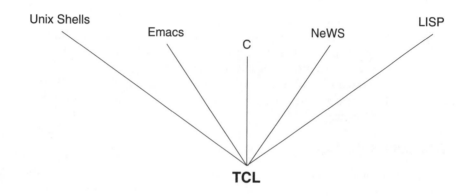

Unix Shells

Emacs

C

NeWS

LISP

TCL

TCL LANGUAGE OVERVIEW

In this chapter, we investigate network programming in the popular scripting language Tcl (otherwise known as the Tool Command Language). Tcl is an interpreted language that was originally designed to be embedded within applications written in other languages (such as C). In addition to its use as an embedded scripting language, Tcl can be used by itself with the simple interpreter. Tcl is both simple and powerful and finds itself in many diverse applications areas. All code for this chapter can be found on the companion CD-ROM at /software/ch14.

ON THE CD

Origin of the Tcl Language

The motivation for creating Tcl was the need for an extension language for hardware design tools at the University of California at Berkeley. In 1990, John Ousterhout observed that some of the power behind Unix and the Emacs editor was that they could be extended by the presence of a programmable command language (shells for Unix and LISP for Emacs) [Ousterhout90]. Rather than create a new language for each design tool that was developed, Tcl provided a library that permitted the tool to be extended (thus the name, Tool Command Language).

Tcl Language Heritage

What makes Tcl most interesting is that it exists as a set of libraries that can be integrated into another application to extend it for scripting capabilities. The Tcl library includes not only the interpreter, but also a set of built-in commands and the ability to extend the language with new commands that are first class (appear as part of the original language).

Tcl has numerous influences that extend beyond the realm of computer languages. One of the major influences on Tcl was the Emacs editor. Emacs is not so much an editor as it is a development environment (a precursor to many Integrated Development Environments, or IDEs, of today). Emacs provided a built-in scripting language to allow users to script capabilities within the editor using LISP. This made Emacs not only extendable, but also very powerful. Tcl also borrowed some aspects of LISP, but also reversed some aspects from LISP such as how expressions are evaluated. Tcl is also similar to the NeWS window system that is based upon the Postscript language. Tcl provides a mechanism to send scripts to a server for execution, similar to NeWS Postscript generation capabilities.

TOOLS

Tcl is managed by the Tcl Core Team (TCT) whose membership is elected by the Tcl community. Like Python, the TCT seeks proposals for enhancements by the development public. Developers can submit TIPs, or Tcl Improvement Proposals, to suggest changes or additions to Tcl [Tcl-TIP]. The version of Tcl used in this chapter (and Chapter 21, Network Code Patterns in Tcl) is version 8.0.

The primary Tcl Web page that provides not only information about Tcl, but also the latest downloads is http://www.tcl.tk.

Interpreter/Tools Used

The standard Tcl distribution is demonstrated here. Tcl operates on a wide range of platforms in a number of different modes. Tcl can be operated via a shell (tclsh) and also as a library that can be integrated into other programs in a large variety of languages. The latest Tcl interpreter can be downloaded at *http://www.tcl.tk*. Tcl operates on Windows, Linux, and numerous other operating systems. One of Tcl's primary strengths is its portability between host architectures and languages.

Tcl's lack of some capabilities makes datagram Sockets programming prohibitive. In order to build datagram and multicast applications, the Tcl-DP package must be used [Tcl-DP03]. Tcl-DP (Tcl-Distributed Programming) is an extension to Tcl that provides a range of services, including datagram sockets, RPC, and TCP and IP connection management. Certain examples in this section require the Tcl-DP package, which may be freely downloaded.

The sample code patterns presented here (and in Chapter 21) and on the CD-ROM will execute on any of the supported environments.

Networking API Used

The Tcl-DP package must be downloaded for some examples presented here. These examples can be identified by their importing of the Tcl-DP package (shown as the line 'package required dp 4.0'). For these examples, the Tcl-DP package must be downloaded at [Tcl-DP03] and installed (using the teki application). The release used here for Tcl-DP is 4.0.

Where to Download

The Tcl interpreter and tools may be downloaded at the source, *http://www.tcl.tk* (under "Latest Releases"). This site includes both source and binary versions of the interpreter. A list of the platforms on which Tcl can currently run is available at [Tcl-Platforms]. The Tcl-DP version 4.0 package can be downloaded at *http://www.cs.cornell.edu/Info/Projects/zeno/Projects/Tcl-DP.html*.

NETWORKING API FOR TCL

Tcl includes a very simple Sockets layer that provides support for TCP stream sockets. We'll intermix the Tcl core Sockets API with the Tcl-DP API, but the Tcl-DP API will be apparent by the dp_ prefix that precedes all Tcl-DP commands.

The Tcl Sockets API is essentially one function, **socket**, but this function returns a channel identifier that can then be used for I/O operations. We introduce these as socket commands for discussion, but it should be noted that they are not actually part of Tcl's Sockets API. The list of commands to be covered for Tcl are shown in Figure 14.1.

Command	Description
`socket [options] host port`	Create a client socket and connect to the peer or create a server socket and await connections.
`fconfigure channel name value`	Configure an option on a given channel.
`read channel [numChars]`	Read from a channel.
`gets channel [varname]`	Read from a channel.
`puts channel [varname]`	Write to a channel.
`close channel`	Close a channel.
`flush channel`	Flush output for a channel.
`fileevent channel [mode] script`	Perform an action when an event has occurred.
`select rd [wr] [ex] [timeout]`	Perform a select operation on a list of socket descriptors with an optional timeout.

FIGURE 14.1 Tcl core socket commands.

The Tcl-DP package includes a number of functions that extend the capabilities of the Tcl core Sockets API. This includes multicast and datagram support in addition to other commands. The Tcl-DP package also includes support for RPC, which is not covered here. The list of commands that are covered for the Tcl-DP API are shown in Figure 14.2.

Command	Description
`dp_connect channel args`	Create a TCP, UDP or multicast client or server socket (channel).
`dp_accept channel`	Accept a new TCP connection from a TCP channel.
`dp_recv channel`	Read from a channel.
`dp_send channel string`	Write to a channel.
`dp_netinfo option args`	Get address or service info.

FIGURE 14.2 Tcl-DP sockets commands.

This is obviously different from the standard BSD API, but as we'll soon see, Tcl provides a complete interface that allows the construction of very complex Sockets applications.

Sockets API Discussion

Let's now walk through the Tcl command and the Tcl-DP package and illustrate how the commands are used.

In Tcl, communication takes place over a channel, so you'll see the word socket and chan-nel used here interchangeably. Tcl provides a number of commands that operate on chan-nels, so if you have socket, the command can be used on it as a channel.

Creating, Destroying, and Connecting Sockets

We'll begin by investigating how Tcl creates and destroys local sockets. Recall that Tcl pro-vides not only socket creation but also connection within the command API. Therefore, in this section, we cover not only socket creation and destruction, but also connection. We'll first look at the Tcl **socket** core command and then the Tcl-DP **dp_connect** command.

The **socket** command within Tcl provides for socket creation and connection, de-pending upon the type of socket being created (client or server). Let's look at the formal arguments for the **socket** command and then dissect it using a number of examples.

The Tcl **socket** command for clients has the following format:

```
socket [options] host port
```

In this basic case, the client **socket** command connects to the host and port and then returns a channel identifier. An example of this is:

```
set clisock [ socket "192.168.1.2" 52000 ]
```

which creates a client socket, and then connects it to the host identified by IP address "192.168.1.2" and port 52000. The resulting socket (channel identifier) is then stored into variable clisock. Note that the brackets shown ([]) mean that this command should be in-voked, with the result being a string (the only data type available in Tcl is the string).

The client version of the **socket** command also supports a number of options, which are shown in the following code in *italics*:

```
socket –myaddr addr –myport port –async host port
```

The first option '-myaddr addr' allows us to specify the host or IP address of the local socket. This option is rarely used for client sockets. The second option '-myport port' al-lows us to specify the port number that we want to bind to for the local socket. Again, this option is rarely used (typically, only servers perform this). Finally, the '-async' option al-lows us to perform the peer connect asynchronously. This means that the **socket** command will return immediately with the client socket channel identifier, but the socket may not have yet connected to the peer. When the client application attempts to use the socket for a read or write operation (**gets** or **puts**), an error may occur identifying that the connect op-eration has not yet completed.

Let's now look at a full example of the client connect, using the options available to us:

```
set clisock [ socket –myaddr 10.0.0.1
              –myport 25000 –async 10.0.0.2 25001 ]
```

In this example, we create a client socket and bind it locally to the interface identified by IP address "10.0.0.1" and port 25000. We also specify that the connect operation should occur asynchronously by the –async option. The client socket should connect to the host identified by IP address "10.0.0.2" and port 25001. Because –async was specified, the call will complete immediately, but the socket will not be usable immediately for **gets** and **puts** commands.

Recall that binding an address to a client socket restricts outgoing connections to that interface. Therefore, in the example previously shown, client connections will be restricted to use only the interface identified by IP address "10.0.0.1".

Let's now look at the server-side of the **socket** command. The Tcl **socket** command for servers has the following format:

```
socket –server command [options] port
```

The basic **socket** server command has the following format:

```
set servsock [ socket –server myaccept 25001 ]
```

This creates a server socket (note the –server option) with the procedure argument that will be called once an incoming connection is accepted for the socket. We bind to IN-ADDR_ANY here (the wildcard) and port 25001. This means that we'll accept connections on any available interface but port 25001.

The server version of the socket command supports a single option, which is shown in the following code in *italics*:

```
socket –server command –myaddr addr port
```

The default configuration for the server **socket** command is that a port must be specified to which the server will locally bind. If the –myaddr option is not specified, then we default to the wildcard and allow incoming connections on any available interface. If the –myaddr option is specified, then we restrict incoming connections from the given interface. A complete example of the **socket** server command is illustrated as:

```
set servsock [ socket –server myaccept
                     –myaddr "10.0.0.2" 25001 ]
```

In this example, the server socket will call the myaccept procedure when an incoming client is connected. We restrict incoming connections to the interface represented by IP address "10.0.0.2" and port 25001.

The Tcl-DP package provides a similar command to Tcl's core **socket** command, but with much more functionality to create a variety of different types of sockets. Let's look at a few examples of the Tcl-DP package for socket creation and connection.

The **dp_connect** command can be used to create client and server sockets for a number of different protocols. The result of **dp_connect** is a new socket (channel identifier). We can create a TCP client socket as:

```
set clisock [ dp_connect tcp -host 192.168.1.1 -port 25001 ]
```

In this case, we use **dp_connect** with the TCP protocol to connect to host 192.168.1.1 and port 25001. To create the server end of this connection, we would do the following:

```
set srvsock [ dp_connect tcp -server true -myport 25001 ]
```

For the srvsock, we use **dp_connect** again specifying our desired protocol of TCP. The −server option specifies that we're creating a server socket, which takes a Boolean. Finally, the −myport option specifies the port to which we'll bind. For **dp_connect**, there is no callback for incoming client connections. This is because the Tcl-DP API includes an **accept** command to service incoming connections.

Creating a UDP protocol socket is similar to the TCP **dp_connect** command:

```
set srvsock [ dp_connect udp -host "10.0.0.1"
                 -port 1999 -myport 2999 ]
```

We specify the UDP protocol identifier after the **dp_connect** command and the host and port to which we'll direct outgoing packets (using the −host and −port options). We also specify the port number to which we'll bind locally (here, 2999). Creating a peer socket for the previous UDP socket definition would be done as:

```
set srvsock [ dp_connect udp -host "10.0.0.2"
                 -port 2999 -myport 1999 ]
```

Note that in the first case, we're directing packets to host 10.0.0.1 port 1999, whereas in the peer case we're directing packets to host 10.0.0.2 port 2199 (and each have bound these ports conversely).

We can also create a multicast socket using the ipm protocol identifier. In this case, we specify the group address that we want to join and the port. An example for multicast is:

```
set msock [ dp_connect ipm -host "239.0.0.2" -myport 48000 ]
```

This allows us to both send and receive multicast datagrams through msock, because all engage in the same communication. We could also set the maximum time-to-live for all datagrams sent using the −ttl option:

```
set msock [ dp_connect ipm -host "239.0.0.2" -myport 48000 -ttl 1 ]
```

When we're finished with a socket, we must close it. To close a previously created socket, clisock, we use the **close** command.

```
close clisock
```

After the **close** command is called, no further communication is possible with the socket. Any data queued for transmission will be given some amount of time to be sent before the connection physically closes.

Socket Addresses

Tcl has no external concept of an Internet or socket address. All address information is specified within the **socket** and **dp_connect** calls. This presents one of the problems with network programming in Tcl. Because addresses are specified within the calls that initiate communication, there is no way to identify the source address of a UDP datagram. For this reason, there's no way to write the datagram server code pattern (in either pure Tcl or Tcl-DP). One caveat is the peer address information returned from the **dp_accept** command, which we see in the next section.

Socket Primitives

In this section, we look at a number of other important socket control commands, relating to the Tcl-DP package. Note that **bind**, **listen**, and **connect** are all implied within the **socket** or **dp_connect** commands and, therefore, no equivalent exists for Tcl. The only primitive that operates in a similar fashion to the BSD API is the **accept** command (**dp_accept**).

dp_accept

The **dp_accept** command is the command used by TCP servers for accepting incoming client connections. The **dp_accept** call blocks until a connection is received, and returns a list identifying the socket for the new client connection and address information about the peer that connected.

Let's look at an example of **dp_accept** in a TCP server (acctest.tcl).

```
package require dp 4.0

set srvsock [ dp_connect tcp —server true —myport 13 ]

set rcvlist [ dp_accept $srvsock ]

set clisock [ lindex $rcvlist 0 ]

set addr [lindex $rcvlist 1 ]

puts "Got a connection from $addr"

close $clisock
close $srvsock
```

In this example, we create our TCP server socket using **dp_connect** and bind it locally to port 13. We then call **dp_accept** on our server socket and return the list result to rcvlist. The first element of this list is the new client socket (channel identifier), and the second is the address from which the peer connected. The lindex command is used to extract these elements from the list. We use the **puts** command at the end to emit who connected to us and then close both sockets with the **close** command.

Sockets I/O

Tcl provides a minimal set of commands to read or write data to a socket. The methods are identical for stream sockets as they are for datagram sockets, so we'll look at examples of the Tcl core API and the Tcl-DP API.

The functions for communicating through a socket are **read**, **gets**, and **puts**, **dp_recv**, and **dp_send**. We look at examples of each of these commands in the following sections.

The Tcl core API utilizes the **read**, **gets**, and **puts** socket methods. Let's now look at some sample code that illustrates the stream-specific methods. The first example illustrates a simple echo server built using stream sockets (acctest2.tcl):

```
proc myaccept { sock addr port }

    fconfigure $sock -buffering line

    while { true } {

        puts $sock [ gets $sock ]

    }

}


proc main{ port } {

    set srvsock [ socket -server myaccept $port ]

    vwait forever

}


main 45000
```

In this example, we create a socket server in our main function and then simply enter Tcl's event loop to await incoming connections (via **vwait**). After a new client connection arrives, Tcl calls the myaccept command with the client socket information. We configure line buffer immediately using **fconfigure** and then enter an infinite loop. Within this loop, we read all data from the socket anonymously using **gets** and then pass this immediately to **puts**. Given this implementation, we could open as many client connections to this server as are allowable by the host machine, each having its own myaccept to process its input.

Now, let's look at the TCP client socket that will connect to the previously defined server, but this time we'll use the Tcl-DP package (testcli.tcl):

```
package require dp 4.0

set clisock [ dp_connect tcp —host localhost —port 45000 ]

fconfigure $clisock —buffering line

dp_send $clisock "Hello\n"

set recv_str [ dp_recv $clisock ]

puts $recv_str

close $clisock
```

After we create our TCP client socket using the **dp_connect** command, we configure it for line buffering using the **fconfigure** command. We issue a **dp_send** command to send our test string and then await a response using **dp_recv**. The received string is then emitted to standard-out using puts and the client socket is closed using the **close** command.

To build datagram applications within Tcl, the Tcl-DP package must be used. Even with Tcl-DP, it's not possible to identify the source of a given datagram (the **dp_recv** or **gets** commands are the only ones available to return data from a socket). Therefore, we can build datagram applications that communicate with one another, but they must have predefined addresses and ports so that each end of the connection can be consistent.

Building a simple datagram application that sends datagrams to a remote peer is simpler than building a stream application because there is no virtual connection between the two endpoints. In the following example, the datagram application simply sends a datagram to the peer and awaits the response (dgramcli.tcl):

```
package require dp 4.0

# Create a datagram socket and point to 127.0.0.1 / 33000
set dsock [ dp_connect udp —host 127.0.0.1 —port 33000 ]

# Configure for line buffering
fconfigure $dsock —buffering line

# Send some data to the remote peer
dp_send $dsock "Some string data..."

# Await the response from the peer
```

```
set line [ dp_recv $dsock ]

# emit the response
puts $line

# Close the socket
close $dsock
```

The datagram client is a fire-and-forget application in that we have no idea if the datagram arrived at the peer. In this particular application, we can know because we expect a response back, but in most cases, we won't know if the datagram was received and no reply was made, or if the datagram was simply lost. The **dp_recv** and **dp_send** commands, as shown in the example, are used in the same way that **send** and **recv** are used in traditional Sockets applications. The **connect** must have been performed before communication may take place so that the Sockets layer can cache the addressing information about the peer. As there is no way to specify the destination information with the packet itself (as is done with Java), Tcl (and Tcl-DP) provide minimal datagram support.

Socket Options

Socket options permit an application to change some of the modifiable behaviors of a socket and change the behavior of the methods that manipulate them. For example, an application can modify the sizes of the send or receive socket buffers or whether input or output commands are blocking or nonblocking.

Tcl provides a minimal set of options that can be manipulated, though they're much easier to deal with than in most other languages. Tcl also provides standard defaults for most options, though the send and receive socket buffers are currently defaulted to 4 KB. Additionally, the SO_REUSEADDR socket option is set by default for server sockets.

Let's look at a simple scalar example first. Let's say that we want to identify the size of the receive buffer for a given socket. This can be done with the following code segment:

```
set srvsock [ socket –server accept_proc 45000 ]

set size [ fconfigure $srvsock –buffersize ]

puts "The SND/RCV buffer sizes are $size"
```

First, we create a new stream server socket using the **socket** command. To get the value of the send and receive buffer size socket option (SO_SNDBUF / SO_RCVBUF), we use the **fconfigure** command. We specify the –buffersize option to the **fconfigure** command, and without specifying any argument, the command returns the value of the option. The disadvantage in Tcl is that there is no way to set the receive buffer size differently than the send buffer size; both must be changed identically.

We can set the buffer sizes for a given socket as the following code snippet demonstrates:

```
set srvsock [ socket —server accept_proc 45000 ]

fconfigure $srvsock —buffersize 8192

set size [ fconfigure $srvsock —buffersize ]

puts "The SND/RCV buffer sizes are $size"
```

The support for manipulating socket options within Tcl is lacking and further degrades its applicability to any network application development other than stream sockets.

Other Miscellaneous Functions

Let's now look at a few miscellaneous commands available within Tcl and the capabilities that they provide. The first method that we discuss provides information about the current host.

Command **info** with the hostname option returns the string name of the current host:

```
set theHost [ info hostname ]

puts theHost
```

The DNS resolver permits us to resolve a host name to an IP address, or vice versa. Command **dp_netinfo** of the Tcl-DP package provides domain name resolution capabilities (as well as other services). To convert a domain name into an IP address, the **dp_netinfo** command can be used with the —address option:

```
set ipAddr [ dp_netinfo —address www.microsoft.com ]

puts ipAddr
```

where the return string ipAddr represents the IP address of the FQDN. What if we wanted to go in the opposite direction, providing an IP address and receiving the FQDN? To achieve this, we use the **dp_netinfo** command, again with the -address option:

```
set hostAddr [ dp_netinfo —address 207.46.249.190 ]

puts hostAddr
```

The domain name of the specified address is contained within the string hostAddr.

Now, let's consider the problem of identifying the port number for a given service. To specifically retrieve the port number associated with a service, we use the **dp_netinfo** command of the Tcl-DP package. Using the —service option, we can specify one of two arguments. We specify either the service name string or the service port number. Consider the following two examples:

```
set line [ dp_netinfo -service http ]

set official_name [ lindex $line 0 ]

set official_port [ lindex $line 1 ]
```

The **dp_netinfo** command with the service option returns a list. The first element of the list is the official name of the service (for http, this is www). The second element of the list is the official port number on which a service can be found. Therefore, upon using the -service option of the **dp_netinfo** command, the lindex command is then used to extract the individual elements.

The **dp_netinfo** command with the –service option can also be used to search for a service based upon the port number. For example:

```
set line [ dp_netinfo -service 80 ]

set official_name [ lindex $line 0 ]

set official_port [ lindex $line 1 ]
```

is synonymous with the previous example, searching for service string http.

Notification

Let's now look at the concept of event notification for sockets. This capability is commonly provided by the **select** primitive. In Tcl, the **select** method is available as part of Extended Tcl (not natively), but Tcl also provides a different capability for event notification that is worthy of investigation.

The **fileevent** command within Tcl provides for event notification on a channel identifier. The **fileevent** requires a channel identifier, mode (readable or writable), and finally a script that will be invoked when the event occurs. Let's now look at an example of the **fileevent** command being used to identify when a socket has data available for read (sel-test.tcl).

```
proc main {} {

    set servsock [ socket -server myaccept 45000 ]

    vwait forever

}


proc myaccept { sock addr port } {

    fconfigure $sock -buffering line
```

```
        fileevent $sock readable [ list myread $sock ]

    }


    proc myread { sock } {

        # Test the socket for eof (peer disconnect)
        if { [ eof $sock ] } {

            puts "Peer disconnected"
            close $sock

        } else {

            # Loopback the data read
            set line [ gets $sock ]
            puts $sock $line

        }

    }


    # Invoke main...
    main
```

We've seen the main procedure before; it's simply setting up a server socket with the callback of myaccept for new incoming connections. When an incoming connection arrives, myaccept is called. The first task is to configure the new client socket for line buffering using **fconfigure**. The **fileevent** command then configures the socket for a callback whenever data is available for read. The script to be called is constructed at the end of **fileevent**. The list command is used to construct a list of the callback function and the channel identifier. This is required by Tcl for a well-formed callback command. When data finally arrives for the new client socket, the myread procedure is called with the channel identifier. We first test the channel for **eof** to see if the peer has disconnected. If so, we close the socket using the **close** command and end the callback. Otherwise, we read the data from the socket using **gets** and then emit this back through the socket using **puts**.

SUMMARY

This short tour of Sockets programming in Tcl provided a brief discussion of the Tcl language in addition to the available socket commands for the network application developer.

Although Tcl is a popular application-glue language, and includes the basic needs for stream sockets development, it lacks native support for datagram sockets and multicast. Tcl does provide some alternative mechanisms for notification, and a very interesting callback architecture for socket events.

REFERENCES

[Ousterhout90] Ousterhout, John K., "Tcl: An Embeddable Command Language," available online at *http://citeseer.nj.nec.com/ousterhout90tcl.html*.

[Tcl-DP03] "Announcing Tcl-DP v. 4.0," available online at *http://www.cs.cornell.edu/Info/Projects/zeno/Projects/Tcl-DP.html*.

[Tcl-Platforms] "Known Compatible Tcl/Tk Platforms," available online at *http://www.tcl.tk/software/tcltk/platforms.html*.

[Tcl-TIP] "Tcl Improvement Proposal Web Site (Tcl Core Team)," available online at *http://www.tcl.tk/community/coreteam/index.html*.

RESOURCES

Tcl FAQs, available online at *http://resource.tcl.tk/resource/doc/faq/*.

Tcl Home Page, available online at *http://www.tcl.tk*.

PART

III

Software Patterns from a Multi-Language Perspective

This third part of the book explores network applications from a multi-language perspective. A number of software patterns are implemented in the target languages for this book as a way to compare and contrast the features that are available in each language.

Each chapter includes implementations of the traditional Daytime service (in both server and client forms) for stream sockets, datagram sockets, multicast sockets, and broadcast sockets. In some cases, the client or server has been implemented more than once in the given language to expose features in the language that are novel or unique. These software patterns provide a quick means to build network applications using the particular paradigm (such as datagram or multicast communication).

Additional patterns implemented in the target languages include a simple HTTP server and SMTP client. These patterns explore more of the features of the target language and can be used to compare and contrast (such as how file input is handled within a given language).

Chapter 15, Software Patterns Introduction, provides an introduction and discussion of the patterns that illustrate the basic design and assumptions with a language-independent treatment.

ON THE CD

All software patterns discussed here are included on the accompanying CD-ROM within the ./software *directory.*

15 Software Patterns Introduction

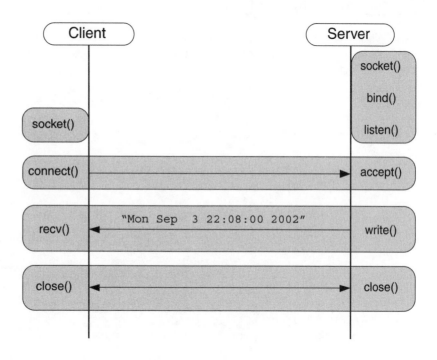

INTRODUCTION

In this chapter, we look at the individual designs of the software patterns covered in this part of the book. Each of the patterns presented in the following sections is implemented in each of the languages, but all follow the same general design. This is because each uses the Sockets API and, therefore, uses the same primitives to achieve the same results. Understanding the design of each pattern will help you to understand what is being achieved in each implementation.

STREAM SERVER/CLIENT DESIGN

The stream server and client utilize a TCP connection to communicate the date and time information from the server to the client. The data flow diagram illustrating the Sockets API calls is shown in Figure 15.1.

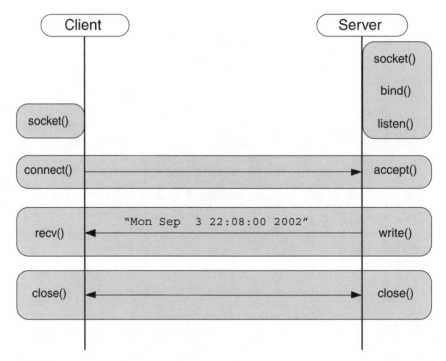

FIGURE 15.1　Stream server/client data flow.

In order to accept client connections, the server must create a socket, **bind** a known name to it, and then use the **listen** call to specify its willingness to accept connections. Similarly, the client must also create a socket using the **socket** function in order to communicate.

Before any communication can occur, the sockets must be connected. From the server perspective, this implies an **accept** call that creates a new client socket at the server through the listening socket. Recall that the listening socket exists only to accept incoming connections. Communication to the client occurs only through the newly created socket returned by the **accept** call. From the client, the **connect** call connects the two socket endpoints together to allow them to communicate.

The **recv** and **write** calls illustrate data communication in this example. These calls could also be **read** and **send**, depending upon the desire of the implementer. In Figure 15.1, we see that the server writes the current time to the client, which is read by the client using the **recv** call.

Finally, the **close** call is used to terminate the connection. Closing the socket is managed by both ends, because each must close their sockets to completely close the conduit between them.

DATAGRAM SERVER/CLIENT DESIGN

The datagram server and client utilize a UDP connection to communicate the date and time information from the server to the client. Recall from the stream server that a connection takes place to create a virtual circuit between the client and server. With datagram communication, this is not the case. Therefore, we must alter the communication in order to make the server aware of the client's request. The data flow diagram illustrating the Sockets API calls is shown in Figure 15.2.

At the server, we set up by creating our socket and then binding a name to it so that the client can locate it. The client similarly creates a socket in order to communicate. Note that the datagram server does not use **listen** or **accept**—these are stream-specific calls that are never used by datagram sockets.

The server then uses a **recvfrom** call to await a client datagram. This is necessary so that the server knows when a client desires the time and date information. Because datagram sockets are unconnected, some way is needed to identify the client request. We've chosen this method because it's simple and provides us with the necessary client information at the server. Once the client sends a NULL message to the server, the server receives this message and, using the **recvfrom** call, can identify the source of the message. This message is then used by the server in the **sendto** call to return the time and date information. Note the symmetry in Sockets API functions for the datagram server.

Finally, both the datagram client and server applications use the **close** function to end the socket session.

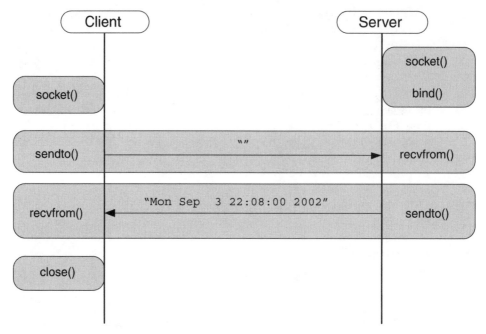

FIGURE 15.2 Datagram server/client data flow.

BROADCAST DATAGRAM SERVER/CLIENT DESIGN

The broadcast datagram server and client utilize a UDP connection to communicate the date and time information from the broadcast server to the client. Because communication is broadcast, there is no physical connection between the client and server. The client need only subscribe to a broadcast port in order to receive broadcast datagrams. The server must use a socket option in order to generate broadcast datagrams, using the **setsockopt** call.

The data flow diagram illustrating the Sockets API calls for the broadcast client and server is shown in Figure 15.3.

The server first creates a socket and then uses **setsockopt** to permit the transmission of broadcast datagrams. In order to receive broadcast datagrams, the client creates a socket and then binds it to the wildcard address (INADDR_ANY) and broadcast port.

The server uses the **sendto** API function to send the date and time datagram, directing it to the broadcast address and port. Because the client is already bound with the address and port information, a **recvfrom** API function is used to receive the broadcast datagram.

The broadcast server then simply acts as a time-based broadcaster of time and date information. The client simply binds itself to the broadcast channel, allowing it to receive any

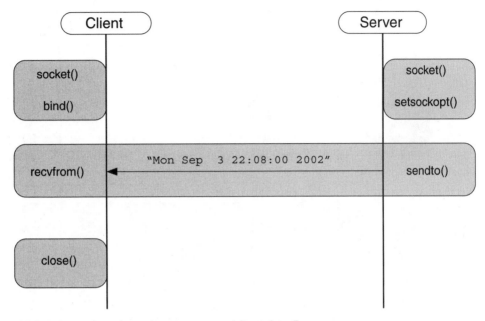

FIGURE 15.3   Broadcast datagram server/client data flow.

datagrams emitted to the broadcast address and port. This differs from our traditional clients and servers, but illustrates the mechanisms to create broadcast clients and servers.

MULTICAST DATAGRAM SERVER/CLIENT DESIGN

The multicast datagram server and client utilize a UDP connection to communicate the date and time information from the multicast server to the client. As in the case of the broadcast server, because communication is multicast, there is no physical connection between the client and server. The client is only required to subscribe to the multicast address and port in order to receive the multicast datagrams. The server may send multicast datagrams without any real configuration. The client must configure itself for membership within the multicast group in order to receive the multicast datagrams. The data flow diagram illustrating the Sockets API calls for the multicast client and server is shown in Figure 15.4.

The server begins by creating a socket and an associated name structure that is used to direct the datagrams. The client is a bit more complicated. After the client socket is created, the SO_REUSEADDR socket option is used to allow the client to bind the multicast address and port number without the potential "port in use" error. The name information is then bound to the socket (multicast group address and port). Finally, the multicast client must

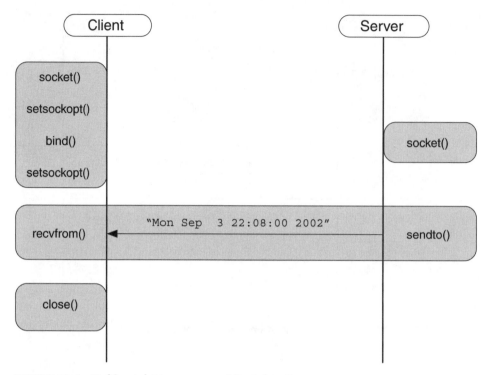

FIGURE 15.4 Multicast datagram server/client data flow.

join the multicast group in order to receive multicast datagrams. This is done using the IP_ADD_MEMBERSHIP socket option, referencing the multicast group address.

The server sends the date and time information to the client using the **sendto** API function. In order to direct the datagrams to the multicast group, the predefined group address and port number are specified within the **sendto** call. Receiving the multicast datagrams is provided by the **recvfrom** API function at the client.

SIMPLE HTTP SERVER DESIGN

The HyperText Transfer Protocol (or HTTP) is a simple ASCII-based protocol that operates on the client/server principle. The protocol is a synchronous request/response in which a client makes an HTTP request through a socket and the server replies with an HTTP response through the same socket. The request and response follow a very simple format that can be simple or very complex depending upon the particular need. An example HTTP request and response is shown in Figure 15.5.

HTTP Request	Get /index.html HTTP/1.1
HTTP Response	HTTP/1.1 200 OK Date: Wed 26 Nov 2002 11:50:59 GMT Connection: close Content-Type: text/html <HTML> <HEAD><TITLE>Your response</TITLE></HEAD> <BODY> Here's the body of the web page. </BODY> </HTML>

FIGURE 15.5 Simple HTTP request and response messages.

The HTTP request shown in Figure 15.5 is the simplest possible—only the request line is provided. This request includes the type of request (GET), the filename being requested (/index.html), and the HTTP version string that is understood by the requester (HTTP/1.1).

The HyperText Transfer Protocol is documented under RFC 2068. This RFC can be freely downloaded and its location is provided in the Resources section of this chapter.

The message flow is based upon a simple stream server and client (as shown in Figure 15.1). HTTP is a stream-based protocol, utilizing the TCP transport. The HTTP data flow showing the utilized Sockets API functions is shown in Figure 15.6. Note that this is a very simple scenario. Production HTTP servers (and clients) provide more robust controls.

As with any other stream socket server, the server begins by creating a socket (using the **socket** function), binding a name to it (using **bind**), and then announcing its willingness to accept client connections (using **listen**). A call to **accept** then permits the server to accept new client connections.

The client (which in this case is a Web browser), creates a socket and then connects to the server using the **connect** API function and an initialized sockaddr_in structure, identifying the server address and port number.

Now that the client is connected to the server, the client can announce its request. To retrieve a file, the GET command is used (recall from Figure 15.5). To specify that the request is complete, the client ends the request message with a blank line. Sending the request is performed using the **write** call (though it could also use **send**; these functions are synonymous). The server receives the request through the **recv** function (but could also be performed using the **read** call).

After the client's request is parsed, the server sends the HTTP response (see again, Figure 15.5). The response is made up of two basic parts, the response message header and the

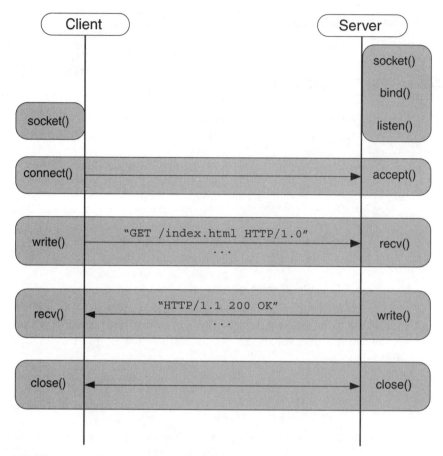

FIGURE 15.6 Simple HTTP server/client data flow.

response message body. The header defines certain information that is used by the protocol, but not commonly seen by the end user. For example, the "Content-Type: text/html" tells the client that the message body should be rendered as an HTML file. The body in this case is the HTML file requested by the user (/index.html). Note again that a blank line separates the message header from the body. The **write** call is shown here to transfer the HTTP response message to the client, who uses the **recv** function to receive the message.

Finally, once the dialog is complete, the **close** function is used to end the session.

The HTTP server implemented by each of the discussed languages provides very basic HTTP services (the HTTP GET method). Two content types will be supported, the "text/plain" type and the "text/html" type. The content type will be determined based upon the file suffix for the file to be served. The server will provide the ability to serve files from the local file system, which are available in the subdirectory where the server was started.

SIMPLE SMTP CLIENT DESIGN

The Simple Mail Transfer Protocol (or SMTP) is another ASCII-based protocol that operates on the client/server principle. The protocol is a synchronous request/response in which a client connects to a server and performs a multistep SMTP dialog to transfer mail from the client to server.

```
220 mailsite.com ESMTP Sendmail 9.8.3/9.9.7; Mon, 15 Nov 2002 22:23:05
HELO mydomain.com
250 mailsite.com Hello IDENT:time@[192.168.1.1], pleased to meet you
MAIL FROM: <me@mydomain.com>
250 <me@mydomain.com>... Sender ok
RCPT TO:<tim@mailsite.com>
250 <tim@mailsite.com>... Recipient ok
DATA
354 Enter mail, end with "." on a line by itself
From: me@mydomain.com
To: tim@mailsite.com
Subject: Test email
Content-Type: text/plain

This is a test email.

.
250 WAA14066 Message accepted for delivery
QUIT
221 mailsite.com closing connection
```

FIGURE 15.7   Sample SMTP dialog between a client and server.

An example SMTP transaction is shown in Figure 15.7; all client commands are shown in bold. After the client connects to the server, a salutation is sent from the server to announce that it's ready for the SMTP dialog (along with some other information that can be simply ignored, such as the mail server version, time, and so on).

Note that all messages received from the mail server are two-part messages. The first part of the message is a numeric status code that identifies the status of the server for this dialog. The second part is a text-readable string that may be ignored. From Figure 15.7, we see that the salutation includes a numeric error code of "220"; this represents normal status and is the indication to the mail client that it may continue with the transaction. The numeric codes differ, depending upon the stage of the SMTP transaction. We'll use a simple subset of the available SMTP commands, and, therefore, don't need to understand the full range of SMTP status codes.

The Simple Mail Transfer Protocol is documented under RFC 821. This RFC can be freely downloaded and its location is provided in the Resources section of this chapter.

The data flow for the SMTP client is a simple socket client with a **send/recv** pair for each SMTP dialog (see Figure 15.8).

After the server-side socket has been established (through **socket/bind/listen/accept**), the client can connect to the server using the **connect** API function. The server immediately emits the salutation string, which is received by the client using the **recv** API function. The client then announces itself to the server using the "HELO" command, which includes the domain of the client. The server responds to the "HELO" command with a hello-response string, but most importantly, a "250" status code. This code tells the client that it is ready for the rest of the e-mail transaction.

The client then specifies the source of the mail and to whom it is destined (using the "MAIL FROM" and "RCPT TO" commands, respectively). The "DATA" command specifies that the body of the e-mail is coming next. The client may then send many lines of text, followed by a '.' on a line by itself to mark the end of the mail. The return numeric status code "250" specifies to the client that the server successfully completed the e-mail transaction and that the e-mail has been accepted for delivery.

The final command to be performed is to end the SMTP transaction. This is done using the "QUIT" command, and causes the server to send a status response and then close the session. The **close** API function is then used on both sockets to physically close the connection.

Each implementation of the simple SMTP client within the target language includes a simple API that permits the developer to specify the recipient of the e-mail, source of the e-mail, subject, and a message body. With this information, the SMTP client will connect to the recipient's mail server and, using the SMTP protocol, deliver the e-mail.

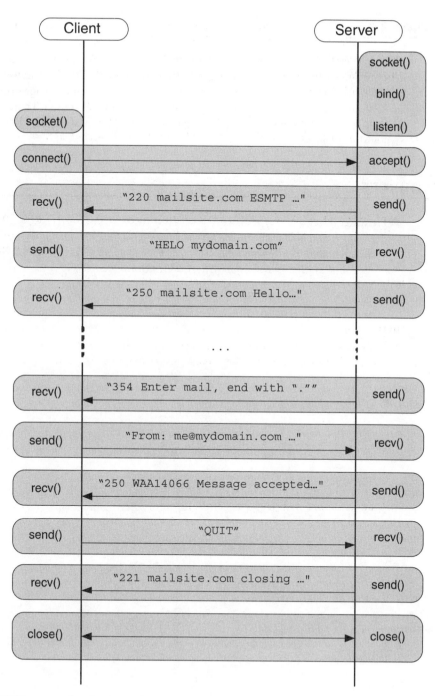

FIGURE 15.8 Simple SMTP client/server data flow.

SUMMARY

This chapter provided design pattern summaries for the sample applications that will demonstrate the features of the languages covered in this book. The patterns were illustrated by data flows outlining the Sockets API functions. Although each language provides a similar set of API functions, each differs in the manner in which it is used. Therefore, by covering the same design pattern in each language, we can better appreciate the offerings of each of the languages.

RESOURCES

Fielding, R., et al., "HyperText Transfer Protocol – HTTP/1.1," RFC 2068, January 1997, available online at *http://www.ietf.org/rfc/rfc2068.txt?number=2068*.

Postel, Jonathan B., "Simple Mail Transfer Protocol," RFC 821, August 1982, available online at *http://www.ietf.org/rfc/rfc0821.txt?number=82*.

16 Network Code Patterns in C

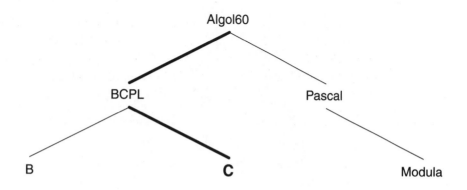

INTRODUCTION

In this chapter, we look at a number of simple socket-based applications in the C language using the BSD Sockets API. These applications can serve as implemented software patterns for the construction of more advanced networking applications. Each of the subsequent language code pattern chapters illustrates the same applications written in the chapter's specified language.

 All software discussed here is also provided on the companion CD-ROM with appropriate Makefiles. Table 16.1 lists the applications and their locations on the CD-ROM.

TABLE 16.1 CD-ROM companion software

Code Pattern	CD-ROM Location
Stream server/client	./software/ch16/stream/
Datagram server/client	./software/ch16/dgram/
Multicast server/client	./software/ch16/mcast/
Broadcast server/client	./software/ch16/bcast
Simple HTTP server	./software/ch16/shttp/
Simple SMTP client	./software/ch16/smtpc/

STREAM (TCP) SERVER/CLIENT

The stream server and client demonstrate the Daytime protocol using stream sockets (reliable TCP-based communication).

Stream Server

The Daytime protocol stream server is shown in Listing 16.1.

Listing 16.1 C language Daytime stream server.

```
1    #include <sys/socket.h>
2    #include <arpa/inet.h>
3    #include <stdio.h>
4    #include <time.h>
5    #include <unistd.h>
6
7    #define MAX_BUFFER            128
8    #define DAYTIME_SERVER_PORT   13
```

```
9
10   int main ( void )
11   {
12     int serverFd, connectionFd;
13     struct sockaddr_in servaddr;
14     char timebuffer[MAX_BUFFER+1];
15     time_t currentTime;
16
17     serverFd = socket(AF_INET, SOCK_STREAM, 0);
18
19     memset(&servaddr, 0, sizeof(servaddr));
20     servaddr.sin_family = AF_INET;
21     servaddr.sin_addr.s_addr = htonl(INADDR_ANY);
22     servaddr.sin_port = htons(DAYTIME_SERVER_PORT);
2
24     bind(serverFd,
25          (struct sockaddr *)&servaddr, sizeof(servaddr));
26
27     listen(serverFd, 5);
28
29     while ( 1 ) {
30
31       connectionFd = accept(serverFd,
32                             (struct sockaddr *)NULL, NULL);
33
34       if (connectionFd >= 0) {
35
36         currentTime = time(NULL);
37         snprintf(timebuffer, MAX_BUFFER, "%s\n",
38                  ctime(&currentTime));
39
40         write(connectionFd, timebuffer, strlen(timebuffer));
41         close(connectionFd);
42
43       }
44
45     }
46
47   }
```

The first item to note about Listing 16.1 is the header files shown in **lines 1 and 2**. These make visible a host of function prototypes, types, and symbolic constants needed for the stream server. The file socket.h (under subdirectory sys) defines the Sockets API (in this case, **socket**, **bind**, **listen**, and **accept**). Header file inet.h provides the INADDR_ANY symbolic, **htonl/htons** macros, and the sockaddr_in type. The remaining include files (**lines 3–5**) make available other system resources such as the time, printf, and write library functions.

Next, we define two symbolic constants in **lines 7 and 8**. These symbolic constants are nothing more than symbols used by the C preprocessor for replacement in the source, and they provide meaningful names in the source rather than ambiguous number constants.

Our sample stream server program begins at **line 10** with the definition of the main program (all C applications have a main, which represents the entry point for the program). Two socket identifiers are declared in **line 12** and our address structure (to represent the server) is declared in **line 13**. We also declare a character buffer (used to send data to the client) in **line 14** and a time variable used to extract the current system time in **line 15**.

We start our server, at **line 17**, with the creation of a new socket to represent our server. The identifier for the sock is returned by the **socket** call. Note that we're specifying three arguments with **socket** to create a stream socket (SOCK_STREAM) for Internet communication (AF_INET). The third argument to socket is defined as 0, and represents a protocol within the family SOCK_STREAM. Because TCP is only recognized here for SOCK_STREAM, no further definition is necessary.

As we're building a server, a name must be bound to the socket so that clients can address it. This naming process occurs in **lines 19–22**. After clearing out the address structure using memset, the protocol family is defined as AF_INET (permitting Internet communication). Next, the address is defined as INADDR_ANY, which means that the server will accept connections on any available interface. Finally, the port for the server is defined as the DAY-TIME_SERVER_PORT (or 13). Note the use of **htonl/htons** in **lines 21 and 22**. The network stack requires that these two elements be in network byte order; therefore, the host-to-network-long/short functions are used to convert these from host to network byte ordering.

In **lines 24 and 25**, we bind the servaddr name created in **lines 19–22** to the socket. This name represents the local name for the socket. After **bind** is performed on an address/port pair, this local name is active and may not be rebound to another socket.

Next, in **line 27**, we permit incoming connections with the server socket using the **listen** call. In this call, we identify the server socket and a backlog. The backlog defines the number of connections that may be queued for the particular server socket. After **listen** is called, incoming connections are permitted for the stream socket.

We begin an infinite loop at **line 29**, where the server accepts and handles incoming client connections. The call to **accept**, **lines 31–32**, blocks until an incoming client connection is accepted. The new client connection, returned from **accept**, is stored into connectionFd. Note that we pass two NULL values to **accept**, which could have been actual pointer references to store the host/port information of the connected client peer. Because we're not interested, we pass NULLs to ignore this information.

Upon verifying that a valid socket was returned from **accept** in **line 34**, we construct a string version of the current system time using the time system function and sprintf to build the string (**lines 36–38**). The **write** function is then used to send the time string (stored in timebuffer) to the client. Note that the connectionFd socket descriptor is used to send the data (as returned by **accept**), referencing both the timebuffer and its length.

Finally, once the data is written through the socket to the client, the newly created socket (represented by connectionFd) is closed using the **close** function (**line 41**). The server socket (serverFd) is not closed, as we'll loop back to the **accept** function at **line 31** to await a new client connection.

Stream Client

Now, let's look at the Daytime stream client (shown in Listing 16.2). The client creates a socket to connect to the server and then awaits a string to be received through the socket. Upon receiving the string, it's printed and the client exits.

Listing 16.2   C language Daytime stream client.

```
1    #include <sys/socket.h>
2    #include <arpa/inet.h>
3    #include <stdio.h>
4    #include <unistd.h>
5    #include <time.h>
6
7    #define MAX_BUFFER          128
8    #define DAYTIME_SERVER_PORT  13
9
10   int main ( )
11   {
12     int connectionFd, in;
13     struct sockaddr_in servaddr;
14     char timebuffer[MAX_BUFFER+1];
15
16     connectionFd = socket(AF_INET, SOCK_STREAM, 0);
17
18     memset(&servaddr, 0, sizeof(servaddr));
19     servaddr.sin_family = AF_INET;
20     servaddr.sin_port = htons(DAYTIME_SERVER_PORT);
21
22     servaddr.sin_addr.s_addr = inet_addr("127.0.0.1");
23
24     connect(connectionFd, (struct sockaddr_in *)&servaddr,
25              sizeof(servaddr));
26
27     while ( (in = read(connectionFd,
28                        timebuffer, MAX_BUFFER)) > 0) {
29       timebuffer[in] = 0;
30       printf("\n%s", timebuffer);
31     }
32
33     close(connectionFd);
34
35     return(0);
36   }
```

Lines 1–14 are identical to the server; refer to the discussion of Listing 16.1 for more information about this code segment.

In **line 16**, we create a socket, identically to the socket created by the server (a stream socket). We also create an address structure (in **lines 18–22**), but in this case, we're not creating the address structure to name the server. In this case, we're creating the address structure to define where the client should connect. Note in **line 22**, that we specify an actual address instead of INADDR_ANY. For the server, the symbolic INADDR_ANY specifies that the server can accept a connection through any interface. In the case of the client, we're specifying an IP address to which we'll connect. This IP address is a special address that represents the loop-back address (127.0.0.1). This means that our connection will be "looped-back" onto the same host (because the client and server, in this example, occupy the same host). This could have been an actual IP address, representing another host on the Internet to which to connect.

We use the **connect** function, at **line 24**, to create a connection between the client and server sockets. Note that we use our previously created socket (connectionFd) and the address structure, servaddr, which represents our intended target server.

With our connection created, we go into a loop to read the time string sent by the server (**line 27 and 28**). Any data received is printed out for view (**line 30**). The call to the **read** function will block, but when the server closes the socket (upon sending the time string), the **read** call will return with an error. This error code will be zero, which allows us to exit out of the loop. Note that this error won't be returned until we've read any data that's available. In the **read** call, we reference the client socket, the timebuffer to store the received string, and the maximum number of bytes to accept in this call.

After we've detected that the server has closed its end of the socket, we close our side at **line 33**. The client is then exited at **line 35** and the process is complete.

DATAGRAM (UDP) SERVER/CLIENT

Now, let's look at a datagram server and client implementation. Recall that datagram communication is not connection-oriented; therefore, each datagram that we send must include the intended recipient with the data.

Datagram Server

The datagram server is illustrated in Listing 16.3.

As discussed with the stream server, **lines 1–5** make visible a number of function prototypes, types, and symbolic constants used by the datagram server. We also define our symbolic constants in **lines 7–8**, to represent the maximum size of the buffer to write (MAX_BUFFER) and the port to which we'll bind for our server socket (DAYTIME_SERVER_PORT).

Listing 16.3   C language Daytime datagram server.

```
1    #include <sys/socket.h>
2    #include <arpa/inet.h>
3    #include <stdio.h>
```

```
 4   #include <time.h>
 5   #include <unistd.h>
 6
 7   #define MAX_BUFFER            128
 8   #define DAYTIME_SERVER_PORT   13
 9
10   int main ( void )
11   {
12     int serverFd, fromlen, msglen;
13     struct sockaddr_in servaddr, from;
14     char timebuffer[MAX_BUFFER+1];
15     time_t currentTime;
16
17     serverFd = socket(AF_INET, SOCK_DGRAM, 0);
18
19     memset(&servaddr, 0, sizeof(servaddr));
20     servaddr.sin_family = AF_INET;
21     servaddr.sin_addr.s_addr = htonl(INADDR_ANY);
22     servaddr.sin_port = htons(DAYTIME_SERVER_PORT);
23
24     bind(serverFd,
25           (struct sockaddr *)&servaddr, sizeof(servaddr));
26
27     while ( 1 ) {
28
29       fromlen = sizeof(from);
30
31       msglen = recvfrom(serverFd, &timebuffer,
32                         MAX_BUFFER, 0, &from, &fromlen);
33
34       if (msglen >= 0) {
35
36         currentTime = time(NULL);
37         snprintf(timebuffer, MAX_BUFFER, "%s\n",
38                  ctime(&currentTime));
39
40         sendto(serverFd, timebuffer, strlen(timebuffer), 0,
41                (struct sockaddr *)&from, fromlen);
42
43       }
44
45     }
46
47     return(0);
48   }
```

Line 10 begins our main program for the datagram server. A number of local variables are declared, each of which is discussed as used in the server.

We create our socket at **line 17**. Note that although we use the domain of AF_INET to create an Internet socket, we specify SOCK_DGRAM as our type (parameter 2). This defines our socket as a datagram socket (compared to a stream socket, using SOCK_STREAM type).

As we're building a server, we need to name our server so that clients can find it. To do this, we create our address structure in **lines 19–22**. After clearing out the structure in **line 19**, we specify the family (socket domain) as AF_INET in **line 20**. The sin_addr element of our address structure (specifically from the server perspective, **line 21**) permits a client to connect through any of the available interfaces (INADDR_ANY). This wildcard could have instead been the address of one interface on the host, which would restrict connections to be allowed only through that interface. Finally, at **line 22**, we specify the port on which this server can be accessed on this host, the Daytime server port (DAYTIME_SERVER_PORT).

To bind the address information to our previously created socket, we use the bind Sockets function at **lines 24–25**. To **bind**, we pass our server socket (serverFd), and the address information (servaddr), including the size of the address structure (computed using the sizeof standard function).

Notably absent from Listing 16.3 are the **listen** and **accept** calls. Datagram sockets aren't connected; therefore, these calls are irrelevant in this paradigm.

Our infinite loop begins at **line 27**. We calculate the size of our address structure at **line 29**, in preparation for a call to **recvfrom** at **line 31**. The **recvfrom** call is used strictly with datagram sockets (and is blocking). This variant of the **recv** function, as can be seen from its name, receives not only a datagram, but also information about the source of the datagram. Recall that with stream sockets, this isn't necessary because all data will arrive from the same source (to which we originally connected). To the **recvfrom** call, we pass our server socket (serverFd), a buffer to receive our datagram (timebuffer), along with the size of the buffer (MAX_BUFFER). The size is important because it defines the maximum size datagram that can be retrieved. The fourth argument (0) represents the flags; in this case, no flags are provided. Finally, we provide an address structure (from) along with a size argument. In this case, we pass fromlen by reference, so that the **recvfrom** call can tell us the size that was returned. Finally, **recvfrom** returns the size of the datagram returned, which is stored within the msglen variable.

At **line 34**, we check the length returned in msglen to ensure that a valid datagram was received. If so (a nonnegative value was returned), then the from address structure is valid and we generate a new time string in timebuffer (**lines 36–38**). The **send** function variant, **sendto**, is used to send a datagram to a specific recipient (**lines 40–41**). Note that we pass our server socket (serverFd), our character buffer (timebuffer), and the length of this buffer using the standard function strlen. We specify no flags, and then the address received in the initial **recvfrom** call (which represents who originally sent us an empty datagram) is used to return the time datagram. Therefore, based upon this code pattern, whoever sent us the empty datagram is to whom we send back the time string.

After the datagram is sent, we loop back around to **line 29** and then await another datagram using the **recvfrom** call. Note that another difference from the stream server is that there is no **close** call for the client socket. Because datagram sockets operate on a message basis, there is no defined connection between any two endpoints. Datagram sockets simply provide the ability to send messages to other datagram sockets.

Datagram Client

The datagram client is shown in Listing 16.4, and corresponds with the datagram server previously shown in Listing 16.3.

Listing 16.4 C language Daytime datagram client.

```
1    #include <sys/socket.h>
2    #include <arpa/inet.h>
3    #include <stdio.h>
4    #include <unistd.h>
5    #include <time.h>
6
7    #define MAX_BUFFER           128
8    #define DAYTIME_SERVER_PORT  13
9
10   int main ()
11   {
12     int connectionFd, in;
13     struct sockaddr_in servaddr;
14     char timebuffer[MAX_BUFFER+1];
15
16     connectionFd = socket(AF_INET, SOCK_DGRAM, 0);
17
18     memset(&servaddr, 0, sizeof(servaddr));
19     servaddr.sin_family = AF_INET;
20     servaddr.sin_port = htons(DAYTIME_SERVER_PORT);
21
22     servaddr.sin_addr.s_addr = inet_addr("127.0.0.1");
23
24     sendto( connectionFd, timebuffer, 1, 0,
25             (struct sockaddr_in *)&servaddr,
26             sizeof(servaddr) );
27
28     in = recv(connectionFd, timebuffer, MAX_BUFFER, 0 );
29     timebuffer[in] = 0;
30
31     printf("\n%s", timebuffer);
32
33     close(connectionFd);
34
35     return(0);
36   }
```

We'll focus our attention now on the differences between the server discussed previously and the client source shown in Listing 16.4. Note in **line 16**, that we create our socket in the same manner as the server. There is no name binding to this socket, as was required with the server. In **lines 18–22**, we create an address structure, but this represents the destination for our datagram. Note the symmetry in this address structure to the one defined for the server. The family and port (AF_INET and DAYTIME_SERVER_PORT) are identical to the server, but in this case, the sin_addr is given an actual address instead of INADDR_ANY (the wildcard). This is because we're defining where to connect, in this case the loopback address (127.0.0.1).

Now that we have a destination set up for our initial datagram, we send a single byte to the server using the **sendto** call at **lines 24–26**. This allows the server to know who we are so that it can send us a response. We provide our client socket descriptor (connectionFd) as well as our temporary string buffer (timebuffer) and a length (1 byte). We provide no flags to **sendto**, but with the servaddr structure, we define the destination for the datagram.

Next, we await a response from the server in **line 28** using the **recv** call. Recall again the symmetry with the server. After the server receives the initial datagram, it uses the source information of this datagram to send back a response to that client. The client uses the **recv** call to receive the server's response. The client socket descriptor is provided along with a temporary string buffer and size. The fourth parameter of **recv** represents the flags argument that is not used here. The return value of **recv** is the number of bytes received in the response datagram (stored in variable in). We assume here that a valid datagram was received, and in **line 29**, we NULL-terminate the timebuffer using the return variable (in) as the index of the end of the string. The string stored in timebuffer is then emitted at **line 31** and the client socket is closed at **line 33**.

Although the client closes its socket, the server has no knowledge of this (compared to stream sockets) because the communication with datagrams is not connected. The client then exits and the dialog between the client and server is complete.

MULTICAST SERVER/CLIENT

The multicast server and client are fundamentally derivatives of the datagram code patterns. This is primarily because multicast communication is based upon the datagram model.

Multicast Server

The C language source code for the multicast server is shown in Listing 16.5. Recall that multicast communication is group-based; therefore, a single message emitted by the server is received by every client that is currently a member of the multicast group.

The necessary function prototypes, types, and symbolic constants are imported in **lines 1–5**. **Lines 7 and 8** define our addressing information. The MCAST_GROUP defines a multicast address, to which all clients and servers will be bound in order to send or receive datagrams. Additionally, a unique port (for UDP) must be specified for the host/port pair.

Listing 16.5   C language Daytime multicast server.

```
1    #include <stdio.h>
2    #include <sys/socket.h>
3    #include <arpa/inet.h>
4    #include <unistd.h>
5    #include <time.h>
6
7    #define MCAST_GROUP      "239.0.0.2"
8    #define MCAST_PORT       45002
9
10   int main ()
11   {
12     int     sock, cnt, addrLen;
13     struct sockaddr_in addr;
14     char    buffer[512];
15
16     sock = socket(AF_INET, SOCK_DGRAM, 0);
17
18     memset(&addr, 0, sizeof(addr));
19     addr.sin_family = AF_INET;
20     addr.sin_port = htons(MCAST_PORT);
21     addr.sin_addr.s_addr = inet_addr(MCAST_GROUP);
22     addrLen = sizeof(addr);
23
24     while (1) {
25
26       time_t t = time(0);
27
28       sprintf(buffer, "%s\n", ctime(&t));
29
30       printf("sending %s", buffer);
31
32       cnt = sendto(sock, buffer, strlen(buffer), 0,
33                    (struct sockaddr *)&addr, addrLen);
34
35       printf("cnt = %d\n", cnt);
36
37       sleep(1);
38
39     }
40
41     return 0;
42   }
```

Our main application begins at **line 10**, and a number of necessary local variables are declared in **lines 12–14**.

At **line 16**, we create our server's socket, specifying AF_INET for Internet communication and SOCK_DGRAM to identify UDP. The return value for the **socket** function is the socket descriptor.

The address structure for the server is initialized in **lines 18–21**. After clearing the structure in **line 18**, the family is set to AF_INET and the port is defined to our multicast port. At **line 21**, we initialize the address for which we'll receive datagrams, in this case our multicast group address (MCAST_GROUP). Recall from Listing 16.3 that our datagram server provided INADDR_ANY for address, which permitted accepting datagrams on any available interface. Here, we specify our multicast address, which represents a special interface. We'll discuss how to restrict these to one or more physical interfaces using the multicast socket options in the Multicast Client section.

At **line 22**, we compute the size of the address structure for use in the infinite loop (because it won't change during the course of execution).

We begin our infinite loop at **line 24**. This multicast server simply emits the current time in string format every second. Because there is no connected state between multicast servers and clients, this arrangement will be used to discuss multicast dialogs.

At **line 26**, we capture the current time (using the standard time function) and then turn it into a string at **line 28** using ctime. We emit this string (contained in buffer) to the console each time it's generated just for debugging purposes.

At **lines 32–33**, we emit the previously constructed time string in buffer using a call to sendto. Note that we use the previously constructed socket descriptor, sock, along with the string buffer and its length as arguments 2 and 3. The fourth argument, flags, is unused in this example. Finally, we specify the address structure (addr) that represents the destination of the datagram and the address structure's length (addrLen). From this single call, the datagram will be multicast to the defined group. This consists not only of applications on other hosts, but also of applications running on the same host that may have subscribed to the multicast group.

Finally, we emit the return value from the **sendto** call (**line 35**), which represents the number of bytes sent through to the multicast group. This should be identical to the string length of buffer. At **line 37**, we sleep one second and then continue the infinite loop at **line 39**.

Multicast Client

The C language source code for the multicast client is shown in Listing 16.6. The multicast client includes a number of new features not previously discussed, but reflects symmetry with the multicast server, as seen with other code patterns.

Lines 1–18 are identical to the server listing in Listing 16.5, with the exception that the time.h header file is not used by the client. **Lines 9–15** begin our C main program and provide some necessary variable declarations (we discuss the variables when used in the client application).

As with all socket-based applications, **line 17** creates a new datagram socket for Internet communication. After the socket is created, we use the SO_REUSEADDR socket option to permit the port to be reused (**lines 19–20**). If this had not been done, the client would not have been able to bind the host/port pair to the socket, because another socket already exists with this port number (the server). By enabling the SO_REUSEADDR socket option, the Sockets API turns off this check and allows us to successfully bind to the port. The enable is defined by the on variable, which has been set to 1 to represent enable (0 represents disable). The SO_REUSEADDR option is a Sockets level option, so the SOL_SOCKET argument is specified.

In **lines 22–25**, we initialize the address structure that we want to bind to the previously created socket. After clearing the structure in **line 22**, we initialize the family for Internet communication and the port to the previously defined multicast port (MCAST_PORT). At **line 25**, the addr field is initialized to the wildcard INADDR_ANY to permit connections from any available interface. This information is then bound to the socket at **line 27**.

Next, in order to receive multicast datagrams, we must add our socket to the multicast membership group for the specific multicast group (**lines 29–32**). This action is performed through a **setsockopt** using the IP_ADD_MEMBERSHIP option. Within the mreq structure, we define the multicast group as the multicast address and the interface as the wildcard INADDR_ANY. This means that we're registering to receive multicast traffic for our defined multicast group, and on any available host interface. We could set the interface to a single interface (if the host were multi-homed) and restrict the receipt of multicast traffic to a single interface. Recall from Listing 16.5 that the server does not perform this socket option. This is because adding a socket to multicast membership is only required to receive multicast datagrams. Multicast datagrams can be sent without membership.

Finally, the client enters an infinite loop in **lines 34–44** to receive multicast datagrams from the server. Datagram receipt is performed using the **recvfrom** call. Although knowing who sent the datagram is not necessary in this example (from the addr structure), we use this information to emit the source along with the time string (**line 42**). Note that the sockaddr_in structure (addr) is reused in this example; the information specified in **lines 23–25** is not relevant to the **recvfrom** call at **line 38**.

Listing 16.6   C language Daytime multicast client.

```
 1   #include <stdio.h>
 2   #include <sys/socket.h>
 3   #include <arpa/inet.h>
 4   #include <unistd.h>
 5
 6   #define MCAST_GROUP      "239.0.0.2"
 7   #define MCAST_PORT       45002
 8
 9   int main ()
10   {
11     int    sock, cnt, addrlen;
12     int    on=1;
13     struct sockaddr_in addr;
14     struct ip_mreq mreq;
15     char   buffer[512];
16
```

```
17      sock = socket(AF_INET, SOCK_DGRAM, 0);
18
19      setsockopt(sock, SOL_SOCKET, SO_REUSEADDR,
20                  (char *)&on, sizeof(on));
21
22      memset(&addr, 0, sizeof(addr));
23      addr.sin_family = AF_INET;
24      addr.sin_port = htons(MCAST_PORT);
25      addr.sin_addr.s_addr = htonl(INADDR_ANY);
26
27      bind(sock, (struct sockaddr *)&addr, sizeof(addr));
28
29      mreq.imr_multiaddr.s_addr = inet_addr(MCAST_GROUP);
30      mreq.imr_interface.s_addr = htonl(INADDR_ANY);
31      setsockopt(sock, IPPROTO_IP, IP_ADD_MEMBERSHIP,
32                  &mreq, sizeof(mreq));
33
34      while (1) {
35
36        addrlen = sizeof(addr);
37
38        cnt = recvfrom(sock, buffer, sizeof(buffer), 0,
39                        (struct sockaddr *)&addr, &addrlen);
40        buffer[cnt] = 0;
41
42        printf("%s : %s\n", inet_ntoa(addr.sin_addr), buffer);
43
44      }
45
46      return 0;
47  }
```

BROADCAST SERVER/CLIENT

The broadcast server and client, like the multicast server and client, are fundamentally derivatives of the datagram code patterns. This is again because broadcast communication is based upon the datagram model.

Broadcast Server

The C language source code for the broadcast server is shown in Listing 16.7. The server is broadcast-based, which means a single message emitted by the server is received by every client that is configured to receive broadcast datagrams (for the given port).

Because much of the discussion of the broadcast server is similar to the datagram server (shown in Listing 16.3), we discuss only some of the elements that make this server broadcast communication specific.

The first item to note is **lines 17–18**, where after the datagram socket is created, we enable the SO_BROADCAST option using setsockopt. This socket option permits a socket to send datagrams to the broadcast address (shown here as 255.255.255.255). This is the limited-broadcast address, because this particular broadcast variant will never cross a router. In this way, datagrams sent with this broadcast address are limited to the LAN of the originating host.

When setting up our address structure (**lines 20–23**), we follow a similar pattern of AF_INET, port and address. Note the use of inet_addr to convert the string address into the 32-bit network-byte-order variant. This address structure defines the recipient of the datagrams sent by the server, in this case the limited broadcast address, port 45003.

Finally, our broadcast server enters an infinite loop, emitting time strings every second. Note **lines 34–35**, in which the datagram is actually emitted. The **sendto** call is datagram-specific and is used to send a buffer to a named endpoint defined by a host and port. In this case, the named recipient was previously defined in our address structure (**lines 20–23**).

Listing 16.7   C language Daytime broadcast server.

```
1   #include <stdio.h>
2   #include <sys/socket.h>
3   #include <arpa/inet.h>
4   #include <unistd.h>
5   #include <time.h>
6
7   #define BCAST_PORT        45003
8
9   int main ()
10  {
11    int    sock, cnt, addrLen, on=1;
12    struct sockaddr_in addr;
13    char   buffer[512];
14
15    sock = socket(AF_INET, SOCK_DGRAM, 0);
16
17    setsockopt(sock, SOL_SOCKET, SO_BROADCAST,
18                &on, sizeof(on));
19
20    memset(&addr, 0, sizeof(addr));
21    addr.sin_family = AF_INET;
22    addr.sin_port = htons(BCAST_PORT);
23    addr.sin_addr.s_addr = inet_addr("255.255.255.255");
24    addrLen = sizeof(addr);
25
26    while (1) {
27
28      time_t t = time(0);
29
```

```
30          sprintf(buffer, "%s\n", ctime(&t));
31
32          printf("sending %s", buffer);
33
34          cnt = sendto(sock, buffer, strlen(buffer), 0,
35                          (struct sockaddr *)&addr, addrLen);
36
37          sleep(1);
38
39      }
40
41  }
```

Broadcast Client

The C language source code for the broadcast client is shown in Listing 16.8. As with the broadcast server, we concentrate on the differences to the standard datagram client.

After the client socket is created at **line 14**, the address structure (addr) is created and initialized in **lines 16–19**. We again define the family for Internet communication, and set the port to our broadcast port. For the client, we set the sin_addr field to the wildcard, INADDR_ANY. This permits the client to receive broadcast datagrams for port BCAST_PORT on any of the available host interfaces. The addressing information and the client socket are joined with the **bind** call at **line 21**.

A difference between the broadcast client and server is that the SO_BROADCAST socket option is not enabled for the client. This option is not used because the client does not send any broadcast datagrams; it only receives them.

Finally, the broadcast client receives datagrams using the **recvfrom** call at **lines 27–28**. The client saves the source address of the datagram, which is used only to print the source with the received buffer at **lines 27–28**.

Listing 16.8   C language Daytime broadcast client.

```
1   #include <stdio.h>
2   #include <sys/socket.h>
3   #include <arpa/inet.h>
4   #include <unistd.h>
5
6   #define BCAST_PORT          45003
7
8   int main ()
9   {
10    int     sock, cnt, addrlen;
11    struct sockaddr_in addr;
12    char    buffer[512];
13
14      sock = socket(AF_INET, SOCK_DGRAM, 0);
```

```
15
16      memset(&addr, 0, sizeof(addr));
17      addr.sin_family = AF_INET;
18      addr.sin_port = htons(BCAST_PORT);
19      addr.sin_addr.s_addr = htonl(INADDR_ANY);
20
21      bind(sock, (struct sockaddr *)&addr, sizeof(addr));
22
23      while (1) {
24
25        addrlen = sizeof(addr);
26
27        cnt = recvfrom(sock, buffer, sizeof(buffer), 0,
28                         (struct sockaddr *)&addr, &addrlen);
29        buffer[cnt] = 0;
30
31        printf("%s : %s\n", inet_ntoa(addr.sin_addr), buffer);
32
33      }
34
35    }
```

SIMPLE HTTP SERVER

 Now, let's look at a simple HTTP server written in the C language. We focus only on the
ON THE CD networking elements; the remaining functions are provided on the CD-ROM.

The first part of our simple HTTP server is the creation and configuration of our
stream server socket. Listing 16.9 shows the start function that is used to start the server.

At **line 7**, we create our stream socket, storing it in serverFd (our server socket descriptor). To ensure that we can restart the server without having to wait two minutes before the address/port combination becomes available, the SO_REUSEADDR socket option is
used. We name our server socket at **lines 16–19**, permitting connections from any interface
(using the INADDR_ANY wildcard address) and the port provided by the caller to start. The
bind call is used at **lines 22–23** to performing the actual name binding.

At **line 26**, we identify our willingness to accept new connections using the **listen** call.
Note that we specify that up to five connections may be outstanding while we process the
current one.

At **line 28**, our infinite loop begins, which represents our simple HTTP server. The
accept call at **line 31** accepts new client connections, and a quick test of the return value
identifies whether an error was returned. If an error was returned, we exit the loop using the
C break keyword (**line 32**). Otherwise, we call the handle_connection function at **line 35**
and pass the client socket descriptor returned by the **accept** API function at **line 31**. Upon
returning from handle_connection, we close the client socket descriptor at **line 38** and
continue with the loop at **line 40**.

If an error had resulted from the **accept** call (detected at **line 32**), control would have resumed at **line 43**, where the server socket would be closed and the server then ended.

Listing 16.9   Simple HTTP server function start.

```
1    static void start( unsigned short port )
2    {
3      int serverfd, clientfd, on=1;
4      struct sockaddr_in servaddr;
5
6      /* Create a new TCP server */
7      serverfd = socket( AF_INET, SOCK_STREAM, 0 );
8
9      /* Make the port immediately reusable after this socket
10      * is close.
11      */
12     setsockopt( serverfd, SOL_SOCKET, SO_REUSEADDR,
13                 &on, sizeof(on) );
14
15     /* Initialize the address structure to which we'll bind */
16     bzero( (void *)&servaddr, sizeof(servaddr) );
17     servaddr.sin_family = AF_INET;
18     servaddr.sin_addr.s_addr = htonl( INADDR_ANY );
19     servaddr.sin_port = htons( port );
20
21     /* Bind the address to the socket */
22     bind( serverfd,
23           (struct sockaddr *)&servaddr, sizeof(servaddr) );
24
25     /* Make the new server visible for incoming connections */
26     listen( serverfd, 5 );
27
28     while ( 1 ) {
29
30       /* Await a connection from a client socket */
31       clientfd = accept( serverfd, NULL, NULL );
32       if (clientfd <= 0) break;
33
34       /* handle the new connection */
35       handle_connection( clientfd );
36
37       /* Close the client connection */
38       close( clientfd );
39
40     }
41
42     /* Close the server socket */
```

```
43    close( serverfd );
44
45    return;
46  }
```

The `handle_connection` function (shown in Listing 16.10) exists to satisfy the HTTP command received by the peer client. The result will be an HTTP response sent back through the same socket. The `handle_connection` function can be logically split into two parts, extracting the HTTP command from the socket and then generating the response.

Lines 9–26 provide the means to read the HTTP command from the socket. Recall from the HTTP server discussion in Chapter 15, Software Patterns Introduction, that an HTTP command is terminated by a single blank line. **Lines 18–20** provide a means to detect this situation and break from the loop of reading from the socket.

The next part of `handle_connection` is providing the response. We first check the type of request at **line 29** and then if it is a GET request, we call the `handle_get_method` function. Note that we also parse the requested filename from the HTTP request prior to calling the `handle_get_method` function.

If the request was not GET, we return an HTTP error message in **lines 39–45**. Note that in this case, a standard HTTP error code is written to the peer client using the **write** function.

Listing 16.10   Simple HTTP server function `handle_connection`.

```
1    static void handle_connection( int sock )
2    {
3      int len, max, ret;
4      char inbuf[1024]={0};
5
6      max = 0;
7
8      /* Read in the HTTP request message from the socket */
9      while ( 1 ) {
10
11       len = read( sock, &inbuf[max], (1024-max) );
12       if (len <= 0) return;
13
14       /* Update the total string size */
15       max += len;
16
17       /* HTTP request message ends with CRLF/CRLF */
18       if ((inbuf[max-4] == 0x0d) && (inbuf[max-3] == 0x0a) &&
19           (inbuf[max-2] == 0x0d) && (inbuf[max-1] == 0x0a)) {
20         break;
21
22       }
23
```

```
24      inbuf[max] = 0;
25
26    }
27
28    /* Determine the request type */
29    if (!strncmp( inbuf, "GET", 3)) {
30
31      char filename[100];
32
33      getFilename( inbuf, filename, 4 );
34
35      handle_get_method( sock, filename );
36
37    } else {
38
39      const char *notimpl=
40                  {"HTTP/1.0 501 Not Implemented.\n\n"};
41
42      printf("Unknown Method %s\n", inbuf);
43
44      /* Unknown file — notify client */
45      write( sock, notimpl, strlen(notimpl) );
46
47    }
48
49  }
```

The handle_get_method function in Listing 16.11 provides the means to satisfy an HTTP GET request. This function performs two basic functions. The first is to identify whether the file requested by the peer client exists on the local file system (**lines 9–17**). If the file is not found, a standard HTTP error message is written to the peer. In this case, it's the HTTP 404 file-not-found error.

If the file was found, then we first determine the type of file we're dealing with (**line 23**). After this is known, we can generate the HTTP response message header (which includes the type of content being sent back to the peer).

Finally, in **lines 30–41**, the contents of the file are read using the **read** function and then written through the client socket to the peer using the **write** function. To close the HTTP transaction, we send a new line representing a blank line (end-of-message) and then close the socket using the **close** function.

Listing 16.11   Simple HTTP server function handle_get_method.

```
1    static void handle_get_method( int sock, char *filename )
2    {
3      int fd, ret;
4      char buffer[255+1];
```

```
5
6      /* Try to open the file requested by the client */
7      fd = open( filename, O_RDONLY );
8
9      if (fd == -1) {
10
11       /* Can't find, emit not found error to the client */
12       const char *notfound=
13                       {"HTTP/1.0 404\n\n File not found.\n\n"};
14
15       write( sock, notfound, strlen( notfound ) );
16
17       printf("File not found.\n");
18
19     } else {
20
21       char *content_type;
22
23       content_type = define_content_type( filename );
24
25       emit_response_header( sock, content_type );
26
27       /* Read and emit the file contents to the client
28        * through the socket.
29        */
30       ret = 1;
31       while ( ret > 0 ) {
32
33         ret = read( fd, buffer, 255 );
34
35         if ( ret > 0 ) {
36
37           write( sock, buffer, ret );
38
39         } else break;
40
41       }
42
43       /* Emit a newline */
44       write( sock, "\015\012", 2 );
45
46       close( fd );
47
48     }
49
50     return;
51   }
```

The final function to consider for the simple HTTP server is the emit_response_header function. This function, shown in Listing 16.12, creates and then sends the HTTP response message header to the client peer through the socket.

The caller provides the client socket and the content_type to the emit_response_header function. Recall that the content_type string may be "text/html" for an HTML page or "application/octet-stream" for a binary stream of data.

The function simply constructs the HTTP response message within the resp_header buffer (using the variable content_type) and then writes this buffer to the peer through the passed client socket (sock).

Listing 16.12 Simple HTTP server function emit_response_header.

```
1    static void emit_response_header( int   sock,
2                                      char *content_type )
3    {
4      char resp_header[256];
5
6      /* Construct the response header (including the variable
7       * content type) and send it to the client through the
8       * socket.
9       */
10     sprintf( resp_header,
11             "HTTP/1.1 200 OK\n"
12             "Server: C shttp\n"
13             "Connection: close\n"
14             "Content-Type: %s\n\n", content_type );
15
16     write( sock, resp_header, strlen(resp_header) );
17
18     return;
19   }
```

SIMPLE SMTP CLIENT

ON THE CD

Now, let's look at a simple SMTP client written in the C language (see Listing 16.13). Again, we focus only on the networking elements; the remaining functions are provided on the CD-ROM.

Lines 8–14 perform the typical client socket creation and specification of the peer address to which we'll connect. **Lines 22–39** provide the ability to translate the mail server address to the raw 32-bit IP address. The source address (in the mailServer string) can either be a dotted-notation IP address (such as "192.168.1.1") or an FQDN (such as "www.microsoft.com"). We first use the **inet_addr** function to attempt to convert the address by assuming it's a dotted-notation IP address string. If we find this was not the case (at **line 22**), then we assume that it's an FQDN and use the gethostbyname API function to resolve the name to a native IP address.

At **line 42**, we connect our client socket to the peer address as defined by the servaddr structure. The result here will be a connected socket to the peer mail server.

Lines 45–89 provide the protocol handling between this client and the SMTP server (as described in Chapter 15, Software Patterns Introduction). Note that the protocol handling is performed within a do/while loop, in which the while has a false conditional. The do/while construct in C provides a minimum of one iteration through the loop, but in this case, because the conditional is false, it will be done only once. The purpose of this construct is to permit a type of goto statement, without the goto statement. If an error occurs in the protocol handling, a break statement is performed. This statement forces a continuation of execution at the end of the loop, thereby providing goto statement functionality.

The actual dialog with the SMTP server is provided by the function dialog (shown in Listing 16.14). This handles not only sending the defined command to the server, but also checking the response status code from the server.

Finally, at **line 91**, we close the client socket and then return a status code to the caller identifying the success or failure of the SMTP transaction.

Listing 16.13   Simple SMTP client sendMail function.

```
1    int sendMail(struct mailHeader *mail)
2    {
3      int connfd, result, ret, goodMsg = 0;
4      struct sockaddr_in servaddr;
5      char mailRcpt[129];
6      char line[256];
7
8      connfd = socket(AF_INET, SOCK_STREAM, 0);
9
10     bzero((void *)&servaddr, sizeof(servaddr));
11     servaddr.sin_family = AF_INET;
12     servaddr.sin_port = htons(25);
13
14     servaddr.sin_addr.s_addr = inet_addr(mailServer);
15
16     /* if the prior inet_addr results in a '-1' (or error),
17      * then we assume that the gateway symbolic is not a
18      * dotted-notation IP address. It must therefore be a
19      * fully-qualified domain name and we use gethostbyname
20      * to resolve it.
21      */
22     if (servaddr.sin_addr.s_addr == 0xffffffff) {
23
24       struct hostent *hptr =
25               (struct hostent *)gethostbyname(mailServer);
```

```
26
27     if (hptr == NULL) {
28       /* Don't know what the mailServer represents... */
29       return(-1);
30     } else {
31
32       struct in_addr **addrs;
33       addrs = (struct in_addr **)hptr->h_addr_list;
34       memcpy(&servaddr.sin_addr,
35         *addrs, sizeof(struct in_addr));
36
37     }
38
39   }
40
41   /* Connect to the SMTP server */
42   result = connect(connfd, (struct sockaddr *)&servaddr,
43                     sizeof(servaddr));
44
45   do {
46
47     /* Look for initial salutation */
48     if ( dialog( connfd, NULL, "220" ) ) break;
49
50     /* Send HELO and await response */
51     if ( dialog( connfd, hello_msg, "250" ) ) break;
52
53     /* Send MAIL FROM and await response */
54     sprintf(line, "MAIL FROM:<%s>\n", mail->sender);
55     if ( dialog( connfd, line, "250" ) ) break;
56
57     /* Send RCPT TO and await response */
58     sprintf(line, "RCPT TO:<%s>\n", mail->recipient);
59     if ( dialog( connfd, line, "250" ) ) break;
60
61     /* Send DATA and await response */
62     if ( dialog( connfd, "DATA\n", "354" ) ) break;
63
64     /* Send out the header first */
65     sprintf(line, "From: %s\n", mail->sender);
66     if ( dialog( connfd, line, NULL ) ) break;
67
68     sprintf(line, "To: %s\n", mail->recipient);
69     if ( dialog( connfd, line, NULL ) ) break;
70
71     sprintf(line, "Subject: %s\n", mail->subject);
72     if ( dialog( connfd, line, NULL ) ) break;
```

```
73
74        if (mail->contentType[0] != 0) {
75          sprintf(line, "Content-Type: %s\n",
76                      mail->contentType);
77          if ( dialog( connfd, line, NULL ) ) break;
78        }
79
80        if ( dialog( connfd, mail->contents, NULL ) ) break;
81
82        /* Send mail-end and await response */
83        if ( dialog( connfd, "\n.\n", "250" ) ) break;
84
85        if ( dialog( connfd, "QUIT\n", "221" ) ) break;
86
87        goodMsg = 1;
88
89      } while (0);
90
91      close(connfd);
92
93      return(goodMsg);
94    }
```

The `dialog` function, as discussed previously, handles the actual communication with the SMTP server (see Listing 16.14). The caller provides the client socket, the command to be sent, and the expected response status code. In some cases, a command is not provided (as we're only expecting a server response). In others, only a command is provided and no server response is expected. The final case provides for sending a command, receiving, and checking a response.

Lines 6–9 provide the command transmission, checking that a command is present before attempting to send using the **write** function. **Lines 11–14** receive and check the response, if expected by the caller.

Listing 16.14   Simple SMTP client `dialog` function.

```
1    int dialog(int sd, char *command, char *resp)
2    {
3      int ret, len;
4      char line[128];
5
6      if (command != NULL) {
7        len = strlen(command);
8        if (write(sd, command, len) != len) return -1;
9      }
10
```

```
11   if (resp != NULL) {
12     ret = read(sd, line, sizeof(line)-1); line[ret] = 0;
13     if (strscan(line, resp, 3) == -1) return -1;
14   }
15
16   return 0;
17  }
```

17 Network Code Patterns in Java

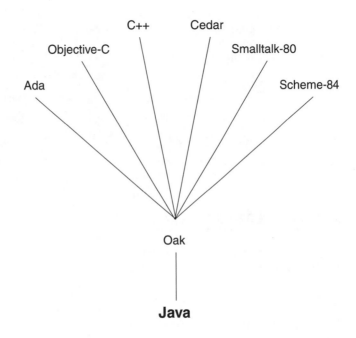

INTRODUCTION

In this chapter, we look at simple socket-based Java applications using the Java Socket APIs and Java's derivative helper classes. These applications can serve as implemented software patterns for the construction of more advanced networking applications.

All software discussed here is also provided on the companion CD-ROM. Table 17.1 lists the applications and their locations on the CD-ROM.

TABLE 17.1   CD-ROM companion software

Code Pattern	CD-ROM Location
Stream server/client	./software/ch17/stream/
Datagram server/client	./software/ch17/dgram/
Multicast server/client	./software/ch17/mcast/
Simple HTTP server	./software/ch17/shttp/
Simple SMTP client	./software/ch17/smtpc/

STREAM (TCP) SERVER/CLIENT

The stream server and client demonstrate the Daytime protocol using stream sockets (reliable TCP-based communication).

Stream Server

The Daytime protocol stream server is shown in Listing 17.1. The first item to note is the importing of a number of Java packages that provide the necessary networking and I/O functionality. The `import` statement is Java's way of identifying the needed external resource classes that will be used (**lines 1–3**). Next, we create a public class (`strmsrv`) to contain our Java application (**line 5**). The `main` method shown is the entry point for the application when started. The `main` acts like a standard C `main`, and accepts a number of arguments, though none are used here. Although only an argument list is shown passed here in `main`, the number of arguments can be retrieved using the `length` attribute of the `args` object (`args.length`). After the declaration of the `main` method of the `strmsrv` class, a number of declarations are made (**lines 9–13**). Rather than discuss them here (because most are self-explanatory), we discuss each as they are utilized in the application.

As in other TCP server applications, we use the ServerSocket constructor to create a new **ServerSocket** object (**line 18**). We pass in our port number variable, so that our new server will be bound to port 13. Next, we enter the server loop to await incoming client connections (**line 20**). The **accept** method is used to accept new connections (**line 23**), for which we store the new **Socket** object in clisock (our client socket object). We don't intend to read any data from our client socket, but we do intend to write the date and time string, so we create a PrintWriter stream over the socket (**lines 28–30**). We use the getOutputStream method on our client socket to get the OutputStream and then pass this anonymously to the PrintWriter constructor to create our PrintWriter object (output). Recall that the second argument (true) is the auto-flush specifier.

Now that we can write simply through our client socket, we create the date and time string (**lines 33–34**). We use the Date() constructor to create a Date object, and then convert this to a string with the toString method. Finally, at **line 37**, we use the println method of the PrintWriter object to write the date and time string to the socket through the PrintWriter stream. We can then close the socket using the **close** method (at **line 41**).

Listing 17.1   Java Daytime stream server.

```
1    import java.net.*;
2    import java.io.*;
3    import java.util.*;
4
5    public class strmsrv {
6
7      public static void main( String[] args ) {
8
9        int port = 13;
10       String line;
11       ServerSocket servsock;
12       Socket clisock;
13       PrintWriter output;
14
15       try {
16
17         // Create a new Server socket
18         servsock = new ServerSocket( port );
19
20         while ( true ) {
21
22           // Await a client connection
23           clisock = servsock.accept();
24
25           try {
26
27             // Create a new PrintWriter for output
28             output = new
29               PrintWriter( clisock.getOutputStream(),
30                            true );
31
```

```
32                // Generate a time string
33                Date now = new Date();
34                String curtime = now.toString();
35
36                // Write the time out to the socket
37                output.println( curtime );
38
39                // Try to close the socket
40                try {
41                  clisock.close();
42                }
43                catch( IOException e ) {
44                  System.out.println( e );
45                }
46
47              }
48              catch( IOException e ) {
49                System.out.println( e );
50              }
51
52            }
53
54          }
55          catch ( UnknownHostException e ) {
56            System.out.println( e );
57            System.exit( -1 );
58          }
59          catch ( IOException e ) {
60            System.out.println( e );
61            System.exit( -1 );
62          }
63
64        }
65
66    }
```

Stream Client

Now, let's look at the Daytime stream client (shown in Listing 17.2). The client creates a socket to connect to the server and then awaits a string to be received through the socket. Upon receiving the string from the server, the message is printed and the client exits.

Listing 17.2 Java Daytime stream client.

```
1    import java.net.*;
2    import java.io.*;
3
```

```
 4   public class strmcli {
 5
 6     public static void main( String[] args ) {
 7
 8       int port = 13;
 9       String server = "localhost";
10       String line;
11       Socket socket;
12       BufferedReader input;
13
14       try {
15
16         // Create a new client socket
17         socket = new Socket( server, port );
18         System.out.println( "Connected to " +
19                             socket.getInetAddress() +
20                             " : " + socket.getPort() );
21
22         try {
23
24           // Build a BufferedReader input stream from
25           // the socket
26           input = new BufferedReader(
27                     new InputStreamReader(
28                       socket.getInputStream() ) );
29
30           // Get a line from the socket and then print it
31           line = input.readLine();
32           System.out.println( line );
33
34           // Try to close the socket
35           try {
36             socket.close();
37           }
38           catch( IOException e ) {
39             System.out.println( e );
40           }
41
42         }
43         catch ( IOException e ) {
44           System.out.println( e );
45         }
46
47       }
48       catch ( UnknownHostException e ) {
49         System.out.println( e );
50         System.exit( -1 );
51       }
```

```
52          catch ( IOException e ) {
53            System.out.println( e );
54            System.exit( -1 );
55          }
56
57        }
58
59    }
```

We begin by making the java.net and java.io package classes available using the im-port statement in **lines 1–2**. Next, we create our strmcli class (**line 4**) and create our ap-plication's main method at **line 6**. A number of variable declarations are made at **lines 8–12**; we discuss each as they arise in the application.

We create our client socket at **line 17** using the **Socket** constructor of the **Socket** class. We specify the server and port in the **Socket** constructor and upon completion of this method, we are connected to the server. We follow socket creation with a simple debug showing the IP address and port number to which we're connected (the remote server) using the **getInetAddress** and **getPort** methods of our client socket.

In order to read from the socket, we create another layered stream to simplify the process (**lines 26–28**). We first get the InputStream of our client socket and anonymously pass this to the InputStreamReader constructor to create a new InputStreamReader ob-ject. This in turn is passed anonymously to the BufferedReader constructor providing us with a BufferedReader object (called input). We use the **readLine** method of our BufferedReader object to get a line from the socket and then emit it to standard-out using the println method. Finally, we close the socket at **line 36** using the **close** method.

DATAGRAM (UDP) SERVER/CLIENT

Now, let's look at a UDP server and client implementation using datagram sockets. Recall that datagram communication is not connection-oriented; therefore, each datagram that we send must also include the intended recipient.

Datagram Server

The datagram server is illustrated in Listing 17.3. It begins in **lines 1–3** by making visible the necessary Java packages that provide the needed functionality (such as the networking classes). Next, the dgramsrv class is created, which will house the Java datagram server ap-plication (**line 5**). Our Java application main method follows next (**line 7**) as well as a set of object declarations in **lines 9–11**.

At **line 12**, we create a byte[] array of 100 bytes in length. This is done in preparation for the creation of the **DatagramPacket** at **line 13**. The **DatagramPacket** object is used to create datagram packets, and in this case, is created to receive the datagram from the client. When creating the new **DatagramPacket** object, we provide to the constructor a byte[] buffer and its length (buffer.length). This differs from the **Socket** and **ServerSocket** classes, because they are stream-oriented and deal simply with payload (recall the Input-Stream / OutputStream discussion). For datagrams, a special class exists to create datagram packets and to manage them.

At **line 19**, we create our **DatagramSocket** object and bind it to the local port (port). We then start our main loop to field datagram requests (**line 22**). The first step in our loop is to receive a datagram from a client. This is done through the receive method of the **DatagramSocket**, specifying the previously created **DatagramPacket** (**line 27**). The purpose of this packet was solely to identify the source of the requester, and at **lines 35 and 36**, we extract the source address and port using the **getAddress** and **getPort** methods on the **DatagramPacket**. We also create our date and time string using the Date() constructor and convert it directly to a string using the toString method. Next, we convert the string into a byte[] object using the getBytes method and store this in the buffer object. A new **DatagramPacket** object is then created at **lines 40–43**, using the new time/date buffer and length as well as the previously extracted client source address (cliadrs) and port (cliport). Finally, we send the packet back to the originator of the original datagram packet using the **send** method at **line 48**.

Listing 17.3   Java Daytime datagram server.

```
1    import java.net.*;
2    import java.io.*;
3    import java.util.*;
4
5    public class dgramsrv {
6
7      public static void main( String[] args ) {
8
9        String server = "localhost";
10       int port = 13;
11       DatagramSocket socket;
12       byte[] buffer = new byte[100];
13       DatagramPacket packet =
14             new DatagramPacket( buffer, buffer.length );
15
16       try {
17
18         // Create the datagram server socket
19         socket = new DatagramSocket( port );
20
21         // The big loop
22         while ( true ) {
23
24           try {
25
26             // Try to receive a datagram packet
27             socket.receive( packet );
```

```
28
29          // Create the time/date string
30          String theDate = new Date().toString();
31          buffer = theDate.getBytes();
32
33          // Extract the address/port information from
34          // the incoming datagram packet.
35          InetAddress cliadrs = packet.getAddress();
36          int cliport = packet.getPort();
37
38          // Create a new datagram packet using the
39          // time/date string.
40          packet = new DatagramPacket( buffer,
41                                       buffer.length,
42                                       cliadrs,
43                                       cliport );
44
45        try {
46
47          // Try to send the response packet to the client
48          socket.send( packet );
49
50        }
51        catch( SocketException e )
52        {
53          System.out.println( e );
54        }
55        catch( IOException e )
56        {
57          System.out.println( e );
58        }
59
60      }
61      catch( IOException e )
62      {
63        // for socket.receive
64        System.out.println( e );
65      }
66
67    }
68
69  }
70  catch( SocketException e )
71  {
72    // for socket creation
73    System.out.println( e );
74  }
```

```
75
76    }
77
78  }
```

Datagram Client

The datagram client is shown in Listing 17.4, and corresponds with the datagram server previously shown in Listing 17.3. After importing the necessary Java packages (**lines 1–2**), we create our class at **line 4** and our main method for our Java application at **line 6**. Next, we create our client socket using the **DatagramSocket** constructor at **line 15** and store our new **DatagramSocket** object in **socket**.

We create a dummy packet at **lines 18–20**, first allocating a byte[] array for the datagram payload and finally creating the **DatagramPacket** at **lines 19 and 20**. At **line 27**, we get an **InetAddress** object for the localhost using the **getByName** method of **InetAddress**, and store this in servaddr. We then use the **setAddress** method of **DatagramPacket** to set the destination address of the packet to the localhost (via the servaddr object) at **line 30**. We also set the port of the packet using the **setPort** method at **line 31**. At **line 36**, we send our dummy datagram to the server using the **send** method, which notifies it that we want to receive a response datagram containing the date and time string.

At **lines 40–41**, we create a new datagram packet, specifying the buffer for the receive payload. At **line 43**, the receive method is used to receive a **DatagramPacket** from the client socket. Upon receipt, the data from the packet is extracted using the **getData** method and anonymously passed into the String constructor to create a String object of the data. This is emitted to standard-out at **line 48**. Finally, the socket is closed at **line 50** using the **close** method.

Listing 17.4 Java Daytime datagram client.

```
1   import java.net.*;
2   import java.io.*;
3
4   public class dgramcli {
5
6     public static void main( String[] args ) {
7
8       String server = "localhost";
9       int port = 13;
10      DatagramSocket socket;
11
12      try {
13
14        // Create a new client socket
15        socket = new DatagramSocket();
16
```

```
17        // Create a dummy packet
18        byte[] buf = new byte[100];
19        DatagramPacket packet =
20            new DatagramPacket( buf, 0 );
21
22        InetAddress servaddr;
23
24        try {
25
26          // Get the address of the current host
27          servaddr = InetAddress.getByName( server );
28
29          // Set the packets destination port and address
30          packet.setAddress( servaddr );
31          packet.setPort( port );
32
33          try {
34
35            // Send the packet to the server
36            socket.send( packet );
37
38            // Await the response packet and then print
39            // the contents (the time string).
40            packet =
41              new DatagramPacket( buf, buf.length );
42
43            socket.receive( packet );
44
45            String received =
46              new String( packet.getData() );
47
48            System.out.println( received );
49
50            socket.close();
51
52          }
53          catch( SocketException e )
54          {
55            System.out.println( e );
56          }
57          catch( IOException e )
58          {
59            System.out.println( e );
60          }
61
62        }
63        catch( UnknownHostException e )
64        {
```

```
65              System.out.println( e );
66          }
67
68      }
69      catch( SocketException e )
70      {
71          System.out.println( e );
72      }
73
74  }
75
76  }
```

MULTICAST SERVER/CLIENT

In many other languages, the multicast server and client code are derivatives of the datagram code patterns. Java differs here because multicast sockets are implemented using its own **MulticastSocket** class. The **MulticastSocket** class is derived from the **DatagramSocket** class, so it inherits the functionality provided there.

Multicast Server

The Java language source code for the multicast server is shown in Listing 17.5. Recall that multicast communication is group-based; therefore, a single message emitted by the server is received by every client that is currently a member of the multicast group.

Because multicast communication has been discussed in previous chapters, we'll forgo discussion of multicast specifics and concentrate solely on the methods that Java provides to achieve multicast communication.

To set up our multicast server (**lines 4 and 5**), we create our mcastsrv class (**line 5**) and the Java application's main method (**line 7**). We create our datagram packet in **lines 13–14**, associated with a byte[] payload of 100 octets. At **line 19**, the multicast socket is created using the **MulticastSocket** constructor. No port is specified with the constructor because we intend not to receive through this socket, but only to send. Next, we create an **InetAddress** object for our multicast group address "239.0.0.2", using the **getByName** method. We then join this multicast group using the **joinGroup** method of the **MulticastSocket** class.

Our setup is complete at **line 31**, and we enter the big loop to distribute the multicast packets containing the date and time string. In each iteration through the loop, we create an anonymous Date object and convert it to a String (**line 34**). We then convert it to a byte[] with getBytes and store it in the buffer. **Lines 39–42** create a new **DatagramPacket** and store not only the date/time string buffer, but also the **InetAddress** of the multicast group and port. At **line 47**, we send the multicast packet using the **send** method. Finally, at **line 51**, we invoke the sleep method of the Thread class to sleep for one second. Upon completion of the sleep, we continue the loop and send another date/time datagram to the multicast group.

Listing 17.5 Java Daytime multicast server.

```java
1    import java.net.*;
2    import java.io.*;
3    import java.util.*;
4
5    public class mcastsrv {
6
7      public static void main( String[] args ) {
8
9        String group = "239.0.0.2";
10       int port = 45002;
11       MulticastSocket socket;
12       byte[] buffer = new byte[100];
13       DatagramPacket packet;
14
15
16       try {
17
18         // Create the multicast socket
19         socket = new MulticastSocket();
20
21         try {
22
23           // Get a binary address for the group
24           InetAddress adrs =
25             InetAddress.getByName( group );
26
27           // Join the multicast group
28           socket.joinGroup( adrs );
29
30           // The big loop
31           while ( true ) {
32
33             // Create the time/date string
34             String theDate = new Date().toString();
35             buffer = theDate.getBytes();
36
37             // Create a new datagram packet using the
38             // date/time string.
39             packet = new DatagramPacket( buffer,
40                                          buffer.length,
41                                          adrs, port );
42
43             try {
44
45               // Try to send the multicast packet to the
```

```
46                    // client(s).
47                    socket.send( packet );
48
49                    // Sleep for one second between sends
50                    try {
51                      Thread.sleep( 1000 );
52                    }
53                    catch( InterruptedException e )
54                    {
55                      // for Thread sleep
56                      System.out.println( e );
57                    }
58
59                  }
60                  catch( IOException e )
61                  {
62                    // for socket send
63                    System.out.println( e );
64                  }
65
66                }
67
68              }
69              catch( IOException e )
70              {
71                // for getByName
72                System.out.println( e );
73              }
74
75            }
76            catch( IOException e )
77            {
78              // for socket creation
79              System.out.println( e );
80            }
81
82          }
83
84    }
```

Multicast Client

The Java source code for the multicast client is shown in Listing 17.6. The multicast client reflects symmetry with the multicast server, as seen with other code patterns.

After making the necessary package classes visible (**lines 1–2**), the Java class and main method for the application are defined (**lines 4 and 6**). The multicast group address is defined at **line 8**, initially as a String, but this will be converted to an **InetAddress** later. We also create our **DatagramPacket** and, using the constructor, initialize it with our byte[ ] buffer.

At **line 18**, we create our multicast socket using the **MulticastSocket** constructor. We also specify our desired multicast port in the constructor. At **lines 23–24**, we convert our multicast group address (in String dotted-notation) to an **InetAddress** object using the **getByName** method. This new object is used at **line 28** to join the multicast group using the **joinGroup** method. We then enter our loop at **line 32** to receive the multicast datagrams. At **line 34**, the **receive** method is invoked with our multicast socket to receive available datagrams. Upon return, we convert the byte[ ] payload of the datagram to a String using the **getData** method. The String is then emitted to standard-out at **line 37**.

Listing 17.6 Java Daytime multicast client.

```
1     import java.net.*;
2     import java.io.*;
3
4     public class mcastcli {
5
6       public static void main( String[] args ) {
7
8         String group = "239.0.0.2";
9         int port = 45002;
10        MulticastSocket socket;
11        byte[] buffer = new byte[100];
12        DatagramPacket packet =
13          new DatagramPacket( buffer, buffer.length );
14
15        try {
16
17          // Create a new client socket
18          socket = new MulticastSocket( port );
19
20          try {
21
22            // Get the address of the current host
23            InetAddress grpaddr =
24              InetAddress.getByName( group );
25
26            try {
27
28              socket.joinGroup( grpaddr );
29
30              try {
31
32                while( true ) {
```

```
33
34                        socket.receive( packet );
35                        String received =
36                          new String( packet.getData() );
37                        System.out.println( received );
38
39                    }
40
41                }
42            catch( SocketException e )
43            {
44              System.out.println( e );
45            }
46            catch( IOException e )
47            {
48              System.out.println( e );
49            }
50
51            }
52        catch( IOException e )
53        {
54          System.out.println( e );
55        }
56
57        }
58    catch( UnknownHostException e )
59    {
60      System.out.println( e );
61    }
62
63    }
64  catch( IOException e )
65  {
66    System.out.println( e );
67  }
68
69  }
70
71  }
```

SIMPLE HTTP SERVER

Now, let's look at a simple HTTP server written in Java. The entire source listing for the Java simple HTTP server is provided in Listing 17.7. Let's now walk through this listing to understand the sample implementation.

Listing 17.7 Java Simple HTTP server source.

```
1   import java.net.*;
2   import java.io.*;
3   import java.util.*;
4
5   public class shttp extends Thread {
6
7     Socket theConnection;
8     static File webRoot;
9     static String indexFile = "index.html";
10
11    public shttp(Socket s) {
12      theConnection = s;
13    }
14
15
16    //
17    //  Determine the content type based upon the
18    //  filename.
19    //
20    public String define_content_type(String name) {
21
22      if (name.endsWith(".html") ||
23          name.endsWith(".htm"))
24        return "text/html";
25      else if (name.endsWith(".txt") ||
26              name.endsWith(".java"))
27        return "text/plain";
28      else if (name.endsWith(".jpg") ||
29              name.endsWith(".jpeg"))
30        return "image/jpeg";
31      else
32        return "text/plain";
33
34    }
35
36
37    //
38    //  Emit the standard HTTP response message header
39    //
40    public void emit_response_header(
41                PrintStream os, String ct ) {
42
43      os.print("Server: Java shttp\r\n");
44      os.print("Connection: close\r\n");
45      os.print("Content-type: " );
```

```
46        os.print( ct );
47        os.print("\r\n\r\n");
48
49    }
50
51
52    //
53    //  HTTP Connection Handler (for GET)
54    //
55    public void handle_connection(
56                   PrintStream os, String file,
57                   String ct ) {
58
59      File theFile;
60
61      try {
62
63        theFile = new File(webRoot,
64                           file.substring(1, file.length()));
65
66        System.out.println("Looking for file " + theFile);
67
68        // Read the data into a byte array
69        FileInputStream fis = new FileInputStream(theFile);
70        byte[] theData = new byte[(int)theFile.length()];
71        fis.read(theData);
72        fis.close();
73
74        os.print("HTTP/1.0 200 OK\r\n");
75        emit_response_header( os, ct );
76
77        os.write(theData);
78        os.close();
79
80      } catch (IOException e) {
81
82          os.print("HTTP/1.0 404 File Not Found\r\n\r\n");
83
84      }
85
86    }
87
88
89    //
90    //  Simple HTTP Server thread
91    //
92    public void run() {
```

```java
93
94     String ct;
95
96     System.out.println("\n--------------------");
97     System.out.println("Request from " +
98        theConnection.getInetAddress().getHostAddress() +
99        " " +
100       theConnection.getInetAddress().getHostName());
101
102    try {
103
104      PrintStream os =
105         new PrintStream(theConnection.getOutputStream());
106      DataInputStream is =
107         new DataInputStream(theConnection.getInputStream());
108      String get = is.readLine();
109      String version = "";
110
111      StringTokenizer st = new StringTokenizer(get);
112      String method = st.nextToken();
113      String file = st.nextToken();
114
115      if (method.equals("GET")) {
116
117        if (file.endsWith("/")) file += indexFile;
118        ct = define_content_type(file);
119        if (st.hasMoreTokens()) {
120          version = st.nextToken();
121        }
122
123        while ((get = is.readLine()) != null) {
124          if (get.trim().equals("")) break;
125        }
126
127        handle_connection( os, file, ct );
128
129        os.close();
130
131      } else { // Method is not GET
132
133        os.print("HTTP/1.0 501 Not Implemented\r\n\r\n");
134        os.close();
135
136      }
137
138    } catch (IOException e) {
139    }
140
```

```
141        try {
142          theConnection.close();
143        } catch (IOException e) {
144        }
145
146    }
147
148
149    //
150    //  Initialization method for the simple HTTP server
151    //
152    public static void main(String[] args) {
153      int thePort;
154      ServerSocket server;
155
156      webRoot = new File(".");
157      thePort = 80;
158
159      try {
160
161        // Create the HTTP server socket
162        server = new ServerSocket(thePort);
163
164        System.out.println(
165              "Accepting connections on port " +
166                          server.getLocalPort());
167        System.out.println("Doc Root: " + webRoot);
168
169        // Start the server accept loop
170        while (true) {
171
172          // Upon accept of new socket, start a new
173          // HTTP thread.
174          shttp t = new shttp(server.accept());
175          t.start();
176
177        }
178      } catch (IOException e) {
179        System.err.println("Server aborted prematurely");
180      }
181
182    }
183
184  }
```

This implementation of the simple HTTP server will be based upon a simple class that provides the HTTP server functionality using Java threads. The only visible method is the Java application's main method, which starts the HTTP server.

The Java HTTP server is made up of six methods: the main method (**lines 152–182**), run (**lines 92–146**), handle_connection (**lines 55–86**), emit_response_header (**lines 40–49**), define_content_type (**lines 20–34**), and the shttp constructor method (**lines 11–13**). We'll look at the methods in their order of use.

The class, shttp, is defined at **line 5** and extends the Thread class (using threads to provide multiple client capabilities). A few local variable objects are created at **lines 7–9**; the client socket connection (theConnection), a String identifying the local file system path (webRoot), and finally the default file that will be loaded if one isn't specified in the HTTP GET request (indexFile).

The first method invoked in the simple HTTP server class is main method (**lines 152–182**). The purpose of the main method is to set up the HTTP server socket and prepare for incoming connections. At **line 156**, the root of the HTTP server's file system is defined as the current path, and stored as a File object into webRoot (we'll see this used in the handle_connection method later). We also define the default port to use for the server as port 80 (thePort) at **line 157**. The serverSocket is created at **line 162** using the **Server-Socket** constructor (passing the desired port to bind to as the single argument). A simple debug message is then emitted to standard-out at **lines 164–167**.

The HTTP server loop begins at **line 170**. At **line 174**, the **accept** method is invoked for the server socket and upon completion of the call (an incoming client connection arrives), the new client socket is passed to the shttp constructor method of the shttp class. Note that a new thread is created here because the shttp class extends the Thread class. The shttp constructor is shown at **lines 11–13**, with the only action being performed that the passed socket is stored into local variable, theConnection. The run method is then invoked for the newly created thread (**lines 92–146**).

The run method is where the HTTP serving functionality begins. Within run, we parse the incoming HTTP GET request and parse it to its basic elements. Some debugging information is emitted identifying the client that has connected (**lines 96–100**) using the **getHostAddress** and **getHostName** methods.

At **lines 104–107**, we create our input and output streams for reading and writing to the client socket. For the output stream, we create a PrintStream over the OutputStream of the socket, and for the input stream, we create a DataInputStream over the InputStream of the socket. The HTTP GET request is then read from the socket using the **readLine** method and the resulting String stored into get. We then create a StringTokenizer object over the HTTP GET request (stored in get) at **line 111**, and then parse out the HTTP request method (storing it in method) at **line 112** and the requested file at **line 113** (storing it in file). The tokenizer simply picks apart the String using space characters as the delimiter, returning the next available token using the nextToken method.

At **line 115**, we test the method String to see if the request was a GET request, and if so we perform the necessary elements for satisfying this HTTP request. We first test the filename to see if it ends with '/' (**line 117**). In this case, the default file is requested, so we append the index filename to the end of this string (index.html). We then call the define_content_type method to determine the type of file that's being requested (and should be returned). We'll discuss this method shortly. We check the tokenizer object again to see if any other tokens are available (hasMoreTokens), and if so, we grab the version from the tokenizer (because it should follow next on the HTTP GET request). We then read the remainder of the HTTP GET request, which we'll ignore (**lines 123–125**). Now that we have the file that's being requested, we call `handle_connection` to handle the specific GET request (**line 127**) and then close the client socket with the **close** method (**line 129**).

Note that if a request other than GET was made to the HTTP server (such as HEAD or POST), we return an error message to the connected client (**line 133**). This is done using the `print` method of the `PrintStream` object. The client socket is then closed with the **close** method (**line 134**).

The `define_content_type` method (**lines 20–34**) is used to identify the type of content being returned based upon the file extension being requested. For example, if the file being requested is an HTML file (identified by the `.html` file extension), then the content type of the HTTP response is "`text/html`". This is necessary for the HTTP client to know how to properly render the content being returned. This method is passed to the `String` filename and returns a new `String` containing the content type. The `endswith` method of the `String` class is used to identify the type of file. We test three unique cases; `.html` and `.htm` result in "`text/html`" (**lines 22–24**), `.txt` and `.java` result in "`text/plain`" (**lines 25–27**), and `.jpg` and `.jpeg` result in "`image/jpeg`" (**lines 28–30**). If the content type can't be determined, then the "`text/plain`" content type is returned by default (**line 32**).

The `handle_connection` method (**lines 55–86**) handles a GET request from a client socket (called by the `run` method). The `handle_connection` method is called with the `PrintStream` object (so that `handle_connection` can send the requested file to the client through the socket), the filename being requested, and the content type. The first task is to open the file that the client is requesting. This is a good example of the exception handling capabilities of Java. At **lines 63–64**, a new `File` object is created, specifying the file path (the current subdirectory in which the Java application was started), and then the filename (ignoring the first character). This new File object creation also attempts to open the file. If the file is not present, an exception is raised and is caught at **lines 80–84** (the `catch` block). In this section of code, we emit an error message back to the client—in this case, an HTTP 404 error code (indicating that the file could not be found). Otherwise, we continue on within the `try` block at **line 69** where we create a `FileInputStream` of the `File` object and create a `byte` array of the size of the file (`theFile.length()`). We then read the file (**line 71**) placing the data read into the previously created `byte` array, and then close the file (**line 72**). Finally, we emit a standard HTTP response header (more on the `emit_response_header` method next) and then write the data array to the peer using the **write** method of the `PrintStream` object (**line 77**). The client socket is then closed with the **close** method and the process is complete (**line 78**).

HTTP response headers are the portion of the response message that indicates to the client what is going to follow. Server responses include first a header that includes the HTTP response code, and then a set of optional lines that provide more information to the client. In `emit_response_header` (**lines 40–49**), we emit some sample information to the client, including the HTTP server name (**line 43**), a status line indicating that the connection should be closed once this request is complete (**line 44**), and, finally, the content type (**lines 45–46**), which tells the HTTP client how to render the data that is contained in the HTTP response message.

SIMPLE SMTP CLIENT

Now, let's look at a simple SMTP client written in Java. The SMTP client is implemented as a class that provides a set of methods to send an e-mail to a specified recipient (Listing 17.8). Two methods exist for the Java SMTP client, smtpc (the class constructor) and send (send the e-mail created by the constructor). One other private method exists internally, which is covered in the smtpc class discussion.

We'll discuss the Java SMTP client class in two parts (Listing 17.8). The first part discusses the test class (smtptest) that we use to test the smtpc class (**lines 4–25**), and the second part discusses the actual smtpc class (**lines 27–193**).

The first class, smtptest (**lines 4–25**) is used to create a new smtpc object and then send a test e-mail using it. Class smtptest is a Java application and provides a main method (at **line 9**). Class smtptest performs only two functions; the first is the creation of a new smtpc object using the smtpc constructor (**lines 12–18**) and the second is the calling of the **send** method of the smtpc class (**line 21**). The smtpc constructor is called with the necessary parameters for the e-mail to send. This includes the subject, sender e-mail address, recipient e-mail address, content type, and, finally, the contents of the e-mail. Note that in the example shown, a text e-mail is sent (as both indicated by the content type "text/html" and the obvious HTML tags within the e-mail contents). Invoking the **send** method then initiates communication with the SMTP server and the configured e-mail is sent.

Now, let's discuss the smtpc class within Listing 17.8 (**lines 27–193**). We discuss the support elements of the class first, and then discuss the primary methods that provide the SMTP functionality. After declaring the class at **line 27**, a set of private variables is created (**lines 30–34**). These variables are used to hold the e-mail data that's passed in to the constructor (**lines 44–54**). Three other objects are created here as well. The first two are the input and output stream objects; a PrintStream is created for output operations through the socket and a BufferedReader is created for input socket operations. These are initialized when the e-mail is to be sent. Finally, a mail server is defined by the String mailServer. This private String is used to identify the IP address of the outgoing SMTP mail server that will be used to send the e-mail.

The smtpc constructor (**lines 44–54**) is used to create a new smtpc object. This method simply stores the information passed by the caller to local variables in the class in preparation for transmission.

The final two support elements of the smtpc class are the creation of a new exception (**lines 60–62**) and the private dialog method, which is used to perform a single transaction with the SMTP server (**lines 68–113**). When an error occurs within a dialog with the SMTP server, rather than return an error message to the method caller, we throw an exception. This allows the calling method to create an exception handler (the catch block) to handle all of the exception cases that are possible with a method that throws the exception. We create a new exception here called MyException that involves creating a new class and method for the exception (**lines 60–62**). This exception can then be thrown and caught, which we'll discuss shortly.

The dialog method (**lines 68–113**) is used to perform a single transaction with the SMTP server. This involves potentially sending a command to the server, then potentially receiving a response, and then verifying that the response was correct (we received what was expected). The first item to note about this method is the declaration (**lines 68–70**). It's a private method that's accessible only to members of the class. It takes two String arguments (the String command and response), and finally it can potentially throw an exception (MyException). We declare our ability to throw this exception early in the method; we'll see shortly where it's actually thrown. The first task is to issue the command, if available. **Lines 73–77** look at the passed command, and if it's non-NULL, we send the command to the SMTP server through the client socket using the print method of our PrintStream. Next, we check to see if a response is expected (**line 82**), and if so, we attempt to read a single line from the socket using the **readLine** method of the BufferedReader object (representing the input stream of our client socket). **Lines 91–93** test to see if the status code (the first numeric identifier returned for SMTP commands) matches what we expect. If so, we simply return. Otherwise, we throw our exception at **line 100**. Note that we create a new exception, as it's an object just like any other in Java. We also throw the new MyException at **line 107**, in the event that there was a problem reading from the socket (such as if a peer closure occurred).

The final method in the smtpc class is the **send** method that is used to actually send the e-mail (**lines 116–191**). This method first creates the client socket (**line 124**) using the **Socket** method and connects it to the mail server (identified by the previously created mailserver String) and the mail server port (25). We then create our input and output streams over the socket to simplify the I/O operations. A BufferedReader is created to read from the socket (**lines 127–129**) and a PrintStream is created to write to the socket (**line 132**).

The actual mail client functionality is implemented in **lines 134–185**. The dialogs with the mail server are contained within a try block in order to catch the MyException that was created to identify errors in communication with the mail server. Note at **lines 181–185**, the catch block exists to catch any MyException exceptions that are thrown within any of the dialogs performed. Because the SMTP command/response functionality was explained in the C chapter, we'll forgo a detailed discussion of it here. Each dialog potentially issues a command, and awaits a valid response before continuing to the next command. When the SMTP dialogs are complete, the **close** method is called (**line 179**) to close the client socket.

The final catch blocks (**lines 188 and 189**) exist to catch exceptions that could be thrown by the **Socket** constructor method call at **line 124**.

Listing 17.8   Java Simple SMTP client source.

```
1    import java.net.*;
2    import java.io.*;
3
4    public class smtptest {
5
6        //
7        // Java main application for the SMTP test.
8        //
9        public static void main( String[] args ) {
```

```
10
11       // Create a new smtpc object
12       smtpc mail = new smtpc(
13           "The Subject",
14           "tim@mtjones.com",
15           "mtj@mtjones.com",
16           "text/html",
17           "<HTML><BODY><H1>This is the mail"+
18           "</H1></BODY></HTML>" );
19
20       // Send the previously created e-mail
21       mail.send();
22
23     }
24
25   }
26
27   class smtpc {
28
29     // Define the local variables
30     private String s_subject;
31     private String s_sender;
32     private String s_recipient;
33     private String s_content_type;
34     private String s_contents;
35     private PrintStream out;
36     private BufferedReader inp;
37
38     // The outgoing SMTP server
39     private String mailserver = "192.168.1.1";
40
41     //
42     //  smtpc class constructor
43     //
44     public smtpc( String subject, String sender,
45                   String recipient, String content_type,
46                   String contents ) {
47
48       s_subject = subject;
49       s_sender = sender;
50       s_recipient = recipient;
51       s_content_type = content_type;
52       s_contents = contents;
53
54     }
55
56     //
57     //  A new exception used to identify SMTP
```

```
58      //   communication errors
59      //
60      class MyException extends Exception {
61        public MyException() { super(); }
62      }
63
64
65      //
66      //   Perform a transaction to the SMTP server
67      //
68      private void dialog( String command,
69                               String expected_response )
70            throws MyException {
71
72        // If a command is available, send it to the server
73        if ( command != null ) {
74
75          out.print( command );
76
77        }
78
79        // If a response is expected, get it and test it
80        // against the desired response.
81
82        if ( expected_response != null ) {
83
84          try {
85
86            // Get a line from the SMTP server through
87            // the socket
88            String line = inp.readLine();
89
90            // Check the status code
91            if (expected_response.equalsIgnoreCase(
92                line.substring( 0,
93                        expected_response.length() ) ) ) {
94
95              return;
96
97            } else {
98
99              // Not the expected response, throw an exception
100             throw new MyException();
101
102           }
103
104         } catch( IOException e ) {
105
```

```
106              // Another error occurred, throw an exception
107              throw new MyException();
108
109        }
110
111      }
112
113    }
114
115
116    public void send () {
117
118      Socket sock;
119
120        try {
121
122          // Create the client socket and connect
123          // it to the server
124          sock = new Socket( mailserver, 25 );
125
126          // Create the input stream
127          inp = new BufferedReader(
128                     new InputStreamReader(
129                           sock.getInputStream() ) );
130
131          // Create the output stream
132          out = new PrintStream( sock.getOutputStream() );
133
134          try {
135
136            // Look for the initial e-mail salutation
137            this.dialog( null, "220" );
138
139            // Send HELO and await response
140            this.dialog( "HELO thisdomain.com\n", "250" );
141
142            // Send the "MAIL FROM" and await response
143            String mailFrom = "MAIL FROM:<" + s_sender + ">\n";
144            this.dialog( mailFrom, "250" );
145
146            // Send the "RCTP TO" and await response
147            String rcptTo = "RCPT TO:<" + s_recipient + ">\n";
148            this.dialog( rcptTo, "250" );
149
150            // Send the DATA command
151            this.dialog( "DATA\n", "354" );
152
153            // Send the e-mail source
```

```
154             String from = "From: " + s_sender + "\n";
155             this.dialog( from, null );
156
157             // Send the e-mail destination
158             String recip = "To: " + s_recipient + "\n";
159             this.dialog( recip, null );
160
161             // Send the subject
162             String subject = "Subject: " + s_subject + "\n";
163             this.dialog( subject, null );
164
165             // Send the Content Type
166             String ct = "Content-Type: " + s_content_type + "\n";
167             this.dialog( ct, null );
168
169             // Send the e-mail body
170             this.dialog( s_contents, null );
171
172             // Await e-mail receipt acknowledgment
173             this.dialog( "\n.\n", "250" );
174
175             // End the mail server dialog
176             this.dialog( "QUIT\n", "221" );
177
178             // Close the client socket
179             sock.close();
180
181         } catch( MyException e ) {
182
183             System.out.println( e );
184
185         }
186
187       }
188     catch( UnknownHostException e ) {}
189     catch( IOException e ) {}
190
191     }
192
193   }
```

18 Network Code Patterns in Python

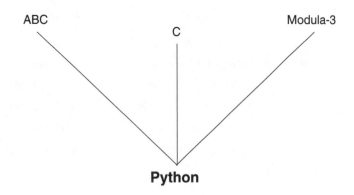

ABC   C   Modula-3

Python

INTRODUCTION

In this chapter, we look at simple socket-based applications in Python using its version of the BSD Sockets API. These applications can serve as implemented software patterns for the construction of more advanced networking applications.

All software discussed here is also provided on the companion CD-ROM. Table 18.1 lists the applications and their locations on the CD-ROM.

TABLE 18.1 CD-ROM companion software

Code Pattern	CD-ROM Location
Stream server/client	./software/ch18/stream/
Datagram server/client	./software/ch18/dgram/
Multicast server/client	./software/ch18/mcast/
Broadcast server/client	./software/ch18/bcast
Simple HTTP server	./software/ch18/shttp/
Simple SMTP client	./software/ch18/smtpc/

STREAM (TCP) SERVER/CLIENT

The stream server and client demonstrate the Daytime protocol using stream sockets (reliable TCP-based communication).

Stream Server

The Daytime protocol stream server is shown in Listing 18.1. As with all Python socket scripts, the **socket** module must be made visible using the import method (**line 1**). We also import the time module, in order to gain access to the current time. We create a new server socket at **lines 4–5** using the **socket** method (of the **socket** module, thus the **socket.socket** name). As with C applications, we specify AF_INET as the address family (an IP protocol) and SOCK_STREAM to represent a TCP socket. These symbols are prefixed by the **socket** module, to identify the source module of the symbol.

We utilize a socket option with method **setsockopt** at **lines 6–7** to make the local address reusable. This is useful if the script is stopped or aborted and needs to be restarted quickly. Otherwise, a two-minute wait period is necessary before the local address is available to be bound again. We use a Sockets level option (identified by SOL_SOCKET) for this purpose, specifying the SO_REUSEADDR option to make it reusable.

At **line 12**, we bind a local name to our socket using the **bind** method. We specify our server socket instance (`servsock`) and provide a single argument to **bind**. This single argument is an address tuple, representing the interface and port from which to accept client connections. In this case, we specify `""` as the host, which is the same as specifying INADDR_ANY, or accept client connections on any interface. The port number specified is 13, which is the well-known port for the Daytime protocol. As a final step in server socket setup, we invoke the **listen** method to enable acceptance of incoming connections. We pass a '1' to the **listen** method, which specifies that we'll permit at most one connection awaiting acceptance on the accept queue.

At **line 18**, we begin the server infinite loop, using the Python while statement. To accept new connections, we use the standard socket **accept** method. This method is used solely by stream sockets and blocks until a client connects. After a client connects to the server, a new client socket is created and returned. At **line 21**, the **accept** method returns the new client socket to the `clisock` object. The **accept** method also returns address information about the client in the second returned parameter, in this case `addr`. This address tuple contains the remote client address and port number.

At **line 24**, we collect the current time using the time module. The time method returns the current time as the number of seconds since a given epoch (January 1st, 1970). The ctime method of the time module converts this quantity into a string representing the local time. This time string (`tm`) is then shipped to the client socket using the **send** method (**line 27**) with the instance of the client socket, `clisock`. The **close** method, at **line 30**, then closes the client socket and control ultimately returns to **line 21**, where we await a new client connection.

Listing 18.1   Python Daytime stream server.

```
1    import socket, time
2
3    # Create a new stream socket
4    servsock = socket.socket( socket.AF_INET,
5                                  socket.SOCK_STREAM )
6
7    # Make the address/port combination reusable
8    servsock.setsockopt( socket.SOL_SOCKET,
9                            socket.SO_REUSEADDR, 1 )
10
11   # Bind our server socket to port 13
12   servsock.bind( ("", 13) )
13
14   # Mark the server's willingness to accept client connections
15   servsock.listen( 1 )
16
17   # The big loop
18   while 1:
```

```
19
20      # Accept a new client connection
21      clisock,addr = servsock.accept()
22
23      # Collect the current time
24      tm = time.ctime( time.time() ) + "\n"
25
26      # Send the time to the client
27      clisock.send( tm )
28
29      # Close the client connection
30      clisock.close()
```

Stream Client

Now, let's look at the Daytime stream client (shown in Listing 18.2). The client creates a socket to connect to the server and then awaits a string to be received through the socket. Upon receiving the string from the server, the message is printed and the client exits.

Listing 18.2   Python Daytime stream client.

```
1      import socket
2
3      # Create a new stream socket
4      sock = socket.socket( socket.AF_INET, socket.SOCK_STREAM )
5
6      # Connect to the daytime port
7      sock.connect( ("localhost", 13) )
8
9      # Receive the time/date information
10     ts = sock.recv( 100 )
11
12     # Emit the received string
13     print ts
14
15     # Close the client socket
16     sock.close()
```

We begin by importing the **socket** module at **line 1** and then creating a new stream socket at **line 4** using the **socket** method from the **socket** module. At **line 7**, we create a connection to the client using the **connect** method. To connect, we pass our address tuple that specifies the localhost interface and the Daytime service port number, 13. After the connection is made, we read the current time through the socket using the **recv** method (**line 10**). We specify the largest size that we'll read from the socket (100 octets), but in fact, we'll actually read much less. The resulting time string is returned and stored into the new object ts. At **line 13**, we print the time using the print statement, and, finally, we close the socket at **line 16** using the **close** method.

DATAGRAM (UDP) SERVER/CLIENT

Now, let's look at a UDP server and client implementation using datagram sockets. Recall that datagram communication is not connection-oriented; therefore, each datagram that we send must also include the intended recipient.

Datagram Server

The datagram server is illustrated in Listing 18.3. After making the socket and time module visible using the import statement, we create our datagram socket using the **socket.socket** method (**line 4**). To create a datagram socket, we specify that the family will be AF_INET (for Internet protocols) and that the socket type will be SOCK_DGRAM (for the datagram protocol). The result of the **socket.socket** method is a new datagram socket, stored in servsock. We use the **setsockopt** method (at **line 7–8**) to permit the local address information to be reused. As a final step in our server setup, the **bind** method is called to bind a local name to the socket (**line 11**). We use the address represented by " " to permit incoming client connections on any interface and specify the Daytime service port, 13. At this stage, our server is ready to accept client connections.

Listing 18.3   Python Daytime datagram server.

```
1    import socket, time
2
3    # Create a new datagram socket
4    servsock = socket.socket( socket.AF_INET, socket.SOCK_DGRAM )
5
6    # Make the address/port combination reusable
7    servsock.setsockopt( socket.SOL_SOCKET,
8                           socket.SO_REUSEADDR, 1 )
9
10   # Bind our server socket to INADDR_ANY and port 13
11   servsock.bind( ("", 13) )
12
13   # The big loop
14   while 1:
15
16     # Receive a client datagram
17     data, addr = servsock.recvfrom( 0 )
18
19     # Collect the current time
```

```
20      tm = time.ctime( time.time() ) + "\n"
21
22      # Send the time to the client
23      servsock.sendto( tm, addr )
```

At **line 14** in Listing 18.3, we start our infinite loop. We await a zero-length datagram from a client at **line 17** using the **recvfrom** method. This method returns two items, the datagram's data buffer (data) and the source address of the datagram (addr). At **line 20**, we create a new time string (identically to the stream server shown in Listing 18.1). Finally, at **line 23**, we send the time string to the client using the **sendto** method. The destination of the time datagram is the source of our initial zero-length datagram (stored as addr).

Datagram Client

The datagram client is shown in Listing 18.4, and corresponds with the datagram server previously shown in Listing 18.3.

After making the socket module visible at **line 1** using the import statement, we create an address tuple that represents the server to which we'll communicate. We create our client socket at **line 7**, specifying SOCK_DGRAM for a datagram protocol socket. As a final setup step, we bind our client socket with a local address using the **bind** method (**line 10**). Although this step is unnecessary, it's done here to demonstrate another element of the **bind** method. We provide the wildcard address (interface) as the first element of the tuple and the second element representing the port number. If an IP address (or host name) had been specified, from the client perspective, this means that our client could communicate only through that interface. This in essence restricts communication to a specific interface, rather than allowing the routing tables to make this decision. Further, the port number specified is 0, which is not a legal port number. The zero port number tells the stack that an ephemeral (dynamic) port should be created for the client. If the script had not performed a **bind**, this exact process would have been done automatically. We provide this example here to illustrate how the **bind** can be used in a client setting.

Listing 18.4   Python Daytime datagram client.

```
1    import socket
2
3    # Create the destination address structure
4    addr = 'localhost', 13
5
6    # Create a new stream socket
7    sock = socket.socket( socket.AF_INET, socket.SOCK_DGRAM )
8
9    # Bind to the INADDR_ANY and the ephemeral port
10   sock.bind( ("", 0) )
11
12   # Send our blank datagram to notify the server that we
```

```
13    # request a time string response.
14    sock.sendto( "", addr )
15
16    # Receive the time/date information
17    ts = sock.recv( 100 )
18
19    # Emit the received string
20    print ts
21
22    # Close the client socket
23    sock.close()
```

To begin the datagram communication, we send an empty datagram to the server at **line 14**. We use the **sendto** method because with datagram communication, we need to specify the destination address of the datagram. The first parameter to **sendto** is our empty datagram and the second is our address tuple (that was created earlier at **line 4**). After the server receives our empty datagram, it sends us a datagram containing the time string. We read this using the **recv** method (**line 17**), specifying a size (the maximum datagram that is to be received). The resulting string is stored into the string object ts. At **line 20**, we emit the time string using the print statement and then close the socket at **line 23** using the **close** method.

MULTICAST SERVER/CLIENT

The multicast server and client are basic derivatives of the datagram code patterns. This is primarily because multicast communication is based upon the datagram model.

Multicast Server

The Python language source code for the multicast server is shown in Listing 18.5. Recall that multicast communication is group-based; therefore, a single message emitted by the server is received by every client that is currently a member of the multicast group.

After making the socket and time modules visible using the import statement (**line 1**), we create our address information that will be used to identify where communication will occur (**lines 3 and 4**). We specify our multicast IP address (noted as group) and our port number (port). Although this is a server, we don't bind this address information to the local socket. Because we're the multicast server, we only send datagrams and never receive. Therefore, the address information defines the group to which we'll send datagrams (to which clients will bind).

We create our datagram socket as we've done in previous datagram applications in **lines 7–8**. We then associate the group and port variables into a single tuple called addr at **line 11**. This tuple will be used as the destination address later to which datagrams will be directed.

Listing 18.5 Python Daytime multicast server.

```
1    import socket, time
2
3    group = '239.0.0.2'
4    port = 45003
5
6    # Create a new datagram socket
7    servsock = socket.socket( socket.AF_INET,
8                              socket.SOCK_DGRAM )
9
10   # Create the destination address structure
11   addr = group, port
12
13   # The big loop
14   while 1:
15
16     # Collect the current time
17     tm = time.ctime( time.time() ) + "\n"
18
19     # Send the time to the client
20     servsock.sendto( tm, addr )
21
22     # Wait one second before sending another time string
23     time.sleep(1)
```

We begin our server's infinite loop at **line 14**, where we'll send out time datagrams at a rate of one per second. At **line 17**, we construct our time string using the time module and store the resulting string into the string variable tm. The sendto method is used at **line 20** to send out the time string to the multicast group. We specify first the message to send (tm) and then the address tuple representing our multicast group (addr). We then sleep for one second at **line 23** and repeat the process of emitting a new time string.

Note that nothing special was required in order to send datagrams to the multicast group. In order to receive from the multicast group, some additional setup is required. We'll now look at these additional steps in the Multicast Client section.

Multicast Client

The Python source code for the multicast client is shown in Listing 18.6. The multicast client includes a number of new features not previously discussed, but reflects symmetry with the multicast server, as seen with prior code patterns.

In addition to importing to the socket module, we also import the string module and the struct module. We discuss the uses of these as they occur in the multicast client. We also define our group and port variables to represent our multicast address at **lines 3–4**. At **line 7**, we create a new datagram socket and make the address/port combination reusable at **lines 11–12** using the SO_REUSEADDR socket option. This option is required because multiple clients may want to bind to the same multicast group/port number; otherwise, the result is one permitted `bind` and subsequent "address in use" errors. We then bind ourselves at **line 15** to the wildcard interface and the defined port number (specified in the `port` variable). Although we've bound ourselves to our predefined multicast port number, we're not able to receive multicast datagrams at this point. For this, we must also join the multicast group.

Let's now look at the process of joining the multicast group. As can be seen from **lines 19–23**, the majority of this code is simply the construction of the `mreq` structure used to join the multicast group. We begin at **line 19** by taking our IP address string and splitting it up into four strings (using the `split` method), representing each of the four numbers of a dotted-notation IP address string. The `map` method that is invoked over the string split converts each of the resulting strings into an integer (as defined by the `int` argument), storing this list in the variable `bytes`.

At **line 21**, we iterate through the `bytes` list and create a new 32-bit integer value of the IP address. For each `byte`, we perform a bitwise 'or' of the `byte` with the new variable `grpAddress` and then shift the entire `grpAddress` left eight bits. Each new iteration shifts the entire `grpAddress` variable eight bits to the left, and then places the new `byte` at the rightmost eight bits. At **line 22**, we create our `mreq` structure that consists of a 32-bit multicast address and a 32-bit interface IP address. Using the `pack` method from the `struct` module, the two 32-bit values are packed together. We specify the '`ll`' template, which defines that those two 32-bit values will be packed into a 64-bit structure. The first 32-bit value is the group address (converted to a network address via the **htonl** method), and the second is the interface address (defined here as the wildcard address, `INADDR_ANY`). This 64-bit structure is stored within the `mreq` variable. As the final setup step (**lines 26–27**), we use the **setsockopt** method with the `IP_ADD_MEMBERSHIP` socket option to join the multicast group. We pass in the `mreq` variable to define the multicast group to which we want to join.

Listing 18.6   Python Daytime multicast client.

```
1    import socket, string, struct
2
3    group = '239.0.0.2'
4    port = 45003
5
6    # Create a new stream socket
7    sock = socket.socket( socket.AF_INET, socket.SOCK_DGRAM )
8
9    # Permit multiple bindings of this address/port
10   # combination
11   sock.setsockopt( socket.SOL_SOCKET,
12                    socket.SO_REUSEADDR, 1 )
13
14   # Bind to the INADDR_ANY and our broadcast port
15   sock.bind( ('', port) )
16
17   # Create the mreq structure for the multicast
18   # binary group
```

```
19   bytes = map( int, string.split( group, "." ) )
20   grpAddress = 0
21   for byte in bytes: grpAddress = (grpAddress << 8) | byte
22   mreq = struct.pack( 'll', socket.htonl(grpAddress),
23                       socket.INADDR_ANY )
24
25   # Use the mreq structure to join the group
26   sock.setsockopt( socket.IPPROTO_IP,
27                    socket.IP_ADD_MEMBERSHIP, mreq )
28
29   # Receive the time/date information
30   ts = sock.recv( 100 )
31
32   # Emit the received string
33   print ts
34
35   # Leave the multicast group
36   sock.setsockopt( socket.IPPROTO_IP,
37                    socket.IP_DROP_MEMBERSHIP, mreq )
38
39   # Close the client socket
40   sock.close()
```

Now that we've joined the multicast group, we're able to receive datagrams that have been sent to it. Using the **recv** method at **line 30**, we read a string of up to 100 bytes and store it into variable ts. This is printed at **line 33** using the print statement. Once printed, we remove ourselves from the multicast group (at **line 36–37**) using the IP_DROP_MEMBERSHIP socket option of the **setsockopt** method. We then close the client socket using the **close** method.

BROADCAST SERVER/CLIENT

The broadcast server and client, like the multicast server and client, are fundamentally derivatives of the datagram code patterns. This is again because broadcast communication is based upon the datagram model.

Broadcast Server

The Python source code for the broadcast server is shown in Listing 18.7. The server is broadcast-based, which means a single message emitted by the server is received by every client that is configured to receive broadcast datagrams (for the given port).

After importing the socket and time modules (**line 1**), we create our datagram socket using the **socket** method from the socket module (**line 4**). At **lines 7–8**, we enable the SO_BROADCAST socket option. This permits us to send datagrams to the broadcast address. Finally, ending the server setup, we use the **bind** method to retrieve an ephemeral port for the socket.

At **line 14**, we create an address tuple that represents the broadcast group to which we'll communicate. The address represents the broadcast address (using the special `<broadcast>` symbol and port 45003). Clients must **bind** this address tuple in order to receive datagrams from the server.

Listing 18.7   Python Daytime broadcast server.

```
1    import socket, time
2
3    # Create a new datagram socket
4    servsock = socket.socket( socket.AF_INET, socket.SOCK_DGRAM )
5
6    # Enable the ability to send to the broadcast address
7    servsock.setsockopt( socket.SOL_SOCKET,
8                         socket.SO_BROADCAST, 1 )
9
10   # Bind our server socket to INADDR_ANY and an ephemeral port
11   servsock.bind( ("", 0) )
12
13   # Create the destination address structure
14   addr = '<broadcast>', 45003
15
16   # The big loop
17   while 1:
18
19     # Collect the current time
20     tm = time.ctime( time.time() ) + "\n"
21
22     # Send the time to the client
23     servsock.sendto( tm, addr )
24
25     # Wait one second before sending another time string
26     time.sleep(1)
```

We begin our infinite server loop at **line 17**. Iteration through the loop begins by constructing a new time string using the `time` module. The time string is stored in the string variable `tm`. At **line 23**, we send the current time string to the broadcast address tuple (created previously at **line 14**) using the **sendto** method. Finally, we use the `sleep` method of the `time` module (**line 26**) to stall for one second and then repeat the process again.

Broadcast Client

The Python source code for the broadcast client is shown in Listing 18.8. As with the broadcast server, we concentrate on the differences to the standard datagram client.

Listing 18.8   Python Daytime broadcast client.

```
1    import socket
2
3    # Create a new stream socket
4    sock = socket.socket( socket.AF_INET, socket.SOCK_DGRAM )
5
6    # Bind to the INADDR_ANY and our broadcast port
7    sock.bind( ("", 45003) )
8
9    # Receive the time/date information
10   ts = sock.recv( 100 )
11
12   # Emit the received string
13   print ts
14
15   # Close the client socket
16   sock.close()
```

The broadcast client is a very simple script that receives a time string datagram from the broadcast server and then exits. After importing the **socket** module at **line 1**, we create our datagram socket at **line 4** (identically to other datagram sockets). At **line 7**, we **bind** the client socket to the broadcast port and the wildcard interface (INADDR_ANY, represented by ""). At **line 10**, we receive the time string datagram using the **recv** method. We could have used the **recvfrom** method, but in this case, we're not interested in the source of the datagram. We specify a maximum of 100 bytes to be received from the server, which, in this case, is much larger than is necessary. At **line 13**, we emit the time string using the print statement and finally close the client socket at **line 16** using the **close** method.

SIMPLE HTTP SERVER

Now, let's look at a simple HTTP server written in Python. The entire source listing for the Python simple HTTP server is provided in Listing 18.9. Let's now walk through this listing to understand the sample implementation.

Listing 18.9   Python simple HTTP server source.

```
1    import socket
2    import string
3    import os
4    import re
5    import time
6
7    class Simple_http_server:
8
```

```
 9      # Define the class variables (host and server socket)
10      host = 'localhost'
11      sock = 0
12
13      #
14      # Define the content type
15      #
16      def define_content_type( self, filename ):
17
18        fileparts = filename.split('.')
19
20        extension = string.lower( fileparts[1] )
21
22        # Based upon the extension, return the content-type string
23        if re.search( 'htm', extension ):
24          return 'text/html'
25        if re.search( 'txt', extension ):
26          return 'text/plain'
27        if re.search( 'py', extension ):
28          return 'text/plain'
29        else:
30          return 'application/octet-stream'
31
32
33      #
34      # Emit the response header
35      #
36      def emit_response_header( self, sock, content_type ):
37
38        # Emit the HTTP response message header
39        sock.send( 'HTTP/1.1 200 OK\n' )
40        sock.send( 'Server: Python shttp\n' )
41        sock.send( 'Connection: close\n' )
42        sock.send( 'Content-Type: ' )
43        sock.send( content_type )
44        sock.send( '\n\n' )
45
46
47      #
48      # Handle an HTTP GET method
49      #
50      def handle_get_method( self, sock, filename ):
51
52        # Remove the leading '/' from the filename
53        if filename[0] == '/':
54          filename = filename[1:len(filename)]
55
56        # If the filename is only '/' or a null string, then
```

```
57      # convert this to the index file.
58      if filename == '/' or filename.__len__() == 0:
59        filename = 'index.html'
60
61    print 'file is', filename
62
63    # Ensure the file is readable
64    if os.access( filename, os.R_OK ):
65
66      # Determine the content type
67      content_type = self.define_content_type( filename )
68
69      print 'content type ', content_type
70
71      self.emit_response_header( sock, content_type )
72
73      # Open the defined file and emit the contents through
74      # the socket.
75      file = open( filename, 'r' )
76      sock.send( file.read() )
77      file.close()
78      sock.send( "\n" )
79
80    else:
81
82      print 'File not found\n'
83
84      sock.send( 'HTTP/1.1 404\n\nFile not found\n\n' )
85
86
87    #
88    # Handle an individual HTTP session (callback).
89    #
90    def handle_connection( self, conn ):
91
92      print 'Handling new connection'
93
94      # Read the HTTP request message
95      line = conn.recv( 200 )
96
97      # Split the HTTP request message into a
98      # space-delimited array
99      elements = string.splitfields( line, ' ' )
100
101     if elements[0] == "GET":
102
103       # If the GET method was requested, call
104       # handle_get_method
```

```
105              self.handle_get_method( conn, elements[1] )
106
107        else:
108
109           print 'Unknown Method ', elements[0], '\n'
110
111           # All other HTTP methods are unimplemented...
112           conn.send( 'HTTP/1.0 501 Method Unimplemented\n' )
113
114
115     #
116     # Initialize an HTTP Server Instance.
117     #
118     def __init__( self, port = 80 ):
119
120        # Create a new server socket and bind it with the host
121        # and port
122        self.sock = socket.socket( socket.AF_INET,
123                                      socket.SOCK_STREAM )
124        self.sock.setsockopt( socket.SOL_SOCKET,
125                                      socket.SO_REUSEADDR, 1 )
126        self.sock.bind( (self.host, port) )
127        self.sock.listen( 1 )
128
129        # The big loop
130        while 1:
131
132           # Await a new client connection
133           conn, (remotehost, remoteport) = self.sock.accept()
134
135           print 'client at ', remotehost, remoteport
136
137           # Call handle_connection with the new client socket
138           self.handle_connection( conn )
139
140           # Bug alert -- close method shuts the socket down
141           # and aborts remaining outgoing data.
142           conn.shutdown(1)
```

This implementation of the simple HTTP server is based upon a simple class that provides the HTTP server functionality. There are no visible methods, but the __init__ method maps directly to the initialization of the creation of a new HTTP server instance. This Python example is different from the others that we've investigated in this chapter because we're creating a simple HTTP server class here, rather than a script. Later in this section, we'll see how to create an instance of the simple HTTP server class.

Lines 118–142 of Listing 18.9 provides the __init__ method for the Simple_http_server class. We use the def statement to define a new method, in this case the special method called __init__ that is invoked when a new instance of the class is created (**line 118**). Note also that the first argument, self, is provided here and in all other methods defined in this class. The self argument represents the instance of this class. Note at **line 11**, we declare the class variable sock. When referenced for the instance of this class, we denote it as self.sock (as in **line 122**). The final item to note at **line 118** is the port argument, which is defaulted here as the value 80. If a value is not provided when the instance is created, the default value of 80 is used. Otherwise, whatever value the caller provides is used.

The purpose of the __init__ method is to create the HTTP socket server object and serve as the big loop for the server. We begin socket creation at **line 122** where we create our socket using the **socket** method. Note the use of self to denote the instance variable. We use the SOCK_STREAM to create a TCP socket (as it's the Transport layer used by HTTP). At **lines 124–125**, we make the local address information reusable with the setsockopt method. At **line 126**, we bind our server socket to the local address tuple. The local address is made up of the instance variable host (defined at **line 10**) and the port argument passed in by the caller. Note that port is not prefixed by self, as port isn't an instance variable but instead a local variable on the stack. Finally, at **line 127**, we invoke the **listen** method to permit the server to accept incoming connections.

The HTTP server loop starts at **line 130** with the while statement. This infinite loop begins with a call to the **accept** method to await a new incoming client connection. The **accept** method returns two items, the client socket and an address tuple representing the peer that connected (**line 133**). We emit this for debug at **line 135** using the print statement. At **line 138**, we invoke the local handle_connection method that is used to handle the current client connection, satisfying its HTTP request. Upon return, we close the client socket. In the current release of Python, using the **close** method when data is waiting to be sent causes this data to be discarded. The workaround is to use the **shutdown** method (specifying the how argument as 1), which permits enqueued data to be sent to the peer. Control then resumes at the **accept** method, awaiting a new client connection.

The handle_connection method (**lines 90–112**) services the HTTP request method received by the simple server. The client socket is passed into the method as the conn variable (**line 90**) along with the self variable representing the object instance. We emit some debugging information at **line 92** to identify that we're servicing a new client connection. We then receive the HTTP request through the client socket using the **recv** method (**line 95**). We specify a maximum of 200 bytes to be received (which gives us at least the first line of the HTTP request, which contains the actual file request). The string result at **line 95** is returned to the line variable. At **line 99**, we split the HTTP request line into a string array using the splitfields method of the string module. Recall that the HTTP request line is made up of the request method (such as GET), the filename for the request, and the HTTP version, all delimited by a space. We test the HTTP request method at **line 101** to see if the client has requested a GET. If so, we call method handle_get_method at **line 105** and include the client socket and the filename from the request (element[1]). Otherwise, if the request method was something other than GET, we emit a debugging message at **line 109** and then send back an unimplemented error message to the client at **line 112** using the **send** method.

Method handle_get_method exists solely to satisfy HTTP GET method requests (**lines 50–84** of Listing 18.9). In addition to the self variable, the method also accepts the client socket (`sock`) and the `filename` being requested. Any leading '/' character is removed in **lines 53–54**, by first testing the `filename` for the character and, if present, removing the first character using the `substring` of the filename excluding the first character. If the filename consisted only of the '/' character, or is now empty, the filename is set to the default 'index.html' (**lines 58–59**). Next, debugging information is emitted at **line 61** to identify the requested file.

After the file being requested has been isolated, we begin the process of serving this file to the client. We begin by identifying whether the `filename` is an actual readable file using the **access** method of the `os` module (**line 64**). If so, we invoke the `define_content_type` method to identify the content type being returned. This string (`content_type`) is emitted for debugging purposes at **line 69** using the `print` statement. The HTTP response message header is then emitted using the `emit_response_header` method (**line 71**). Finally, the actual file is sent to the client. The file is first opened at **line 75** using the `open` method. At **line 76**, the entire file is read using the **read** method, and immediately sent to the client using the **send** method. After closing the file using the **close** method, we emit a final carriage-return through the client socket and the client request is complete.

If the file had not been accessible on the file system (as determined at **line 64**), a debugging message would have been emitted at **line 82** and an HTTP error message emitted to the client using the **send** method at **line 84**.

The `emit_response_header` method (**lines 36–44**) sends the HTTP response message header to the client through the socket. The response message header consists of two items, the status line (the first line emitted at **line 39**) and then a set of optional headers (**lines 40–43**). The response message header is emitted through to the client socket using the **send** method. At the end of the response message header is a blank line (**line 44**), which is required to separate the optional headers from the message body (emitted in method `handle_get_method`).

The final method in the `Simple_http_server` class is the `define_content_type` method. This method is used to identify the content type based upon the suffix of the filename to be served. For example, if an HTML file is to be served (such as 'index.html'), then the content type is 'text/html'. If a text file is to be served (such as 'file.txt'), then the content type is 'text/plain'. The only other content type that is served by the server is 'application/octet-stream'. This type is commonly stored by the HTTP client in a file, because it does not know how to deal with the content.

Method `define_content_type` is shown in **lines 16–30** of Listing 18.9. The `filename` is split into two parts (the filename and the suffix) using the `split` method. The split occurs on the '.' character, which is used to separate the actual filename from the suffix. The extension is then converted to lowercase using the `lower` method of the `string` module, storing the result in variable `extension`. We then use the `search` method of the `re` (regular-expression) module to identify which particular suffix appears, and based upon a successful match, the `content-type` is returned. If the suffix is not matched, the 'application/octet-stream' content type is returned to the caller. This content type is then transferred to the client in the HTTP response message header as the value of the 'Content-Type' key.

The final item to discuss with the `Simple_http_server` class is how an application can create a new `Simple_http_server` object. As shown in Listing 18.10, a new

Simple_http_server object is created by using the class name as a method and providing any arguments that may be necessary. In this case, we provide the port (80), though for this default value, we could have passed nothing. The result of **line 5** is actually the invocation of the __init__ method, as shown in Listing 18.9. If the __init__ method returned, the next step (at **line 7**) is destroying the object using the del statement.

Listing 18.10   Instantiating a new Python HTTP server.

```
1   #
2   # Create a new HTTP server and start it on port 80
3   #
4
5   shttp = Simple_http_server( 80 )
6
7   del shttp
```

SIMPLE SMTP CLIENT

Now, let's look at a simple SMTP client written in Python. The SMTP client is implemented as a class that provides a set of methods to send an e-mail to a specified recipient (see Listing 18.11).

Four methods are provided by the Mailer class for e-mail transport. These are the __init__ method (though abstracted by the class name Mailer), dialog, send, and finish. Only method dialog is used internally by the Mailer class to implement the actual SMTP protocol, all others are used as API functions for transporting e-mail. We'll investigate each of the methods in their order of appearance within a sample use. One item to note before the methods is the creation of global class variables. In this case (**lines 8–13**), we define a set of instance variables and initialize them each to 0. As we'll see in Listing 18.11, referencing these class variables requires the use of the self modifier.

The Python language Mailer class is shown in Listing 18.11. It begins by making the **socket** module visible using the import statement at **line 1**. We define the Mailer class at **line 3**, and then, as defined in the previous paragraph, we create our class global variables.

Listing 18.11   Python language SMTP client source.

```
1   import socket
2
3   class Mailer:
4
5     #
6     # Define the class variables (host and server socket)
7     #
8     subject = 0
9     sender = 0
```

```
10      recipient = 0
11      content_type = 0
12      contents = 0
13      sock = 0
14
15
16      #
17      # Class initialization method
18      #
19      def __init__( self, subject, sender, recipient,
20                    content_type, contents ):
21
22        self.subject = subject
23        self.sender = sender
24        self.recipient = recipient
25        self.content_type = content_type
26        self.contents = contents
27        self.sock = 0
28
29
30      #
31      # Perform a dialog with the SMTP server
32      #
33      def dialog( self, command, expected_response ):
34
35        if command != 0:
36
37          self.sock.send( command )
38
39        if expected_response != 0:
40
41          line = self.sock.recv( 100 )
42
43          if line.startswith( expected_response ) == 0:
44
45            raise IOError
46
47
48      #
49      # Send the mail based upon the initialized parameters
50      #
51      def send( self ):
52
53        mail_elements = self.recipient.split('@')
54
55        self.sock = socket.socket( socket.AF_INET,
56                                   socket.SOCK_STREAM )
57
```

```
58          self.sock.connect( (mail_elements[1], 25) )
59
60          # Look for initial e-mail salutation
61          self.dialog( 0, "220" )
62
63          # Send HELO and await response
64          self.dialog( "HELO thisdomain.com\n", "250" )
65
66          # Send "MAIL FROM" command and await response
67          string = "MAIL FROM: " + self.sender + "\n"
68          self.dialog( string, "250" )
69
70          # Send "RCPT TO" command and await response
71          string = "RCPT TO: " + self.recipient + "\n"
72          self.dialog( string, "250" )
73
74          # Send "DATA" to start the message body
75          self.dialog( "DATA\n", "354" )
76
77          # Send out the mail headers (from/to/subject)
78          string = "From: " + self.sender + "\n"
79          self.dialog( string, 0 )
80
81          string = "To: " + self.recipient + "\n"
82          self.dialog( string, 0 )
83
84          string = "Subject: " + self.subject + "\n"
85          self.dialog( string, 0 )
86
87          # Send the Content Type
88          string = "Content-Type: " + self.content_type + "\n"
89          self.dialog( string, 0 )
90
91          # Send the actual message body
92          self.dialog( self.contents, 0 )
93
94          # Send the end-of-email indicator
95          self.dialog( "\n.\n", "250" )
96
97          # Finally, close out the session
98          self.dialog( "QUIT\n", "221" )
100
101
102     #
103     # Close out the smtp session by closing the socket
104     #
105     def finish( self ):
106
```

```
107        self.sock.shutdown(1)
```

Let's now look at the methods created within the `Mailer` class. The first method used by mail users is the `__init__` method that is accessed by the class name as a method (abstracted by `Mailer`). The application creates an instance of the `Mailer` class using the `Mailer` method. With the new instance, the `Mailer` (`__init__`) method is called to specify the `subject`, `sender`, `recipient`, `content_type`, and message body (`contents`). As shown in **lines 22–26**, these variables are copied from the application into the class instance variables and are bound to this instance of the class. The final `__init__` method step is initializing the local `sock` variable, in this case to zero (**line 27**).

Next, we look at the **send** method (**lines 51–98**). The first step at **line 53** is to identify the mail server to which we'll connect to send our e-mail. We can identify this mail server by taking the FQDN from the recipient (such as `mtjones.com` from the e-mail address `mtj@mtjones.com`). The `split` method is used to separate the username from the domain name, with the resulting two-element array stored in `mail_elements` (**line 53**). A stream socket is created next at **lines 55–56**, which will be used as our conduit to the mail server. At **line 58**, we use the **connect** method to connect to the mail server. We utilize the second element of the `mail_elements` array as the mail server to which we're connecting, and specify the standard SMTP port number, 25.

Lines 61–98 of the **send** method from Listing 18.11 implement the basics of the SMTP client side. Using the `dialog` method (provided within this class), transactions with the SMTP server are performed. We do not discuss the details of the SMTP protocol here; however, a discussion of SMTP and the design pattern is provided in Chapter 15, Software Patterns Introduction.

The `dialog` method (shown in **lines 33–45** of Listing 18.11) provides the ability for the client to perform communicative transactions with the SMTP server. The `dialog` method accepts three arguments, the `self` variable, the `command` (the string to send to the server), and the `expected_response` (the string error code that is expected back for a successful transaction). The `dialog` function can send a command and receive a response, send a command without expecting a response, or send no command but receive a response. This behavior is implemented simply by checking the `command`/`expected_response` for their presence and acting accordingly. At **line 35**, we check the presence of a command (`command` is not zero), and if present, the command is sent using the **send** method at **line 37**. At **line 39**, we check whether the caller provided an `expected_response`. If so, a response is expected, so we read up to 100 octets from the socket using the **recv** method at **line 41**. Using the `startswith` method, we check to see if the `expected_response` is the first set of characters in the server response (stored in `line`). If the expected response is not found, we raise an `IOError`. This error causes us to abort the dialog and raises the exception back to the user application, which in this case deals with the exception.

Finally, to complete the SMTP session, the `finish` method must be called (**lines 105–107**). This method simply closes the client socket of the `Mailer` object using the **shutdown** method.

Now, let's look at a sample application of the `Mailer` class. Listing 18.12 shows a sample usage of the `Mailer` class. At **line 6**, a new `Mailer` object is created by calling the __init__ method of the Mailer class (implicit in the Mailer call). We specify the subject (**line 6**), source and destination e-mail addresses (**lines 7 and 8**), the content type (**line 9**), and finally the mail message (shown as two split strings at **lines 10 and 11**). Note that, in this case, we're showing an HTML encoded e-mail being sent, as is illustrated by the HTML content in the mail message.

Now that the mail object has been created and initialized, we can send the e-mail using the **send** method (at **line 15**). Note the try statement at **line 14** that bounds the **send** method. For methods that extend the possibility of raising an exception, the try method is used to note this. If an exception is raised within the bounds of the try block, the except statement identifies which exceptions to handle and how to handle them. In this case (**line 16**), we're handling the IOError exception, and we handle it by simply emitting a debug message using the print statement (**line 17**). If an exception was not raised within the try block, execution continues at **line 20** where we use the finish method to complete the SMTP transaction and close the socket. Finally at line 23, the mail object is deleted using the del statement.

Listing 18.12   Instantiating a new SMTP client.

```
1   #
2   # Create a new mailer and send an e-mail
3   #
4
5   # Create a new Mailer instance
6   mail = Mailer( "The Subject",
7                  "tim@mtjones.com",
8                  "mtj@mtjones.com",
9                  "text/html",
10                 "<HTML><BODY><H1>This is the mail"
11                 "</H1></BODY></HTML" )
12
13  # Try to send the mail
14  try:
15    mail.send()
16  except IOError:
17    print "Couldn't send mail"
18
19  # Finish up the mail session
20  mail.finish()
21
22  # Remove the instance
23  del mail
```

19 Network Code Patterns in Perl

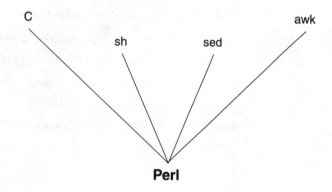

INTRODUCTION

In this chapter, we look at simple socket-based applications using Perl. These applications can serve as implemented software patterns for the construction of more advanced networking applications.

ON THE CD *All software discussed here is also provided on the companion CD-ROM. Table 19.1 lists the applications and their locations on the CD-ROM.*

TABLE 19.1 CD-ROM companion software

Code Pattern	CD-ROM Location
Stream server/client	./software/ch19/stream/
Datagram server/client	./software/ch19/dgram/
Multicast server/client	./software/ch19/mcast/
Broadcast server/client	./software/ch19/bcast
Simple HTTP server	./software/ch19/shttp/
Simple SMTP client	./software/ch19/smtpc/

STREAM (TCP) SERVER/CLIENT

The stream server and client demonstrate the Daytime protocol using stream sockets (reliable TCP-based communication).

Stream Server

The Daytime protocol stream server is shown in Listing 19.1. Because all networking applications must have access to the socket module, we first make the socket module visible by including the socket module using the use statement (**line 1**). At **lines 4–5**, we specify the host and port variables that define to what we'll bind (how we'll advertise the server). At **line 8**, we create our socket using the **socket** function. This function looks very similar to the standard function in C, except for the fact that our return server socket is a return argument within the function rather than a return from the function. For the stream socket, we specify the AF_INET domain and the SOCK_STREAM type.

*The die function in Perl (for example, Listing 19.1 **line 10**) is a simple way to raise an exception. In Listing 19.1, the die function is used in a number of cases to halt script execution if the preceding function fails. Upon failure, the $! symbol is replaced with the string error return, which is emitted to standard-out.*

NOTE

At **line 13**, we create a packed address, which is synonymous with the C sockaddr_in structure. Note that here in Perl, the function to convert an address and a port into the sockaddr_in structure is called **sockaddr_in**. This function takes a port and host address and returns the packed address. This is used at **line 21** to bind the server socket to the local address defined by the packed address (the address we want our server to be advertised as) using the **bind** function. At **line 16**, we use the **setsockopt** function to make the local address reusable, to avoid "address in use" errors. Next, at **line 22**, we enable our willingness to accept client connections using the **listen** function.

Our big loop begins at **line 26**, where we accept client connections and then emit the current date and time to them. At **line 29**, we accept new client connections using the **accept** function. The first parameter is the returned client socket descriptor and the second argument is the server socket from which we'll be accepting connections. For a successful **accept** (true return), we emit the current date and time to the client using the **print** function (**line 32**). The current time is created using scalar localtime, and a newline is appended. The client socket is then closed using the **close** function (**line 35**).

Listing 19.1 Perl Daytime stream server.

```
1    use Socket;
2
3    # Create the host/port information
4    $host = INADDR_ANY;
5    $port = 13;
6
7    # Create a new stream socket
8    socket( servSock, Socket::AF_INET,
9            Socket::SOCK_STREAM, 0 )
10     or die "socket: $!";
11
12   # Create a packed address
13   $paddr = sockaddr_in( $port, $host );
14
15   # Make the local address reusable
16   setsockopt( servSock, Socket::SOL_SOCKET,
17               Socket::SO_REUSEADDR, 1 );
18
19   # Bind the local address and enable incoming
20   # connections
21   bind( servSock, $paddr ) or die "bind: $!";
22   listen( servSock, Socket::SOMAXCONN ) or
23     die "listen: $!";
24
25   # The big loop
26   while(1) {
27
```

```
28    # Accept a client connection
29    if ( accept( cliSock, servSock ) ) {
30
31      # Emit the current time
32      print cliSock scalar localtime, "\n";
33
34      # Close the client socket
35      close cliSock;
36
37    }
38
39  }
40
41  # Close the server socket
42  close servSock;
```

Stream Client

Now, let's look at the Daytime stream client (shown in Listing 19.2). The client creates a socket to connect to the server and then awaits a string to be received through the socket. Upon receiving the string from the server, the message is printed and the client exits.

Listing 19.2   Perl Daytime stream client.

```
1   use Socket;
2
3   # Create the host/port information for the server
4   $host = 'localhost';
5   $port = 13;
6
7   # Create a stream socket
8   socket( mySock, Socket::AF_INET,
9           Socket::SOCK_STREAM, 0 ) or
10    die "socket: $!";
11
12  # Create a packed address
13  $addr = inet_aton( $host );
14  $paddr = sockaddr_in( $port, $addr );
15
16  # Connect to the server
17  connect( mySock, $paddr ) or die "connect: $!";
18
19  # Receive the time string
20  recv( mySock, $buffer, 100, 0 );
21
22  # Print the received time string
```

```
23   print $buffer;
24
25   # Close the client socket
26   close( mySock ) || die "close: $!";
```

We begin by importing the Socket module at **line 1** using the use function. The server to which we want to connect is then identified at **lines 4–5** (address and port number). At **line 8**, we create our client socket using the **socket** command, specifying the AF_INET domain and SOCK_STREAM type. We then convert our address to a binary address using the **inet_aton** function. This value is then used at **line 14** to create the packed address (using the binary address, addr, and the predefined port). This address is used at **line 17** to connect to the server, which returns the new client socket called mySock.

To retrieve the current date and time string, the **recv** function is used at **line 20** to get a line from the socket. This function looks identical to the standard C BSD function. We provide the socket, a buffer to store the data, and an associated length (maximum amount of data to return). The final argument, zero, represents the flags (none are specified). At **line 23**, the string returned by **recv** is emitted to standard-out using the print function. Finally, the client socket is closed at **line 26** using the **close** function.

DATAGRAM (UDP) SERVER/CLIENT

Now, let's look at a UDP server and client implementation using datagram sockets. Recall that datagram communication is not connection-oriented; therefore, each datagram that we send must also include the intended recipient.

Datagram Server

The datagram server is illustrated in Listing 19.3. After importing the Socket module, we create our port variable at **line 4**, which represents the port to which we'll bind (as a server socket). **Lines 7–9** represent socket creation, using SOCK_DGRAM as the type for a datagram socket. We use the **sockaddr_in** function at **line 12** to create our socket address, using the predefined port variable and the INADDR_ANY constant. Recall that INADDR_ANY specifies that, for a server, we'll accept incoming connections from any available interface. At **line 15**, we bind to the previously created packed address structure ($paddr). In this case, there is no call to **listen**, because this is required only of stream sockets.

The infinite server loop begins at **line 17** where we await incoming datagram packets using the **recv** function (**lines 20–21**). The packet received is simply a dummy datagram from the client, which gives us the source address of a client that wants to receive the date/time string. We store the return value from **recv** to $from, which represents the source of the datagram, and then use it at **line 27** with the **send** function. This is a special case of the **send** function, where the fourth argument represents the destination to which we want to direct the datagram. The datagram was constructed at **line 24** using sprintf to construct the string and command scalar localtime to retrieve the date and time in string format.

Listing 19.3 Perl Daytime datagram server.

```perl
1    use Socket;
2
3    # Create the host/port information for the server
4    $port = 13;
5
6    # Create a datagram socket
7    socket( srvSock, Socket::AF_INET,
8            Socket::SOCK_DGRAM, 0 )
9      or die "socket: $!";
10
11   # Convert the address and port into a packed address
12   $paddr = sockaddr_in( $port, INADDR_ANY );
13
14   # Bind the server
15   bind( srvSock, $paddr ) or die "bind: $!";
16
17   while (1) {
18
19     # Receive the dummy datagram from the client
20     $from = recv( srvSock, $line, 1, 0 ) or
21       die "recv: $!";
22
23     # Get the time and convert it to a string
24     $line = sprintf( "%s\n", scalar localtime );
25
26     # Write the date/time string to the client
27     send( srvSock, $line, 0, $from ) or
28       die "send: $!";
29
30   }
31
32   # Close the client socket
33   close( srvSock ) || die "close: $!";
```

Datagram Client

The datagram client is shown in Listing 19.4, and corresponds with the datagram server previously shown in Listing 19.3.

After importing the Socket module at **line 1** using the use statement, we specify the host and port information of the server with which we want to communicate (**lines 4–5**). Using the **socket** function (**lines 8–10**), we create a new datagram socket. At **line 13**, we convert the destination host address to a numeric binary address using **inet_aton**, storing the result into $srvaddr. At **line 16**, we use $srvaddr and our predefined port to create a packed address using **sockaddr_in**.

Given our new packed address ($paddr), we send our dummy datagram to the server using the **send** function. Upon the receiving this datagram, the server should respond with the date/time string datagram, which we'll read here using the **recv** function. The resulting datagram payload (date and time) is stored into $line at **line 23**, which is emitted to standard-out at **line 26**. Finally, we close the client socket at **line 29** using the **close** function.

Listing 19.4   Perl Daytime datagram client.

```
1    use Socket;
2
3    # Create the host/port information for the server
4    $host = 'localhost';
5    $port = 13;
6
7    # Create a datagram socket
8    socket( mySock, Socket::AF_INET,
9            Socket::SOCK_DGRAM, 0 ) or
10     die "socket: $!";
11
12   # Convert the host/address to a numeric address
13   $srvaddr = inet_aton( $host );
14
15   # Convert the address and port into a packed address
16   $paddr = sockaddr_in( $port, $srvaddr );
17
18   # Send a dummy packet to identify ourself with
19   # the server
20   send( mySock, " ", 0, $paddr ) or die "send: $!";
21
22   # Receive the date/time string from the server
23   recv( mySock, $line, 128, 0 ) or die "recv: $!";
24
25   # Print the received time string
26   print $line, "\n";
27
28   # Close the client socket
29   close( mySock ) || die "close: $!";
```

MULTICAST SERVER/CLIENT

The multicast server and client are fundamentally derivatives of the datagram code patterns, though we use another API because Perl does not natively support sending multicast packets.

Multicast Server

The Perl language source code for the multicast server is shown in Listing 19.5. Recall that multicast communication is group-based; therefore, a single message emitted by the server is received by every client that is currently a member of the multicast group.

To send datagrams, no multicast support is actually needed. All we do is send datagrams to an address that is a multicast address. In order to receive multicast datagrams, we must join the specific multicast group.

At **line 1**, we import the standard Socket module using the use statement. Next, we create the group and port variables at **lines 4–5**. Note that instead of a host address, we define a group variable, which contains a multicast address. At **lines 8–10**, we create our socket that is a standard datagram socket (we'll note a difference between this and the multicast client in Listing 19.6). In order to bind to the address and port, we use the SO_REUSEADDR socket option to avoid the common "address in use" error. At **line 16**, we create the packed address using **sockaddr_in** and then bind ourselves to this address (not completely necessary in this example).

To direct our datagrams, we need a packed address structure containing the destination address (group multicast address) and port. This is accomplished with **lines 20 and 22**, where we first create a numeric version of our group address using **inet_aton**, and then use **sockaddr_in** to create a packed address with the port number.

Our infinite loop begins at **line 24**. We create a time string at **line 26**, and then send the string to the group as a datagram payload using the **send** function. Note that we specify not only our server socket and date/time string buffer, but also the packed address representing the multicast group ($paddr). Finally, at **line 30**, we sleep for one second and then repeat the process.

Listing 19.5 Perl Daytime multicast server.

```
1    use Socket;
2
3    # Create the host/port information for the server
4    $group = '239.0.0.2';
5    $port = 45002;
6
7    # Create a new multicast socket
8    socket( srvSock, Socket::AF_INET,
9            Socket::SOCK_DGRAM, 0)
10     or die "socket: $!";
11
12   # Make the local address reusable
13   setsockopt( srvSock, Socket::SOL_SOCKET,
14               Socket::SO_REUSEADDR, 1 );
15
16   $myaddr = sockaddr_in( $port, INADDR_ANY );
17
```

```
18    bind( srvSock, $myaddr );
19
20    $srvaddr = inet_aton( $group );
21
22    $paddr = sockaddr_in( $port, $srvaddr );
23
24    while( 1 ) {
25
26      $line = sprintf( "%s\n", scalar localtime );
27
28      send( srvSock, $line, 0, $paddr ) or die "send: $!";
29
30      sleep( 1 );
31
32    }
33
34    close( srvSock ) or die "close: $!";
```

Multicast Client

The Perl source code for the multicast client is shown in Listing 19.6. The multicast client includes a number of new features not previously discussed, but reflects symmetry with the multicast server, as seen with other code patterns.

In the case of the multicast client, the Multicast API must be used. We import not only the standard Socket module at **line 1**, but also the Multicast module at **line 2** (also called IO::Socket::Multicast). In **lines 5–6**, we specify our group multicast address and port to which we want to subscribe. At **line 9**, we create a new multicast socket using the **new** function of the Multicast module. We specify the port to which we want to bind (LocalPort), the protocol of the socket (UDP) and type (SOCK_DGRAM), and finally an option that we want enabled (Reuse, or SO_REUSEADDR). This is an alternate mechanism for socket creation that can simplify a number of operations into a single call.

Now that we have our multicast datagram socket, we need to join the multicast group in order to receive multicast datagrams. At **line 17**, we use the **mcast_add** function to join the group. This is synonymous with the IP_ADD_MEMBERSHIP socket option. After this function completes, we're subscribed to the multicast group and can receive multicast datagrams. We call the **recv** function to receive a datagram, at **line 20**, and the result is stored in $line. We emit this to standard-out at **line 23** using the print function and, finally, drop ourselves from the multicast subscription using the **mcast_drop** function. This command is the same as using the IP_DROP_MEMBERSHIP socket option in the standard BSD API.

Listing 19.6   Perl Daytime multicast client.

```
1    use IO::Socket;
2    use IO::Socket::Multicast;
3
```

```
4    # Create the host/port information for the server
5    GROUP = '239.0.0.2';
6    PORT = '45002';
7
8    # Create a new multicast socket
9    my $sock = IO::Socket::Multicast->new(
10                                 LocalPort=>PORT,
11                                 Proto=>'udp',
12                                 Type=> SOCK_DGRAM,
13                                 Reuse => 1 )
14           or die "new mcast socket $!";
15
16   # Join the predefined multicast group
17   $sock->mcast_add( GROUP ) or die "mcast_add $!";
18
19   # Receive the date/time string
20   $sock->recv( $line, 128 );
21
22   # Emit it to standard-out
23   print $line;
24
25   # Remove ourselves from the multicast group
26   $sock->mcast_drop( GROUP ) or die "mcast_drop $!";
```

BROADCAST SERVER/CLIENT

The broadcast server and client, like the multicast server and client, are fundamentally derivatives of the datagram code patterns. This is again because broadcast communication is
based upon the datagram model.

Broadcast Server

The Perl source code for the broadcast server is shown in Listing 19.7. The server is broadcast-based, which means a single message emitted by the server is received by every client
that is configured to receive broadcast datagrams (for the given port).

At **line 1**, we import the **Socket** module and then define the host and port to which
we'll direct our datagrams (**lines 4–5**). Note that the $host variable is a broadcast address,
but this could be further constrained given the subnet in which it's executed (for example,
"192.168.1.255"). At **line 7**, we create our datagram socket using the **socket** function and
then enable sending of broadcast datagrams at **lines 13–14** using the **setsockopt** function
with the SO_BROADCAST option.

At **line 18**, we use **inet_aton** to convert the broadcast address from dotted-string notation to a binary address. This is then stored, with the port, into a packed address using the
sockaddr_in function. The remainder of this pattern is similar to the datagram server, except that we don't wait for dummy datagrams from clients (although, it is identical to the

multicast server pattern). In the big loop (**lines 23–32**), we create a new date/time string using sprintf and scalar localtime, and then emit it out using the **send** function. Any socket currently bound to the broadcast address and port will then receive the datagram.

Listing 19.7 Perl Daytime broadcast server.

```
1    use Socket;
2
3    # Create the host/port information for the server
4    $host = '255.255.255.255';
5    $port = 45003;
6
7    # Create a datagram socket
8    socket( srvSock, Socket::AF_INET,
9            Socket::SOCK_DGRAM, 0 ) or
10     die "socket: $!";
11
12   # Make the socket broadcast-send capable
13   setsockopt( srvSock, Socket::SOL_SOCKET,
14               Socket::SO_BROADCAST, 1 ) or
15     die "setsockopt: $!";
16
17   # Convert the host/address to a numeric address
18   $srvaddr = inet_aton( $host );
19
20   # Convert the address and port into a packed address
21   $paddr = sockaddr_in( $port, $srvaddr );
22
23   while ( 1 ) {
24
25     $line = sprintf( "%s\n", scalar localtime );
26
27     send( srvSock, $line, 0, $paddr ) or
28       die "send: $!";
29
30     sleep( 1 );
31
32   }
33
34   # Close the client socket
35   close( srvSock ) || die "close: $!";
```

Broadcast Client

The Perl source code for the broadcast client is shown in Listing 19.8. As with the broadcast server, we concentrate on the differences to the standard datagram client.

The broadcast client pattern is identical to the datagram client, except for the difference in host addresses (in this case, the broadcast address). We begin at **line 1** by importing the Socket module and then defining the host and port variables with the broadcast address and selected port. At **line 8**, we create the datagram socket using the socket function (type SOCK_DGRAM).

Next, we create a binary address of the dotted-notation string IP address using **inet_aton** (**line 13**) and then create our packed address structure with the binary address ($srvaddr) and the port to create $paddr (**line 16**). This packed address is then used at **line 19** with **bind** to bind the socket to the local address.

Recall that broadcast datagrams are received by all sockets that are bound to the broadcast address. Therefore, when we call **recv** at **line 22**, we'll receive any broadcast datagram just like any other client bound to the same address (on the same host, or different host on the subnet). We emit the received date/time string ($line) to standard-out using the print function and then close the socket at **line 28** using the **close** function.

Listing 19.8   Perl Daytime broadcast client.

```perl
1    use Socket;
2
3    # Create the host/port information for the server
4    $host = '255.255.255.255';
5    $port = 45003;
6
7    # Create a datagram socket
8    socket( cliSock, Socket::AF_INET,
9              Socket::SOCK_DGRAM, 0 ) or
10      die "socket: $!";
11
12   # Convert the host/address to a numeric address
13   $srvaddr = inet_aton( $host );
14
15   # Convert the address and port into a packed address
16   $paddr = sockaddr_in( $port, $srvaddr );
17
18   # Bind to the broadcast address and port
19   bind( cliSock, $paddr ) or die "bind: $!";
20
21   # Retrieve the time/date string from the socket
22   recv( cliSock, $line, 128, 0 ) or die "recv: $!";
23
24   # Emit the line to standard-out
25   print $line;
26
27   # Close the client socket
28   close( cliSock ) || die "close: $!";
```

SIMPLE HTTP SERVER

Now, let's look at a simple HTTP server written in Perl. The entire source listing for the Perl simple HTTP server is provided in Listing 19.9. Let's walk through this listing to understand the sample implementation.

The simple HTTP server in Perl is made up of five subroutines, as shown in Listing 19.9. These subroutines range from the main subroutine, Simple_Http_Server (**lines 124–163**), handle_connection (**lines 92–118**), handle_get_method (**lines 46–86**), emit_response_header (**lines 29–40**), and finally define_content_type (**lines 6–23**). We'll walk through each of these subroutines, in their order of use.

The first subroutine is Simple_Http_Server, which is the initialization routine for the server (**lines 124–163**). This subroutine is responsible for setting up the server socket and awaiting incoming client connections. Upon receiving a new connection, the handle_connection subroutine is called to service it. At **line 124**, we declare our new subroutine and then assign our incoming arguments using the shift function. Note the similarity here to standard shell languages. The shift operator is used to shift out the contents of a list (starting from the left). The default parameter to shift (if no argument is specified) is @ARGV, the standard function call parameter list. Therefore, the first shift gets the first argument (in this case the $addr) and the second shift extracts the second argument ($port). If no argument was present for $port, the die operator would be called, causing the script to exit.

At **line 131**, we create our server socket of type SOCK_STREAM. To ensure reusability of the local address, we call **setsockopt** with the SO_REUSEADDR option. We then construct the local address to which we'll bind. At **line 140**, we take the address passed in by the caller ($addr) and convert this to a numeric address (from dotted-string notation). Next, this numeric address and the user-defined $port are used with **sockaddr_in** to create our packed address (**line 143**). This parameter is then used at **line 146** to **bind** our server socket to the local address defined by $paddr. The final setup element is shown at **line 149**, where we permit incoming client connections using the **listen** call.

The final portion of Simple_Http_Server is the server loop (**lines 152–161**). Using the **accept** call (**line 154**), we accept new incoming client connections, storing the new client socket in CLISOCK. We then call handle_connection with the new client socket to service its request (**line 156**). Upon return at **line 157**, we close the client socket using the **close** function.

The handle_connection subroutine is used to service an individual client connection (**lines 92–118**). We first shift out the client socket passed in by the caller ($sock) and then call the **recv** function to get the HTTP request from the client (**line 98**). Using the split function, we extract the first three elements of the $line string. These three elements are space delimited and represent the HTTP method (GET, HEAD, etc.), the object requested (filename), and the HTTP version. At **line 104**, we check the request, and if it's the GET method, then we call the handle_get_method subroutine (**line 107**). Otherwise, we emit an error to the client (unimplemented method) using the **print** function (**line 114**).

As the simple HTTP server supports only the GET method, the meat of the HTTP server is the handle_get_method subroutine (**lines 46–86**). At **lines 48–49**, we shift out the

$sock and $fname (filename) arguments and then use a regular expression at **line 53** to remove any '/' characters from the filename. The '=~' operator forces the regular expression to be applied to the lefthand side of the expression (here $fname). This regular expression consists of a command (s/, for substitution), the character we want to substitute (\/, which means the '/' character), and, finally, the character that we want to replace it with (because nothing comes between the / separator and the final / terminator, this means no character). Finally, the 'g' means globally replace all instances. After this operation is complete, we test the resulting filename with an empty string at **line 56**, and if an empty string did result from the regular expression operation, we replace the filename with the default (index.html).

At **line 60**, we call the define_content_type subroutine with the filename to identify the type of content being returned to the client. Note that the my statement here simply means that the following variable ($content_type) is scoped to the current block (this subroutine). The subroutine define_content_type returns a string representing the content-type string, which is then stored into $content_type. We next call emit_response_header at **line 62**, providing the $content_type variable (because it will use this to declare our content type for our response message).

We then test the existence of the file being requested using '-e' at **line 65**. If not present, we report an error to the client at **line 82** (error code 404, file not found). Otherwise, we open the file at **line 70**, and read the contents of the file into the list @thefile. This list is then written to the client socket at **line 72** using the **print** function. A final newline is also written to the client socket at **line 73** before the file is closed at **line 75** using the **close** function.

The emit_response_header subroutine (**lines 29–40**) is used to send a standard HTTP message response back to the client. It sends an HTTP status line and then an optional set of headers. The subroutine first shifts out the client socket handle (argument 1) and then the content type string (stored into $ct) at **lines 31–32**. The success status line is emitted to the client at **line 34** using the **print** function and then a set of optional headers (**lines 35–37**). Note the specification of content type to the client at **line 37**. This uses the previously created content_type string returned from the define_content_type subroutine. Finally, two newlines are emitted at **line 38** to the client to separate the HTTP message response header from the body.

The final subroutine in the simple HTTP server is the define_content_type subroutine. This is used to identify the type of content being returned to the HTTP client. Although there are a variety of ways to do this, the simplest involves looking at the file extension and defining the content type based upon it. For example, the content type "text/html" is required when returning HTML content. Therefore, if we look at the file extension and find ".htm", then we can assume it's an HTML file and set the content type accordingly. After shifting the filename out of the argument list, we test it against a list of prospective extensions (**lines 11–19**). We first check (using a regular expression) to see if the string '.htm' is present in the filename. If so, we return the content type 'text/html' to the caller. For '.txt' and '.pl' extensions, we return 'text/plain'. Finally, for extensions that are unknown, we return 'application/octet-stream'. This is the default content type for unknown types. At **line 21**, we return the content type to the caller.

Finally, to start the simple HTTP server (**line 169**), we call the `Simple_Http_Server` subroutine and provide it with the host and port on which we want to bind. At this point, connections to the server are possible.

Listing 19.9 Perl simple HTTP server source.

```
1    use Socket;
2
3    #
4    #   Determine the content type based upon the filename
5    #
6    sub define_content_type {
7
8      my $name = shift
9        or die 'usage is define_content_type( $filename )';
10
11     if      ($name =~ /.htm/) {
12       $string = "text/html";
13     } elsif ($name =~ /.txt/) {
14       $string = "text/plain";
15     } elsif ($name =~ /.pl/) {
16       $string = "text/plain";
17     } else {
18       $string = "application/octet-stream";
19     }
20
21     return( $string );
22
23   }
24
25
26   #
27   #   Emit the standard HTTP response message header
28   #
29   sub emit_response_header {
30
31     my $sock = shift;
32     my $ct = shift;
33
34     print $sock "HTTP/1.1 200 OK\n";
35     print $sock "Server: Perl shttp\n";
36     print $sock "Connection: close\n";
37     print $sock "Content-Type: ", $ct;
38     print $sock "\n\n";
39
40   }
```

```perl
41
42
43    #
44    #   HTTP 'GET' Method Handler
45    #
46    sub handle_get_method {
47
48      my $sock = shift;
49      my $fname = shift or
50        die 'usage is handle_get_method( $sock, $filename )';
51
52      # Remove any '/' characters
53      $fname =~ s/\///g;
54
55      # If filename is now empty, convert it to the default
56      if ($fname eq "") {
57        $fname = "index.html";
58      }
59
60      my ($content_type) =  define_content_type( $fname );
61
62      emit_response_header( $sock, $content_type );
63
64      # Test the extistence of the requested file
65      if (-e $fname) {
66
67        # Open the file and emit it through the socket
68        open( INFILE, $fname );
69
70        @theFile = <INFILE>;
71
72        print $sock @theFile;
73        print $sock "\n";
74
75        close( INFILE );
76
77      } else {
78
79        print "File not found.\n";
80
81        # Unknown file -- notify client
82        print $sock "HTTP/1.1 404\n\nFile not found.\n\n";
83
84      }
85
86    }
87
88
```

```
89    #
90    #   HTTP Connection Handler
91    #
92    sub handle_connection {
93
94      my $sock = shift or
95        die 'usage is handle_connection( $sock )';
96
97      # Get the request line from the client
98      recv( $sock, $line, 1024, 0 );
99
100     # Split the GET request into its parts
101     ($Request, $Filename, $Version) = split( ' ', $line, 3 );
102
103     # Check the request — we handle only GET requests
104     if ( $Request eq "GET" ) {
105
106       # Call our GET method handler
107       handle_get_method( $sock, $Filename );
108
109     } else {
110
111       print "Unknown Method ", $Filename, "\n";
112
113       # Unknown method — notify client
114       print $sock, "HTTP/1.1 501 Unimplemented Method\n\n";
115
116     }
117
118   }
119
120
121   #
122   #   Initialization function for the simple HTTP server
123   #
124   sub Simple_Http_Server {
125
126     my $addr = shift;
127     my $port = shift or
128       die 'usage is Simple_HTTP_Server( $addr, $port )';
129
130     # Create a new TCP Server socket
131     socket( SRVSOCK, Socket::AF_INET,
132             Socket::SOCK_STREAM, 0 )
133       or die "socket: $!";
134
135     # Make the local address reusable
136     setsockopt( SRVSOCK, Socket::SOL_SOCKET,
```

```
137                          Socket::SO_REUSEADDR, 1 );
138
139      # Conver the dotted IP string to a binary address
140      my $baddr = inet_aton( $addr );
141
142      # Create a packed address for the server
143      my $paddr = sockaddr_in( $port, $baddr );
144
145      # Bind the local address to the server
146      bind( SRVSOCK, $paddr ) or die "bind: $!";
147
148      # Enable incoming connections
149      listen( SRVSOCK, Socket::SOMAXCONN ) or
150        die "listen: $!";
151
152      while (1) {
153
154        if ( accept( CLISOCK, SRVSOCK ) ) {
155
156          handle_connection( CLISOCK );
157          close( CLISOCK );
158
159        }
160
161      }
162
163    }
164
165
166    #
167    #  Create a new HTTP server and start it on port 80
168    #
169    Simple_Http_Server( '127.0.0.1', 80 );
```

SIMPLE SMTP CLIENT

Now, let's look at a simple SMTP client written in Perl (see Listing 19.10). The Perl implementation is made up of two subroutines, Mail_Send (**lines 45–108**) and dialog (**lines 6–39**). The Mail_Send subroutine is the main entry point for sending e-mails, whereas the dialog is a support function that provides the ability to send commands and receive and verify responses from the SMTP server, thus simplifying the overall complexity of Mail_Send.

We discuss the dialog subroutine first, and then look at the Mail_Send subroutine that actually provides the meat of the SMTP client protocol.

Subroutine dialog (**lines 6–39**) provides the means to perform a single dialog with the SMTP server. Recall that SMTP is a command and response protocol, so dialog provides the mechanism to send a command and then conditionally await and check an associated response. Subroutine `dialog` is provided the client socket (`$sock`), the optional command to send (`$command`), and the optional response code that should be matched against the potential response (`$exp_resp`). The `shift` operator is used to gather the arguments from the `@ARGV` list (**lines 8–10**). The `$command` is tested at **line 13** to see if a command does exist, and if so, the **send** function is used to send this command through the client socket to the SMTP server (**line 15**). The `$exp_resp` is tested at **line 21** to see if the caller is expecting a response. This variable contains a three-digit SMTP status code that will be used to verify that the SMTP server accepted and processed our optional `$command` properly. If the caller expects a response, the response is read from the socket at **line 24** using the **recv** function. At **line 27**, we grab the first three digits from the line (which represent the status code from the SMTP server) and then at **line 31**, we test this with our `$exp_resp` variable. If they don't match (an error has occurred), we cause an exception using the `die` operator with a simple error message. Otherwise, we simply return and allow control to continue at the caller (`Mail_Send`).

The `Mail_Send` subroutine (**lines 45–108**) implements a very simple version of the SMTP client protocol. Six arguments are provided to `Mail_Send` that completely identify both the e-mail to send and the destination SMTP server (Mail Transfer Agent) to which the e-mail will be transported. The first argument (`$mail_server`) defines the destination address of the SMTP server to which we'll communicate (otherwise known as the Mail Transfer Agent, or MTA). The `$subject` is the subject line of the e-mail, and `$sender` and `$recipient` are the source and destination e-mail addresses, respectively. The `$content_type` is the string content type that is used to identify how the receiving e-mail client (Mail User Agent, or MUA) should render the body of the e-mail. If we're sending an e-mail that contains HTML tags, we'll specify a content type of "`text/html`"; otherwise, we'll use the default "`text/plain`". Finally, the `$contents` variable is used to specify the e-mail body itself. These arguments are shifted from the `@ARGV` list at **lines 48–53**.

At **line 56**, the client socket is created using a type of `Socket::SOCK_STREAM` (TCP, stream socket). In order to connect to the SMTP server, we must next create an address structure representing the address and port to which we'll connect. We first take the mail server address represented by `$mail_server` (**line 61**) and convert this into a binary, 32-bit address (`$addr`) using **inet_aton**. This new address, along with the SMTP port 25 is passed to **sockaddr_in** to create our packed address, `$paddr` (**line 62**). With this new packed address, we can connect to the remote SMTP server using the **connect** function (**line 65**).

The remainder of `Mail_Send` is a very simple version of the SMTP client protocol. The `dialog` subroutine is used to perform the command/response transactions with the SMTP server. As this is discussed in Chapter 15, Software Patterns Introduction, we discuss only a couple of dialogs here to illustrate the Perl functionality. At **line 68**, we await the SMTP salutation from the SMTP server. No command is sent, because the salutation should be emitted immediately after **connect**. Therefore, we provide no command, and simply await the "220" response status code. The next example, **line 71**, begins the client side of the session by issuing a `HELO` command. This command is used to identify us to the SMTP server. We send the `HELO` command with a bogus domain (should be the domain from which we connect), and expect back a status response code of "250".

The remainder of this function represents the SMTP client for sending a simple e-mail (**lines 73–103**). Recall from the dialog subroutine that if an incorrect status response code is received, the client will die, severing the connection to the server and, therefore, the SMTP session. If we successfully complete the dialogs in Mail_Send, we close the client socket at **line 106** using the **close** function.

The final element of the SMTP client is the sample client invocation at **lines 114–120**. We invoke the Mail_Send subroutine, specifying the mail_server and the e-mail that we want to send. Note at **lines 119 and 120**, the period ('.') is used to concatenate the contents string argument together.

Listing 19.10   Perl language SMTP client source.

```
1    use Socket;
2
3    #
4    #  Perform a dialog with the SMTP server
5    #
6    sub dialog {
7
8      $sock = shift;
9      $command = shift;
10     $exp_resp = shift;
11
12     # Send the command if the user provided one
13     if ( $command ne "" ) {
14
15       send( $sock, $command, 0);
16
17     }
18
19     # Only check for a response if the user defined the
20     # expected response code.
21     if ( $exp_resp ne "" ) {
22
23       # Get a line from the connection
24       recv( $sock, $line, 128, 0 );
25
26       # Parse the status code from the line
27       $stscode = substr( $line, 0, 3 );
28
29       # Check the status code from the server with the
30       # response code that's expected.
31       if ( $stscode ne $exp_resp ) {
32
33         die "Error sending mail";
34
```

```
35        }
36
37      }
38
39   }
40
41
42   #
43   #   Send the mail based upon the defined parameters
44   #
45   sub Mail_Send {
46
47      # Grab the arguments
48      $mail_server = shift;
49      $subject = shift;
50      $sender = shift;
51      $recipient = shift;
52      $content_type = shift;
53      $contents = shift;
54
55      # Create a stream socket
56      socket( SOCK, Socket::AF_INET,
57              Socket::SOCK_STREAM, 0 ) or
58        die "socket: $!";
59
60      # Create a packed address
61      $addr = inet_aton( $mail_server );
62      $paddr = sockaddr_in( 25, $addr );
63
64      # Connect to the server
65      connect( SOCK, $paddr ) or die "connect: $!";
66
67      # Look for the initial e-mail salutation
68      dialog( SOCK, "", "220" );
69
70      # Send HELO and await response
71      dialog( SOCK, "HELO thisdomain.com\n", "250" );
72
73      # Send "MAIL FROM" command and await response
74      $tempstr = "MAIL FROM: " . $sender . "\n";
75      dialog( SOCK, $tempstr, "250" );
76
77      # Send "RCPT TO" command and await response
78      $tempstr = "RCPT TO: " . $recipient . "\n";
79      dialog( SOCK, $tempstr, "250" );
80
81      # Send "DATA" to start the message body
```

```
82     dialog( SOCK, "DATA\n", "354" );
83
84     # Send out the mail headers (from / to / subject)
85     $tempstr = "From: " . $sender . "\n";
86     dialog( SOCK, $tempstr, "" );
87     $tempstr = "To: " . $recipient . "\n";
88     dialog( SOCK, $tempstr, "" );
89     $tempstr = "Subject: " . $subject . "\n";
90     dialog( SOCK, $tempstr, "" );
91
92     # Send the content type
93     $tempstr = "Content-Type: " . $content_type . "\n";
94     dialog( SOCK, $tempstr, "" );
95
96     # Send the actual message body
97     dialog( SOCK, $contents, "" );
98
99     # Send the end-of-email indicator
100    dialog( SOCK, "\n.\n", "250" );
101
102    # Finally, close out the session
103    dialog( SOCK, "QUIT\n", "221" );
104
105    # Close the client socket
106    close( SOCK );
107
108  }
109
110
111  #
112  #  Create a new HTTP server and start it on port 80
113  #
114  Mail_Send ("192.168.1.1",
115           "The Subject",
116           "tim\@mtjones.com",
117           "mtj\@mtjones.com",
118           "text/html",
119           "<HTML><BODY><H1>This is the mail" .
120           "</H1></BODY></HTML>" );
```

20 Network Code Patterns in Ruby

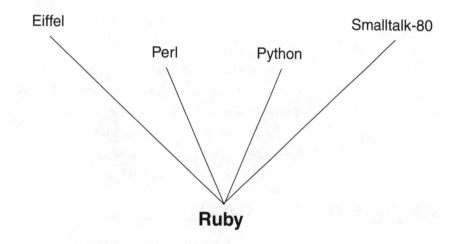

INTRODUCTION

In this chapter, we look at simple socket-based applications in Ruby using the BSD Sockets API and Ruby's derivative helper classes. These applications can serve as implemented software patterns for the construction of more advanced networking applications.

ON THE CD

All software discussed here is also provided on the companion CD-ROM. Table 20.1 lists the applications and their locations on the CD-ROM.

TABLE 20.1   CD-ROM companion software

Code Pattern	CD-ROM Location
Stream server/client	./software/ch20/stream/
Datagram server/client	./software/ch20/dgram/
Multicast server/client	./software/ch20/mcast/
Broadcast server/client	./software/ch20/bcast
Simple HTTP server	./software/ch20/shttp/
Simple SMTP client	./software/ch20/smtpc/

STREAM (TCP) SERVER/CLIENT

The stream server and client demonstrate the Daytime protocol using stream sockets (reliable TCP-based communication).

Stream Server

The Daytime protocol stream server is shown in Listing 20.1. Because all networking applications must have access to the socket classes, we first make the socket classes visible by including the socket library using require (**line 1**). At **line 2**, we create our server socket using the **TCPServer** class. We bind our new server to all interfaces (as we've left the interface spec out) and the daytime port (which will resolve to 13). We could achieve the same result by specifying the host as "0.0.0.0", which is synonymous with INADDR_ANY (from the C language perspective).

With our server socket defined in servsock, we emit some data for debugging purposes (to show the address of the server). The address printed will be "0.0.0.0", as discussed previously. Next, we enter the infinite loop awaiting connections from external client sockets.

To accept a client connection, we use the **accept** instance method. This method returns a socket instance representing our connection to the client (**line 13**). We again emit some debugging information to show that a client successfully connected, and some information about it (**lines 16 and 17**).

In **lines 20 and 21**, we emit the time through the client socket. Note the use of `Time::new`, which returns a string representing the current time. We anonymously create the time string (returned by `Time::new`) and send it to the client using the **write** method. Because the time string contains no newline, we follow with another **write** (**line 21**) to ensure that the client receives a newline for proper printing. Because we're now finished communicating with the client socket, we close it at **line 24**. This stops all communication with the client, except that any data remaining to be sent will be sent before the socket is physically closed.

Finally, at **line 26**, we terminate our `while` loop so processing begins again at **line 13**, awaiting a new client connection.

Listing 20.1   Ruby Daytime stream server.

```
1    require 'socket'
2
3    # Create a new TCP Server using port 13
4    servsock = TCPserver::new("daytime")
5
6    # Debug data -- emit the server socket info
7    print("server address : ",
8          servsock.addr::join(":"), "\n")
9
10   while true
11
12     # Await a connection from a client socket
13     clisock = servsock.accept
14
15     # Emit some debugging data on the peer
16     print("accepted ", clisock.peeraddr::join(":"), "\n")
17     print(clisock, " is accepted\n")
18
19     # Emit the time through the socket to the client
20     clisock.write( Time::new )
21     clisock.write( "\n" )
22
23     # Close the client connection
24     clisock.close
25
26   end
```

Before we move on to discuss the stream client, let's discuss the method used to emit the debugging data at **lines 8 and 16**. The instance methods **addr** and **peeraddr** return information about the local socket and peer socket, respectively, in array form. The purposes of the `join` method are to create a new string by joining the elements together of the initial string and to separate them by the string argument to `join` (in this case, a ':'). This permits us to view the contents of the given array. For example, **line 8** generates:

```
server address : AF_INET:13:0.0.0.0:0.0.0.0
```

The contents of the new array are "AF_INET" (the address family), 13 (the port), the resolved address ("0.0.0.0", which is actually unresolved), and, finally, the IP address attached to the interface ("0.0.0.0"). Recall that because we specified no host in the creation of the TCP server socket, it was bound to the wildcard address so that connections could be accepted on any available interface (including localhost).

Stream Client

Now, let's look at the Daytime stream client (shown in Listing 20.2). The client creates a socket to connect to the server and then awaits a string to be received through the socket. Upon receiving the string from the server, the message is printed and the client exits.

Listing 20.2   Ruby Daytime stream client.

```
1    require 'socket'
2
3    # Create a new client socket to the daytime port
4    mysock = TCPSocket::open("localhost", "daytime")
5
6    # Read a line from the socket
7    line = mysock.gets
8
9    # Print the line
10   print line
11
12   # Close the socket
13   mysock.close
```

We begin by making the socket library visible using the require method in **line 1**. Next, we create our stream using the **TCPSocket** class and the **open** method (**line 4**). We specify "localhost" as the host to connect, which is the loopback interface on the current host. We also specify the "daytime" port, which Ruby will resolve to 13. Recall that the **TCPSocket::open** method not only creates the client socket, but also connects it to the server as defined by the arguments. Therefore, when this method completes, we'll be ready to communicate with the server.

To receive the time string from the server, we use the **gets** method with our previously created **TCPsocket** instance (mysock). The result of the **gets** method is a single line of input from the socket, which is stored in the line variable (**line 7**). We emit this string at **line 10** using the print method and finally close the socket at **line 13** using the **close** method.

Before we leave stream clients, let's look at a simplified version of the Ruby stream client (shown in Listing 20.3). This client performs the socket creation followed by a **gets** method to retrieve the time string from the server. The entire line is preceded by the print method that emits the response from the socket methods.

Listing 20.3 Simplified Ruby Daytime stream client.

```
1    require 'socket'
2
3    # Short version -- open, read, close, and print in one line
4    print TCPSocket::open("localhost", "daytime")::gets
```

Scripts built with this style of method invocation are limited, but as is illustrated by Listing 20.3, can still be very useful.

DATAGRAM (UDP) SERVER/CLIENT

Now, let's look at a UDP server and client implementation using datagram sockets. Recall that datagram communication is not connection-oriented; therefore, each datagram that we send must also include the intended recipient.

Datagram Server

The datagram server is illustrated in Listing 20.4. After making the socket classes visible using the required method, we create our UDP socket using the **UDPSocket::new** method (**line 4**). We specify only that the UDP socket be of the AF_INET family. This is actually an optional argument, because the default is automatically Socket::AF_INET. The return of **UDPSocket::new** is the server socket.

In order to be visible to external clients, we name our UDP socket using the **bind** method (**line 7**). We bind to the "localhost" interface and the datagram port (13). Once bound, we emit the server address information using the **addr::join** method pair (as discussed previously with the stream server).

At **line 13**, we begin our server's infinite loop to accept client datagrams and emit to them the current time in string form. At **line 16**, we await a datagram from the client using the **recvfrom** method. This method is used because we need to know the source of the datagram in order to know where to send the time string. We specify that we're expecting a datagram of size 0 (no payload data) and we specify no flags. When a datagram does arrive, the **recvfrom** method returns two values. The first is the actual datagram payload (there won't be one in this example) and the second is the source of the datagram.

At **line 19**, we respond to the client using the **send** method. The **send** method used here is a variant of the traditional **send** using a number of optional parameters to specify the destination of our datagram. The first parameter in **send** is the string we're going to send (our datagram payload). This is made up of a time string returned by the Time::new::asctime method appended with a newline. The second parameter is the flags settings (0). The third and fourth parameters are the host and port to which the datagram will be sent. We use the previously received from parameter that contains the host in the third element (from[2]) and the port in the second element (from[1]).

Finally, we continue the infinite loop from **line 22** and await another client datagram at **line 16**.

Listing 20.4 Ruby Daytime datagram server.

```ruby
1   require 'socket'
2
3   # Create a new UDP Socket
4   servsock = UDPSocket::new(Socket::AF_INET)
5
6   # Bind the server socket to the daytime port
7   servsock.bind("localhost", "daytime")
8
9   # Debug data -- emit the server socket info
10  print("server address : ", servsock::addr::join(":"), "\n")
11
12  # The big loop
13  while true
14
15    # Receive a datagram from a client
16    reply, from = servsock.recvfrom(0, 0)
17
18    # Emit the time through the socket to the client
19    servsock.send( Time::new::asctime+"\n", 0,
20                      from[2], from[1] )
21
22  end
```

Datagram Client

The datagram client is shown in Listing 20.5, and corresponds with the datagram server previously shown in Listing 20.4.

Listing 20.5 Ruby Daytime datagram client.

```ruby
1   require 'socket'
2
3   # Create a new UDP Socket
4   clisock = UDPSocket::new(Socket::AF_INET)
5
6   # Send a zero-length datagram to the server on the daytime port
7   clisock.send("", 0, "localhost", "daytime")
8
9   # Await receipt of the time from the server
10  print clisock.recv(100)
11
12  # Close the socket
13  clisock.close
```

For the client, we create our datagram socket at **line 4** using the **UDPSocket::new** method. Though the default, we specify the address family for the socket as `Socket::AF_INET`. At **line 7**, we send our empty datagram using the **send** method to let the server know that we'd like to receive a time string. This datagram serves only to notify the server of our address for the return datagram containing the time. We specify four parameters to the **send** method, an empty string representing the zero-length payload, optional flags, destination for the datagram (in this case, the "`localhost`" interface), and the port ("`daytime`", or 13).

We then await the response datagram containing the time string at **line 10**. We use the **recv** call here instead of **recvfrom** because we're not interested in the source address information. We print the string out immediately upon return rather than storing it first in a temporary string. Finally, at **line 13**, we close the socket using the **close** method.

MULTICAST SERVER/CLIENT

The multicast server and client are fundamentally derivatives of the datagram code patterns. This is primarily because multicast communication is based upon the datagram model.

Multicast Server

The Ruby language source code for the multicast server is shown in Listing 20.6. Recall that multicast communication is group-based; therefore, a single message emitted by the server is received by every client that is currently a member of the multicast group.

As multicast communication has been discussed in previous chapters, we'll forgo discussion of multicast specifics and concentrate solely on the methods that Ruby provides to achieve multicast communication. To set up our multicast server (**lines 4 and 5**), we first define the multicast group to which we belong and the port (in this case, "`239.0.0.2`", 45002). Clients wanting to communicate with the server must also utilize this same group and port combination. Next, we create a typical datagram socket, at **line 8**, using the **UDPSocket::new** class method. We utilize the `SO_REUSEADDR` socket option (**lines 11 and 12**) in order to ensure that other clients (on the same host) can bind to this same group and port combination. Then at **line 16**, using the previously defined port number, we bind to the port number and `Socket::INADDR_ANY` (the wildcard interface).

Next, we enter the big loop, starting at **line 22**. Within this loop, we simply emit the current time at one second intervals to the multicast group (**lines 25–26**). Any clients having membership to this group and port will receive this time string. We emit the time string as would be done with any datagram server, specifying the string to send (`Time::new:asctime`), any flags, and, finally, the destination IP address and port (in this case, the multicast group and port number). The `sleep` method is used to wait for one second, and the process begins again by sending another time-string datagram. Finally, at **line 34**, we close the server socket using the **close** method.

Listing 20.6 Ruby Daytime multicast server.

```
1   require 'socket'
2
3   # Multicast Group and Port
4   mcast_group = "239.0.0.2"
5   group_port = 45002
6
7   # Create a new UDP Socket
8   servsock = UDPSocket::new(Socket::AF_INET)
9
10  # Make the address/port association immediately reusable
11  servsock.setsockopt( Socket::SOL_SOCKET,
12                       Socket::SO_REUSEADDR, 1 )
13
14  # Bind the server socket to the daytime port and
15  # any interface
16  servsock.bind( Socket::INADDR_ANY, group_port )
17
18  # Debug data -- emit the server socket info
19  print("server address : ", servsock.addr::join(":"), "\n")
20
21  # The big loop
22  while true
23
24    # Emit the time through the socket to the client
25    servsock::send( Time::new::asctime+"\n", 0,
26                    mcast_group, group_port )
27
28    # Wait one second
29    sleep 1
30
31  end
32
33  # Close the server socket
34  servsock.close
```

Multicast Client

The Ruby source code for the multicast client is shown in Listing 20.7. The multicast client includes a number of new features not previously discussed, but reflects symmetry with the multicast server, as seen with other code patterns.

After creating our UDP socket at **line 8** (using the **UDPSocket::new** method), we make the address/port pair immediately reusable with the SO_REUSEADDR socket option (**lines 11 and 12**). At **line 16**, we use the **bind** method to permit our socket to accept datagrams from any port, and our previously defined multicast port (specified at **line 5**).

The next step is for the server to physically join the multicast group. This is performed using a socket option (IP_ADD_MEMBERSHIP), but before doing this, we must create a special address structure. The first step is creating a binary image of the IP address for which we want to join. Recall that our multicast group is "239.0.0.2", so we create a binary array using the Array class and the pack method to create the binary equivalent (**line 19**). **Lines 20 and 21** append another binary string representing the IP address of the current host. We first identify the host name for the host on which we're executing using the **Socket::gethostname** method. This is used as an argument to **Socket::gethostbyname**, which communicates with the local resolver to resolve the host name to an IP address. Therefore, the variable mreq will represent a binary string containing eight bytes, four of the multicast group and four of the current host's IP address. We then use this structure at **lines 22 and 23** to actually join the multicast group using the **setsockopt** method.

Now that we've subscribed to the multicast group, we use the **recv** method at **line 26** to receive a datagram from the group. Note that we use **recv** rather than **recvfrom** because we're not interested in the source of the datagram (we know from whom it's coming). We immediately emit the response datagram using the print method.

Finally, before we exit the client, we must remove ourselves from the multicast group. We use the **setsockopt** method again with IP_DROP_MEMBERSHIP to remove ourselves from the group (**lines 29 and 30**). The client socket is then closed using the **close** method (**line 33**).

Listing 20.7 Ruby Daytime multicast client.

```
1    require 'socket'
2
3    # Multicast Group and Port
4    mcast_group = "239.0.0.2"
5    group_port = 45002
6
7    # Create a new UDP Socket
8    clisock = UDPSocket::new(Socket::AF_INET)
9
10   # Make the address/port association immediately reusable
11   clisock.setsockopt( Socket::SOL_SOCKET,
12                       Socket::SO_REUSEADDR, 1 )
13
14   # Bind the client socket to the daytime port and
15   # any interface
16   clisock.bind( Socket::INADDR_ANY, group_port )
17
18   # Join the multicast group
19   address = [239, 0, 0, 2].pack('CCCC')
20   mreq = address +
21         Socket::gethostbyname(Socket::gethostname)[3]
22   clisock.setsockopt( Socket::IPPROTO_IP,
23                       Socket::IP_ADD_MEMBERSHIP, mreq )
24
25   # Await receipt of the time from the server
```

```
26    print clisock.recv(100)
27
28    # Leave the multicast group
29    clisock.setsockopt( Socket::IPPROTO_IP,
30                            Socket::IP_DROP_MEMBERSHIP, mreq )
31
32    # Close the client socket
33    clisock.close
```

BROADCAST SERVER/CLIENT

The broadcast server and client, like the multicast server and client, are fundamentally derivatives of the datagram code patterns. This is again because broadcast communication is based upon the datagram model.

Broadcast Server

The Ruby source code for the broadcast server is shown in Listing 20.8. The server is broadcast-based, which means a single message emitted by the server is received by every client that is configured to receive broadcast datagrams (for the given port).

Broadcast communication is datagram-specific, and, therefore, we begin at **line 4** by creating a datagram socket using **UDPSocket::new**. In order to send broadcast datagrams, we use the **setsockopt** method and enable the SO_BROADCAST socket option (**lines 8 and 9**). For debugging purposes only, we emit the server address information at **lines 11 and 12**.

We begin our infinite loop at **line 15**, where the broadcast server will emit broadcast datagrams once per second. We begin at **line 18** by creating a new time string using the Time::new::asctime method. This returns a string representing a human-readable date and time string, to which we append a newline. We then use the **send** method at **line 21** to emit our datagram. We specify our data (ts), our flags (0), the limited broadcast address ("255.255.255.255"), and our chosen broadcast port, 45003. Note that this is the limited broadcast address because it is limited to the local LAN and won't be forwarded. Finally, at **line 24**, we sleep for one second and then start the time emission process all over again.

Listing 20.8   Ruby Daytime broadcast server.

```
1    require 'socket'
2
3    # Create a new UDP Socket
4    servsock = UDPSocket::new(Socket::AF_INET)
5
6    # Allow sending of broadcast datagrams.
7    servsock::setsockopt( Socket::SOL_SOCKET,
8                            Socket::SO_BROADCAST, 1 )
9
```

```
10    # Debug data -- emit the server socket info
11    print("server address : ",
12          servsock.addr::join(":"), "\n")
13
14    # The big loop
15    while true
16
17      # Create a time string with the current time
18      ts = Time::new::asctime + "\n"
19
20      # Emit the time through the socket to the client
21      servsock.send( ts, 0, "255.255.255.255", 45003 )
22
23      # Sleep for 1 seconds
24      sleep 1
25
26    end
```

Broadcast Client

The Ruby source code for the broadcast client is shown in Listing 20.9. As with the broad-cast server, we concentrate on the differences to the standard datagram client.

We begin again by creating our datagram socket at **line 4** and then bind the broadcast address and port to it using the **bind** method at **line 7**. We bind our client socket to the broadcast address and port because the client must receive broadcast datagrams. Note that the server does not use **bind**. If the server were to receive broadcast datagrams, it would be required to bind as the client does. At **line 10**, we receive the broadcast datagram using the **recvfrom** method. We could have used the **recv** method here to simplify the client. At **line 13**, we emit the time string received by the server using the print method and then, finally, at **line 16**, we close the client socket using the **close** method.

Listing 20.9   Ruby Daytime broadcast client.

```
1    require 'socket'
2
3    # Create a new UDP Socket
4    clisock = UDPSocket::new(Socket::AF_INET)
5
6    # Bind the server socket to the daytime port
7    clisock.bind("255.255.255.255", 45003)
8
9    # Receive a datagram through the socket
10   reply, from = clisock.recvfrom(100, 0)
11
12   # Print the received datagram payload
13   print reply
```

```
14
15    # Close the client socket
16    clisock.close
```

SIMPLE HTTP SERVER

Now, let's look at a simple HTTP server written in Ruby. The entire source listing for the Ruby simple HTTP server is provided in Listing 20.10. Let's now walk through this listing to understand the sample implementation.

Listing 20.10 Ruby simple HTTP server source.

```
1    require 'socket'
2
3    class Simple_http_server
4
5      #
6      #  Determine the content type based upon the filename
7      #
8      def define_content_type( filename )
9
10       fileparts = filename.split(".")
11
12       case fileparts[1]
13         when "html","htm", "HTML", "HTM"
14           return "text/html"
15         when "txt", "TXT"
16           return "text/plain"
17         when "rb"
18           return "text/plain"
19         else
20           return "application/octet-stream"
21       end
22
23     end
24     protected:define_content_type
25
26
27     #
28     #  Emit the standard HTTP response message header
29     #
30     def emit_response_header( sock, content_type )
31
32       sock.write( "HTTP/1.1 200 OK\n" )
33       sock.write( "Server: Ruby shttp\n" )
```

```
34        sock.write( "Connection: close\n" )
35        sock.write( "Content-Type: " )
36        sock.write( content_type )
37        sock.write( "\n\n" )
38
39      end
40      protected:emit_response_header
41
42
43      #
44      # HTTP 'GET' Method Handler
45      #
46      def handle_get_method( sock, filename )
47
48        filename = filename.delete("/")
49
50        if File::exist? filename then
51
52          print "Requested filename is "
53          print filename + "\n"
54
55          content_type = define_content_type( filename )
56
57          emit_response_header( sock, content_type )
58
59          # Emit the file through the socket
60          file = open( filename, "r" )
61          sock.write( file.read )
62          file.close
63          sock.write( "\n" )
64
65        else
66
67          print "File not found.\n"
68
69          # Unknown file -- notify client
70          sock.write( "HTTP/1.0 404\n\n File not found\n\n" )
71
72        end
73
74      end
75      protected:handle_get_method
76
77
78      #
79      # HTTP Connection Handler
80      #
```

```ruby
81    def handle_connection( sock )
82
83       print "Handling new connection\n"
84
85       # Grab a line from the client
86       line = sock.gets
87
88       # Removing any leading/trailing whitespace and
89       # trailing newline
90       line.strip!.chomp!
91
92       # Split the line into three elements
93       # (command, filename, version)
94       elements = line.split
95
96       # Check the request -- we handle only GET requests
97       if elements[0] =~ "GET"
98
99          # Call our GET method handler
100         handle_get_method( sock, elements[1] )
101
102      else
103
104         print "Unknown Method " + elements[0] + "\n"
105
106         # Unknown method -- notify client
107         sock.write("HTTP/1.0 501 Unimplemented Method\n\n")
108
109      end
110
111   end # handle_connection
112   protected:handle_connection
113
114
115   #
116   # Initialization function for the simple HTTP server
117   #
118   def start( port = 80 )
119
120      # Create a new TCP Server using port 80
121      servsock = TCPserver::new( port )
122
123      # Make the port immediately reusable after this
124      # socket is closed
125      servsock.setsockopt( Socket::SOL_SOCKET,
126                           Socket::SO_REUSEADDR, 1 )
127
```

```
128        # Debug data -- emit the server socket info
129        print("server address : ",
130              servsock.addr::join(":"), "\n\n")
131
132        # The big loop
133        while true
134
135          # Await a connection from a client socket
136          clisock = servsock.accept
137
138          # Emit some debugging data on the peer
139          print("Accepted new connection",
140                clisock.peeraddr::join(":"), "\n")
141
142          # handle the new connection
143          self::handle_connection( clisock )
144
145          print "\n"
146
147          # Close the client connection
148          clisock.close
149
150        end
151
152      end # start
153
154    end
```

This implementation of the simple HTTP server is based upon a simple class that provides the HTTP server functionality. The only visible method is the start method, which starts a new HTTP server using a given port. Because the start method is the entry point to our simple HTTP server class, we'll start here first.

The start method accepts a single argument, the port number to use for the HTTP server (**line 118**). If the caller provides no port number, the default port number of **80** is used. Next, as with all stream servers, we create a new TCP server socket using the **TCPServer::new** method. We provide our port variable, which would have been defaulted if not specified. With our new server socket, stored in servsock, we make the address/port combination reusable using the SO_REUSEADDR socket option (**lines 125 and 126**). Our server socket is now ready to accept connections, so we emit some debugging data using the **addr** method to identify some information about the server.

At **line 133**, we start our server loop. This infinite loop awaits and then handles incoming HTTP sessions. At **line 136**, we accept new client connections using the **accept** method and the server socket (servsock). When a client connects, a new socket will be returned representing the socket to use for this session. This will be stored in clisock. To identify that a new client connection has been received, we use the **peeraddr** function. This returns information identifying the peer end of the new client socket, which we emit to the display. Next, at **line 143**, we call the **handle_connection** method to service the given HTTP request. This method is a protected method within our Simple_http_server class. After the session has been serviced using handle_connection, we close the client socket at **line 148** using the **close** method and start the server loop again awaiting a client connection with the **accept** method.

The handle_connection method of our class is responsible for handling the client's HTTP request (**lines 81–112**). We begin getting a line of text from the socket, which represents the actual request (the first line defines the HTTP message type and file of interest). The gets method is used on the client socket and results in a string being stored in line. We use the strip! to remove leading and trailing whitespace from the line (**line 90**). We then use the chomp! method on line to remove the trailing record separator (in this case, '\n'). Certain Ruby methods utilize the '!' modifier, which operates directly on the particular variable. In the case of strip and chomp, the given variable is modified instead of returning a new variable.

At **line 94**, we use the split method to split the line into N separate strings. The new string array is stored in variable elements. At **line 97**, we check the first element to see that it's the 'GET' string (representing the HTTP GET method). In this simple HTTP server, we handle only the GET method. If we have received a GET request from the client, we use the method handle_get_method, to service the actual client request. If the server has received some other type of request, we specify an error to the client using the **write** method at **line 107**. We emit a standard HTTP error, which is split into three parts, the HTTP version number, the error code, and a human-readable string identifying the error. The particular error we emit here is the unimplemented error, or 501.

One final point to note about the method is **line 112**. The protected:handle_connection specifies that the handle_connection method is protected to this class (Simple_http_server). Therefore, the method can only be invoked by methods within this class, and not by external means. The only method that isn't protected within the Simple_http_server class is the start method, which is used to start the server.

Method handle_get_method, called from handle_connection, is invoked only when a request is received containing the HTTP GET method (**lines 46–75**). Method handle_get_method is called with two arguments representing the client socket and the filename (parsed from the GET request). We begin by stripping the '/' character, which will be present at the beginning of the filename (**line 48**). This character specifies the root file system (directory portion of the filename), but isn't useful when we're trying to locate the file using file access methods. We test whether the file actually exists using the File::exist? method at **line 50**. If the exist? method returns false, we emit another HTTP response method using the **write** method. This method is error code 404, or file-not-found (**line 70**).

If the file is found (in the directory where the server was started), we identify the content type of the file using the internal method define_content_type (at **line 55**). The string response from define_content_type is stored in the local variable content_type. We next emit an HTTP response header using emit_response_header, specifying our client socket and the content type that we're serving (**line 57**).

We emit the file following the HTTP response header in **lines 60–63**. We open the file for read, storing a file descriptor in the `file` variable. Next, we use the socket **write** method to emit the contents of the file returned by `file.read` (**line 61**). The entire file will be read and emitted by this single line. Next, we close the file using the **close** method and then write a final newline to the socket using the **write** method at **line 63**.

Let's now look at the final two methods that are used by the `Simple_http_server`. At **lines 30–40**, the HTTP response message header is served through the socket. This consists of a basic number of message headers, including the status response (first line) and the content type being served (through the previously created `content_type` variable). The **write** method is used in each case to emit the headers. Note that in the end, two newlines are emitted, which are required by HTTP to separate the HTTP message header from the message body (a blank line must separate them).

The `define_content_type` method is used to identify the content type of the content requested based upon the suffix of the `filename` (**lines 8–24**). We split the `filename` into two parts at **line 10** using the `split` method. The first element of the array will contain the filename and the second element will contain the file extension (suffix). **Lines 12–21** provide a `case` statement to return a string based upon the file extension found. The string that's returned is used solely in the construction of the response header. The `content_type` informs the client how it should treat the response message body.

That completes the source for the `Simple_http_server` class. Now, let's look at how a user might use the class. Listing 20.11 provides a sample illustration of its use. At **line 5**, we create a new HTTP server using the new method (on the `Simple_http_server` class). We then call the `start` method of our new class instance and specify that it should occupy port 80 (**line 9**). That's it! In a threaded design, we could specify another HTTP server by simply instantiating the `Simple_http_server` to another class instance, and start it on another port.

Listing 20.11   Instantiating a new HTTP server.

```
1    #
2    # Create a new HTTP server and start it on port 80
3    #
4
5    server = Simple_http_server::new
6
7    print "Starting new HTTP server on port 80\n"
8
9    server.start 80
```

SIMPLE SMTP CLIENT

Now, let's look at a simple SMTP client written in Ruby. The SMTP client is implemented as a class that provides a set of methods to send an e-mail to a specified recipient (see Listing 20.12).

Four methods are provided by the `Mailer` class for e-mail transport. These are `init`, `dialog`, `send`, and `finish`. Only method `dialog` is hidden in the class (a protected method), and is utilized by the `Mailer` class to implement the actual SMTP protocol. We investigate each of these methods in their order of appearance within a sample use. One item to note before the methods is the creation of global class variables. In this case (**lines 8–13**), we define a set of instance variables and initialize them each to nil. We know these to be instance variables because each is preceded by a '@' character. These variables are global to the class, and retain their value statically between method calls. This is important because we want to be able to define the socket variable within the `init` method, and still have it available when we later call the **send** method. Other types of Ruby variables include class variables, preceded by "@@" and pure global variables, preceded by "$".

Listing 20.12 Ruby language SMTP client source.

```
1    require 'socket'
2
3    class Mailer
4
5      #
6      # Globals defined within the init method.
7      #
8      @subject = nil
9      @sender = nil
10     @recipient = nil
11     @content_type = nil
12     @contents = nil
13     @sock = 0
14
15
16     #
17     # Initialize the internal mail variables
18     #
19     def init( subject, sender, recipient,
20               content_type, contents )
21
22       @subject = subject
23       @sender = sender
24       @recipient = recipient
25       @content_type = content_type
26       @contents = contents
27
28     end
29
30
31     #
32     # Perform a dialog with the SMTP server
```

```
33     #
34     def dialog( command, expected_response )
35
36        # Only send the command if the caller provided one.
37        if command != nil
38
39           @sock.write( command )
40
41        end
42
43        # Only check for a response if the user defined the
44        # expected response code.
45        if expected_response != nil
46
47           # Get a single line from the connection
48           line = @sock.gets
49
50           # Check the received line with the expected
51           # response code
52           if line[0..expected_response.length-1] !=
53              expected_response
54
55              # Not what was expected, raise an exception.
56              raise
57
58           end
59
60        end
61
62     end
63     protected:dialog
64
65
66     #
67     # Send the mail based upon the initialized parameters
68     #
69     def send
70
71        mail_server = @recipient.split('@')[1]
72
73        @sock = TCPSocket::open(mail_server, "mail")
74
75        # Look for initial e-mail salutation
76        self::dialog( nil, "220" )
77
78        # Send HELO and await response
79        self::dialog( "HELO thisdomain.com\n", "250" )
80
```

```
81      # Send "MAIL FROM" command and await response
82      string = "MAIL FROM: " + @sender + "\n"
83      self::dialog( string, "250" )
84
85      # Send "RCPT TO" command and await response
86      string = "RCPT TO: " + @recipient + "\n"
87      self::dialog( string, "250" )
88
89      # Send "DATA" to start the message body
90      self::dialog( "DATA\n", "354" )
91
92      # Send out the mail headers (from/to/subject)
93      string = "From: " + @sender + "\n"
94      self::dialog( string, nil )
95
96      string = "To: " + @recipient + "\n"
97      self::dialog( string, nil )
98
99      string = "Subject: " + @subject + "\n"
100     self::dialog( string, nil )
101
102     # Send the content type
103     string = "Content-Type: " + @content_type + "\n"
104     self::dialog( string, nil )
105
106     # Send the actual message body
107     self::dialog( @contents, nil )
108
109     # Send the end-of-email indicator
110     self::dialog( "\n.\n", "250" )
111
112     # Finally, close out the session
113     self::dialog( "QUIT\n", "221" )
114
115   rescue
116
117     puts "Couldn't send mail"
118     @sock.close
119
120   end
121
122
123   def finish
124
125     @sock.close
126
127   end
```

```
128
129
130    end
```

The first function used by mail users is the `init` method. The application will create an instance of the `Mailer` class using `Mailer::new`. With the new instance, the `init` method is called to specify the `subject`, `sender`, `recipient`, `content_type`, and message body (contents). As shown in **lines 19–28**, these variables are copied from the application into the class instance variables and are bound to this instance of the class.

The send method is the next method invoked by the user (**lines 69–120**). This method implements SMTP and performs all communication with the SMTP server. The first step, at **line 71**, is to extract the domain name to which we're connecting (the target mail server). This is taken from the `recipient` mail address using the `split` method, using the '@' character to split the `recipient` username from the target domain name. For example, "`mtj@mtjones.com`" is split into an array of two elements, "`mtj`" and "`mtjones.com`". Note that at **line 71**, we use only the second element, and store this to the `mail_server` local variable. Variable `mail_server` is then used at **line 73** to create the TCP socket and connect it to the server using **TCPSocket::open**. The remainder of the **send** method is SMTP-specific communication to the mail server. This was discussed previously in Chapter 15, Software Patterns Introduction.

One item to note in the **send** method is the use of the `rescue` statement. If a method within the **send** method's calls-chain raises an error (using the `raise` statement), then execution continues immediately at the `rescue` section. In this section (at **line 115**), we close the socket to end the session using the **close** method.

The `dialog` method (**lines 34–63**) is a protected method that performs the communication transactions with the SMTP server. Recall that SMTP commanding is command/response-based. If the caller provides a `command`, it is communicated to the SMTP server using the **write** method (**lines 37–41**). If a `response` is expected (**lines 45–60**), then it is retrieved using the **gets** method. The `dialog` method also checks the `response` from the expected numeric code (**lines 52 and 53**) as defined by the caller (`expected_response`), and if a different code is found, an exception is raised (**line 56**). Otherwise, the `dialog` method returns and the **send** method continues with the SMTP transactions.

The final method in the `Mailer` class (see Listing 20.12) is the `finish` method. This method is a cleanup method that closes the client socket using the **close** method. The `finish` method is the final method to be called once the SMTP session is complete.

Let's now look at a sample usage of the `Mailer` class. In Listing 20.13, a simple application that utilizes the `Mailer` class to send an e-mail is shown. At **line 4**, we create a new instance of the `Mailer` class using `Mailer::new`. We store this new instance in a variable called `mail`. At **lines 6–10**, we call `init` and initialize the instance variables with our set of values to prepare for sending a new e-mail. We specify (in order) the `subject`, `sender`, `recipient`, `content_type`, and `contents`. Note that in this instance, we specify the content_type as "`text/html`". Note at **line 10**, we specify an HTML-encoded e-mail. Therefore, the content_type must be "`text/html`" to permit the receiving mail client to properly render the e-mail. If we were sending pure text rather than HTML, we'd use "text/plain" as the content_type. At **line 11**, we call the **send** method to send the e-mail, and, finally, at **line 13**, we close out the SMTP session by calling the finish method.

Listing 20.13   Instantiating a new SMTP client.

```
1    #
2    # Create a new mailer and send an e-mail
3    #
4    mail = Mailer::new
5
6    mail.init( "The Subject",
7              "tim@mtjones.com",
8              "mtj@mtjones.com",
9              "text/html",
10             "<HTML><BODY><H1>:Your mail</H1></BODY></HTML>" )
11   mail.send
12
13   mail.finish
```

21 Network Code Patterns in Tcl

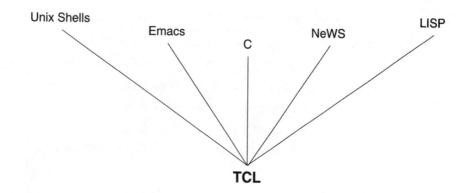

Unix Shells

Emacs

C

NeWS

LISP

TCL

INTRODUCTION

In this chapter, we look at simple socket-based applications in Tcl. These applications can serve as implemented software patterns for the construction of more advanced networking applications.

All software discussed here is also provided on the companion CD-ROM. Table 21.1 lists the applications and their locations on the CD-ROM.

TABLE 21.1 CD-ROM companion software

Code Pattern	CD-ROM Location
Stream server/client	./software/ch21/stream/
Datagram server/client	./software/ch21/dgram/
Multicast server/client	./software/ch21/mcast/
Broadcast server/client	./software/ch21/bcast
Simple HTTP server	./software/ch21/shttp/
Simple SMTP client	./software/ch21/smtpc/

STREAM (TCP) SERVER/CLIENT

The stream server and client demonstrate the Daytime protocol using stream sockets (reliable TCP-based communication).

Stream Server

The Daytime protocol stream server is shown in Listing 21.1. We begin the stream Daytime server with the Stream_Server procedure (**lines 4–13**). The procedure begins with the declaration and the requirement of a port argument (the port number on which the server should be bound on the current host). At **line 8**, a new socket is created as a server (-server) with the definition of a callback procedure ServerAccept to call for accepting new client connections. We also present the port number that was passed in by the caller. The new socket is stored into the s variable, but as can be noted, this variable will no longer be referenced. The final line in the Stream_Server procedure is the vwait command. This command invokes the Tcl event loop, which is used to await new events from the underlying system. The vwait command references a variable, which upon change will cause a return and a continuance. The forever variable is not declared (and will never be changed), so this procedure stalls at this line indefinitely.

The event loop (as invoked by the vwait call in Listing 21.1) is required in this application for proper operation (under Tcl). If this application had utilized the wish interpreter, the event loop would have been automatic, and, therefore, the vwait command would not have been necessary.

The ServerAccept procedure (**lines 19–40**) is the callback for the server socket that is called when an incoming connection arrives. The callback includes the new client socket (created as with the standard BSD **accept** function), the address of the client that was accepted (addr), and the port of the remote client socket (port). The first task is to emit some debugging data identifying the new client socket and from where the client is connecting (**line 22**). Next, the client socket is configured for buffering. This allows us to send data without having to perform a flush operation and so that data for read is available right away. We specify the socket for **fconfigure** and the type of configuration we're requesting—in this case line, buffering (**line 26**). In **lines 29 and 32**, we create our date and time string by first reading the current clock value in seconds (using the clock command with parameter seconds) and storing this in clock_val (**line 29**). Then, we convert this to a string using the clock command with the option format and the time value stored in clock_val (**line 32**). Finally, we emit the newly created date and time string (date_str) at **line 35** using the **puts** command referencing the client socket and the date_str and then close the socket at **line 38** using the **close** command.

The stream server is started at **line 46** by invoking the newly created Stream_Server procedure, specifying the port number on which we want to bind (in this case, port 13).

Listing 21.1   Tcl Daytime stream server.

```
 1   #
 2   # Stream server setup
 3   #
 4   proc Stream_Server { port } {
 5
 6      # Create the server socket and set the accept
 7      # event callback.
 8      set s [socket -server ServerAccept $port]
 9
10      # Wait for an event to occur
11      vwait forever
12
13   }
14
15
16   #
17   # Event callback for the socket accept
18   #
19   proc ServerAccept { sock addr port } {
20
```

```
21      # Emit some debugging information
22      puts "Accept $sock from $addr port $port"
23
24      # Configure the socket so that each puts results
25      # in a socket send.
26      fconfigure $sock -buffering line
27
28      # Get the current time in seconds
29      set clock_val [ clock seconds ]
30
31      # Convert the seconds time into string
32      set date_str [ clock format $clock_val ]
33
34      # Emit the date/time string to the socket
35      puts $sock $date_str
36
37      # Close the socket
38      close $sock
39
40    }
41
42
43  #
44  # Start the stream server on port 13
45  #
46    Stream_Server 13
```

Stream Client

Now, let's look at the Daytime stream client (shown in Listing 21.2). The client creates a socket to connect to the server and then awaits a string to be received through the socket. Upon receiving the string from the server, the message is printed and the client exits.

The client activity in Tcl is very simple, as is illustrated in Listing 21.2. First, the client socket is created at **line 2** using the **socket** command, specifying the host to which we'll connect (in this case, localhost) and the port (13). The new client socket is then stored in variable sock. At **line 5**, we read a line from the socket using the **gets** command, specifying the client socket and storing the result in line. Finally, at **line 8**, we emit the line read from the socket to standard-out using the **puts** command.

Listing 21.2   Tcl Daytime stream client.

```
1    # Create the client socket and connect it to the server
2    set sock [socket localhost 13]
3
4    # Receive a line of text through the socket
5    gets $sock line
6
```

```
7   # Emit the line of text
8   puts $line
```

Another potential solution to the Daytime stream client using the Tcl-DP package is shown in Listing 21.3. Although no simpler than the version using the standard Tcl libraries, it represents another way to handle streams.

Listing 21.3   Tcl Daytime stream client using Tcl-DP.

```
 1   package require dp 4.0
 2
 3   # Connect via TCP to port 13 of the local host
 4   set chan [dp_connect tcp -host localhost -port 13]
 5
 6   # Read a line from the channel
 7   set line [dp_recv $chan]
 8
 9   # Emit the line
10   puts $line
```

In Listing 21.3, **line 1** begins by importing the package **dp**. The package command specifies that the following script requires the package (via the require keyword), the package to be imported (**dp**), and, finally, the version of the package that is desired (in this case, 4.0). **Line 4** uses the **dp_connect** command to connect to the defined host and port using the stream protocol, TCP. The result is a new channel for the newly connected socket, which we store in chan. At **line 7**, we read a line from the socket using the **dp_recv** command, specifying the channel from which we'll read (chan) and the variable to which we'll store the result (line). Finally, we emit the line read from the socket to standard-out using the **puts** command.

DATAGRAM (UDP) CLIENT

Now, let's look at a UDP client implementation using datagram sockets. This implementation uses the Tcl-DP package that can be used to simplify the development of datagram-based applications. Given that it's not possible to identify the source of a datagram through Tcl (nor the Tcl-DP package), the datagram server cannot be written in Tcl without modifications to the Tcl interpreter. The client will be presented, which successfully works with other datagram servers implemented in other languages in this book.

Datagram Client

The datagram client is illustrated in Listing 21.4. We begin by importing our **dp** (Tcl-DP package) for the script using the package require command. Next, we create a procedure for the datagram client at **line 3**, calling the new procedure Datagram_Client. This procedure takes two arguments, the address and port to which we want the client to connect (represented by addr and port).

Within Datagram_Client, we create a new datagram socket and connect it to the requested peer using the **dp_connect** command (**line 7**). Note that we specify the protocol desired (udp) as well as the host and port arguments passed in by the caller. The newly created datagram socket is stored in the variable dsock. Using the **fconfigure** command, we configure the new socket for line buffering, which means that we'll read or write lines of text as soon as they're available (**line 10**). In order to identify ourselves to the remote datagram server, we send a dummy datagram packet to it, which is performed here using the **dp_send** command, referencing our socket (dsock) and the data to send (here, an 'a' character). After the server receives our dummy datagram, it will send us a date/time string in a new datagram packet, using the source address of the previously received packet as the destination of the new datagram. We read this new datagram using the **dp_recv** command, storing the result in variable line (**line 16**). Finally, we emit the received line to standard-out using the puts command (**line 19**) and close the socket using the **close** command at **line 22**.

The final task for our datagram client is to start it. We call the Datagram_Client at **line 31** specifying the host to which we want to connect (localhost) and the port number (13).

Listing 21.4   Tcl Daytime datagram client.

```
1    package require dp 4.0
2
3    proc Datagram_Client { addr port } {
4
5      # Create a UDP client socket and connect to the defined
6      # host and port.
7      set dsock [dp_connect udp -host $addr -port $port]
8
9      # Configure for immediate return of data
10     fconfigure $dsock -buffering line
11
12     # Send a dummy packet to the peer
13     dp_send $dsock "a"
14
15     # Read a line from the channel
16     set line [dp_recv $dsock]
17
18     # Emit the line
19     puts $line
20
21     # Close the client socket
22     close $dsock
23
24   }
25
26
```

```
27   #
28   # Start the datagram client
29   #
30
31   Datagram_Client localhost 13
```

MULTICAST SERVER/CLIENT

The multicast server and client are basic derivatives of the datagram code patterns, using the Tcl-DP package. This is primarily because multicast communication is based upon the datagram model and because Tcl does not support multicast (or datagrams) natively.

Multicast Server

The Tcl language source code for the multicast server is shown in Listing 21.5. Recall that multicast communication is group-based; therefore, a single message emitted by the server is received by every client that is currently a member of the multicast group.

Listing 21.5   Tcl Daytime multicast server.

```
1    package require dp 4.0
2
3
4    proc sleep { value } {
5      set temp 0
6      after $value {set temp 1}
7      vwait temp
8    }
9
10
11   #
12   #  Multicast Server setup
13   #
14   proc Multicast_Server { addr port } {
15
16     # Create the multicast socket
17     set mcsock [dp_connect ipm -group $addr -myport $port]
18
19     # Configure the socket so that each puts results
20     # in a socket send.
21     fconfigure $mcsock -buffering line
22
23     while (1) {
24
25       # Get the current time in seconds
26       set clock_val [ clock seconds ]
```

```
27
28          # Convert the seconds time into string
29          set date_str [ clock format $clock_val ]
30
31          # Emit the date/time string to the socket
32          puts $mcsock $date_str
33
34          sleep 1000
35
36      }
37
38  }
39
40
41  #
42  # Start the multicast server
43  #
44
45  Multicast_Server 239.0.0.2 45002
```

The multicast server, like the datagram client, is written as a simple procedure that is called with the address and port for which to register as a server. At **line 1**, we import the Tcl-DP package so that we can create multicast datagram sockets.

A sleep procedure is constructed using the after and vwait commands. The after command specifies that we'll delay the number of milliseconds specified in the input value to the procedure, and upon completion of that delay, perform the command that follows (in this case, set the temp value to 1). The procedure continues to the vwait command, where we turn control over to Tcl's event loop. The vwait command awaits a change of the temp variable, which is set by the after command (after the delay completes). Therefore, once the variable temp changes, the vwait command completes and permits execution to continue.

The Multicast_Server procedure is created next (**line 14**), with two arguments specifying the host and port to which the server must locally bind. At **line 17**, we create the multicast socket using the **dp_connect** command, specifying the protocol type (ipm, for IP multicast), the group address, and the port. After the multicast server socket is created, the resulting socket is stored in mcsock. The **fconfigure** command is then called to allow data to be immediately sent upon **puts** (**line 21**). We then begin our infinite loop at **line 23**, and generate the date and time string at **lines 26 and 29** (resulting in the date_str). At **line 32**, we use the **puts** command to write our date_str to the socket to be multicast to anyone else happening to have joined the multicast group. Our previously discussed sleep procedure is then called, with an argument of 1000 ms (1 second).

*The usage of **dp_connect** here shouldn't be confused with the standard **connect** call of the BSD API. In this usage, the **dp_connect** call provides the **bind** functionality, binding the local group address and port to the newly created socket. This functionality could also be used to create a datagram server socket.*

Finally, at **line 45**, we call the `Multicast_Server` procedure and pass to it the group address we want to join (239.0.0.2) and the port (45002).

Multicast Client

The Tcl source code for the multicast client is shown in Listing 21.6. This source illustrates symmetry with the previously discussed multicast server, shown in Listing 21.5.

Listing 21.6　Tcl Daytime multicast client.

```
1    package require dp 4.0
2
3    #
4    #   Multicast Client setup
5    #
6    proc Multicast_Client { addr port } {
7
8      # Create the multicast socket
9      set mcsock [dp_connect ipm -group $addr -myport $port]
10
11     # Configure the socket so that each puts results
12     # in a socket send.
13     fconfigure $mcsock -buffering line
14
15     # Read a line from the channel
16     set line [dp_recv $mcsock]
17
18     # Emit the date/time string to the socket
19     puts $line
20
21     # Close the client socket
22     close $mcsock
23
24   }
25
26
27   #
28   # Start the multicast client
29   #
30
31   Multicast_Client 239.0.0.2 45002
```

At **line 1**, we import the Tcl-DP package for access to multicast sockets. At **line 6**, we create our Multicast_Client procedure that accepts two arguments, the host and port to which we'll bind (or in this case, the group to which we'll join). This is done using the

dp_connect call with the ipm argument specifying that we're creating an IP multicast socket. We configure our socket for line buffering at **line 13** using the **fconfigure** command and then read in a line of text from the socket at **line 16** using **dp_recv**. This line of text (the date and time string) are emitted to standard-out using the **puts** command (**line 19**), and, finally, the socket is closed at **line 22** using the **close** command.

The multicast client is started at **line 31** with a call to the Multicast_Client procedure. As shown at **line 6**, the procedure takes two arguments, the group address and port to which we'll bind. At **line 31**, we specify our group address as 239.0.0.2 (a multicast IP address) and port 45002.

BROADCAST (UDP) CLIENT

Now, let's look at a UDP broadcast client implementation using datagram sockets. The broadcast client, like the multicast client, is fundamentally a derivative of the datagram code patterns. This is again because broadcast communication is based upon the datagram model. Only the broadcast client is shown because the broadcast server cannot be written in Tcl, or the Tcl-DP package. This is because the APIs lack the ability to set the broadcast socket option that makes it possible to send broadcast datagrams. An application is still able to receive broadcast datagrams without this option.

The broadcast client in Tcl is shown in Listing 21.7. As with all of the Tcl-DP scripts, **line 1** begins with the importing of the Tcl-DP package. We then declare our Broadcast_Client procedure, which accepts an address and port to which the procedure will bind (similar to the multicast join). We use the **dp_connect** command with the udp protocol at **line 7** to create the client socket, and bind it to the broadcast IP address and port number defined by the caller (addr and port). The final setup element utilizes **fconfigure** to enable line buffering on the client socket (**line 10**).

The **dp_recv** command is used to read a line from the socket, which is stored in variable line (**line 13**). At **line 16**, the line is emitted to standard-out using the puts command. The broadcast datagram socket is then closed at **line 19** using the **close** command.

The broadcast datagram client is started at **line 28**. This is done by calling the Broadcast_Client procedure with the broadcast address and the port number from which we want to receive datagrams.

Listing 21.7 Tcl Daytime broadcast client.

```
1   package require dp 4.0
2
3   proc Broadcast_Client { addr port } {
4
5     # Create a UDP client socket and connect to the defined
6     # host and port.
7     set dsock [dp_connect udp -myaddr $addr -myport $port]
8
```

```
 9      # Configure for immediate return of data
10      fconfigure $dsock -buffering line
11
12      # Read a line from the channel
13      set line [dp_recv $dsock]
14
15      # Emit the line
16      puts $line
17
18      # Close the client socket
19      close $dsock
20
21    }
22
23
24    #
25    # Start the datagram client
26    #
27
28    Broadcast_Client 255.255.255.255 45003
```

SIMPLE HTTP SERVER

Now, let's look at a simple HTTP server written in Tcl. The entire source listing for the Tcl simple HTTP server is provided in Listing 21.8. Let's now walk through this listing to understand the sample implementation.

The Tcl simple HTTP server is made up of five procedures, with a general design mimicking that of all other simple HTTP servers presented in this book. In this discussion, we cover each of the procedures in their order of use, which in this case means last procedure first (from Listing 21.8).

The http_server procedure is the main entry for the simple HTTP server (**lines 125–134**). The http_server procedure accepts a single argument, which is the port on which the server should become resident (**bind**, in essence). At **line 128**, the **socket** command is used with the –server option to create the server socket. Immediately following the –server option at **line 128** is the procedure that will be called upon accept of a new client socket (http_accept). The last argument for the **socket** command is the port, which we'll be bound to. The new server socket is stored into servsock (which is not used again here). At **line 132**, we allow the Tcl event loop to take over by calling vwait on an unused variable.

The http_accept procedure (**lines 83–119**) handles a new client connection and determines the type of request being made from the client. The http_accept procedure (a callback for the **socket** function) accepts the new client socket (sock) and the address and port from which the new client connection came (addr and port). We emit a debug message to standard-out at **line 85** and then configure the new client socket for line buffering using **fconfigure** (**line 87**). A line of text is then read from the client using the **gets** command and stored in line (**line 90**). Recall that an HTTP request is made up of the request type (GET, etc.), the object desired (file), and the HTTP version. These are separated by spaces, so we use the split command at **line 93** to split up the request into the three separate elements. **Lines 97–99** extract the newly split elements from the list and store them to the individual variables using the lindex command. At **line 102**, we test the request against "get" (using string compare), and, if it matches (**line 105**), we call the handle_get_method procedure. Otherwise, an error message is emitted to standard-out (**line 112**) and to the client (**line 113**) using the **puts** command. Finally, at **line 117**, the client socket is closed using the **close** command.

The handle_get_method procedure (**lines 49–77**) handles the connection from an HTTP GET message perspective and is the meat of the simple HTTP server. The procedure accepts as arguments the client socket (sock) and the filename being requested (filename). The first step is cleaning up the filename that was requested by the Web client. We trim the '/' characters from the beginning and end of the filename using the string trim command (**line 52**). After this is complete, we check to see if the filename is now empty (for example, if a '/' was requested indicating the default file). If the filename is now zero length (as identified by string compare with a null string at **line 55**), we set the filename to the default file (index.html).

At **line 62**, we open the filename defined by the client using the open command. We then call define_content_type to determine the type of content being returned to the user (**line 65**). We pass the filename to define_content_type, as the content type is determined here by the file extension. We then emit the HTTP response header (**line 70** using a call to emit_response_header) and then emit the contents of the file through the socket using the **puts** command at **line 73**. Note that we read the file anonymously using the **read** command and pass the data read immediately to the socket using the **puts** command. Finally, we close the file opened at **line 62** with the **close** command at **line 75**.

The final two procedures to be discussed are support procedures called from the handle_get_method procedure. The first is emit_response_header (**lines 32–43**) and define_content_type (**lines 4–26**).

Procedure emit_response_header is used to emit the HTTP message response header to the client. It accepts as arguments the client socket and the content type. The **puts** command is used to write the response header to the client, and specifies a message header with status code (**line 37**) followed by a number of optional elements. An important element to notice is the content-type line (**line 40**), which tells the Web client how to render the associated message body. A blank line is also emitted (**line 41**), which is used to identify the end of the response message, and the beginning of the HTTP response.

Procedure define_content_type is used to identify the type of content being returned to the client, based upon the extension of the filename (**lines 4–26**). The filename is split into a list at **line 7** using the split command, and the extension extracted from the list using lindex at **line 10**. A series of if commands are then performed (**lines 14–24**) to identify the supported content types (returned as strings). If an unsupported content type is found, the default is returned ("application/octet-stream").

Listing 21.8   Tcl simple HTTP server source.

```
1   #
2   # Determine the content type based upon the file extension
3   #
4   proc define_content_type { filename } {
5
6     # Split the filename into filename / extension
7     set fileparts [split $filename {"."}]
8
9     # Grab the extension part
10    set extension [lindex $fileparts 1]
11
12    # Test the extensions and return the appropriate
13    # content-type string
14    if      { [string compare $extension "html"] == 0 } {
15      return "text/html"
16    } elseif { [string compare $extension "htm"] == 0 } {
17      return "text/html"
18    } elseif { [string compare $extension "txt"] == 0 } {
19      return "text/html"
20    } elseif { [string compare $extension "tcl"] == 0 } {
21      return "text/plain"
22    } else {
23      return "application/octet-stream"
24    }
25
26  }
27
28
29  #
30  # Emit the standard HTTP response message header
31  #
32  proc emit_response_header { sock content_type } {
33
34    # Emit the header through the socket with one
35    # blank line to identify the header / message
36    # separation.
37    puts $sock "HTTP/1.1 200 OK"
38    puts $sock "Server: TCL shttp"
39    puts $sock "Connection: close"
40    puts $sock "Content-Type: $content_type"
41    puts $sock "\n"
42
43  }
44
45
```

```
46    #
47    #  HTTP Get Method Handler
48    #
49    proc handle_get_method { sock filename } {
50
51      # Trim leading and trailing '/'
52      set newfile [string trim $filename "/"]
53
54      # If new file is blank, replace with 'index.html'
55      if { [string compare $newfile "" ] == 0 } {
56
57        set newfile "index.html"
58
59      }
60
61      # Open the requested file
62      set f [open $newfile r]
63
64      # Call define_content_type and get the content type
65      set ct [ define_content_type $filename ]
66
67      puts "The content type is $ct"
68
69      # Emit the response header
70      emit_response_header $sock $ct
71
72      # Emit the contents of the file
73      puts $sock [read $f]
74
75      close $f
76
77    }
78
79
80    #
81    #  HTTP Connection Handler
82    #
83    proc http_accept { sock addr port } {
84
85      puts "Handling new connection from $addr port $port"
86
87      fconfigure $sock -buffering line
88
89      # Grab a line from the client
90      gets $sock line
91
92      # Split the command out into request / file / version
```

```
93      set message [split $line {" "}]
94
95      # Split out the elements of the HTTP request into
96      # separate strings.
97      set request [ lindex $message 0 ]
98      set filename [ lindex $message 1 ]
99      set version [ lindex $message 2 ]
100
101     # Test the request
102     set get_test [string compare -nocase $request "get"]
103
104     # If 'GET', then call the get method handler
105     if { $get_test == 0 } {
106
107       handle_get_method $sock $filename
108
109     } else {
110
111       # Emit the unimplemented error
112       puts "Unknown method $request"
113       puts $sock "HTTP/1.0 501 Unimplemented Method"
114
115     }
116
117     close $sock
118
119   }
120
121
122   #
123   # Initialization function for the simple HTTP server
124   #
125   proc http_server { port } {
126
127     # Create the TCP Server using the user-defined port
128     set servsock [ socket -server http_accept $port ]
129
130     # Wait indefinitely (allow the event-loop to take
131     # over)
132     vwait forever
133
134   }
```

The simple Tcl HTTP server is started as shown in Listing 21.9. In this listing, we simply call the http_server procedure (**line 7**) and specify the port number on which we want the server to bind.

Listing 21.9 Starting the Tcl simple HTTP server.

```
1   #
2   # Start the HTTP server on port 80
3   #
4
5   puts "Starting the HTTP server on port 80"
6
7   http_server 80
```

SIMPLE SMTP CLIENT

The simple Tcl SMTP client (**lines 1–99**) in Listing 21.10 is made up of only two procedures, the main procedure for sending the e-mail (Mail_Send, **lines 42–99**) and the dialog procedure (dialog, **lines 4–36**) for communicating dialog transactions with the SMTP server.

Let's look at the SMTP client in the reverse order. We first discuss the dialog procedure and then its user, Mail_Send. The dialog procedure (**lines 4–36**) provides the ability to send a command and receive a response from the SMTP server. The caller may optionally specify no command, or no response. If a response is specified, the response is expected from the server and is verified with what the server actually returned.

Line 4 begins with the definition of the dialog procedure. The dialog procedure expects three arguments from the caller: the client socket, string command, and string response. At **line 7**, we check to see if the caller specified a command to send by testing the length of the command using the string command with the length option. If the length is greater than zero, a command is present and is then sent through the socket at **line 9** using the **puts** command. Note that the -nonewline option is specified. We'll allow the caller to add its newline if needed for the command.

At **line 15**, we begin checking for a response. If a response is expected (again identified by the string length of the exp_response argument), we get a line of text from the socket using the **gets** command at **line 18**. Recall that the first three characters of an SMTP server response represent the response status code. At **line 21**, the first three characters are extracted from the line read through the socket and stored in stscode. We then compare the response code extracted from the server response to the response code that the caller expected at **line 25** using the string compare command. The result is stored in check, and then tested at **line 27** (non-zero means that the strings did not compare successfully). If the response codes were not equal, we issue an error, otherwise we simply return. The error command is a mechanism to raise errors to calling procedures, which we discuss in the Mail_Send procedure.

The Mail_Send procedure (Listing 21.10, **lines 40–99**) performs the actual SMTP client protocol. It uses the dialog procedure to perform the actual command transactions with the server, and is, therefore, very simple because the input and output operations are abstracted away.

The Mail_Send procedure accepts a number of arguments for defining the location of the outgoing mail server (mail_server) as well as the e-mail to be sent. The e-mail is defined by the subject (the subject to be shown in the e-mail), the sender (who this e-mail will be shown as arriving from), the recipient (the e-mail address for whom this e-mail is directed), the content-type (how the body of the e-mail should be rendered by the receiving client), and, finally, the e-mail body itself (contents).

We begin at **line 46** by creating the client socket (using the **socket** command) and connecting it to the mail_server defined by the caller and the standard SMTP server port (25). Once the socket is connected, we configure the socket for line buffering (**line 49**) using the **fconfigure** command. At **line 52**, we use the catch command. The catch command catches errors that may occur in the script block within the catch command. If an error is raised in the script, the script is exited and a 1 is returned to indicate that an error occurred. Otherwise, a 0 is returned indicating that no error occurred. The catch block, therefore, allows us to catch any errors that are raised for any of the dialogs, and immediately exit the script within the catch block. The error is stored within variable err (as shown at **line 52**).

Within the catch block (**lines 55–86**), we perform our dialogs with the SMTP server. These won't be detailed here because they are discussed in detail in Chapter 15, Software Patterns Introduction. Each dialog potentially issues a command and in most cases awaits a valid response from the server. Upon completion of the dialogs, we check our error status (**line 91**) and if an error occurred, we simply issue a warning to standard-out using the puts command to notify the user. The socket is then closed at **line 97** using the **close** command.

The final step in Listing 21.10 is calling the Mail_Send procedure at **line 105** to send an e-mail. This step specifies the outgoing mail server, subject line, sender and recipient, content type, and, finally, the message body.

Listing 21.10   Simple Tcl SMTP client source.

```
1    #
2    # Perform a dialog with the SMTP server
3    #
4    proc dialog { sock command exp_response } {
5
6      # Send the command if the user provided one
7      if { [string length $command] > 0 } {
8
9        puts -nonewline $sock $command
10
11     }
12
13     # Only check for a response if the user defined the
14     # expected response code
15     if { [string length $exp_response] > 0 } {
16
```

```
17          # Get a single line from the connection
18          set line [ gets $sock ]
19
20          # Parse the status code from the line
21          set stscode [ string range $line 0 2 ]
22
23          # Check the status code from the server with the
24          # response code that's expected
25          set check [ string compare $stscode $exp_response ]
26
27          if { $check != 0 } {
28
29            # Not what was expected, raise an error
30            error "bad dialog"
31
32          }
33
34        }
35
36    }
37
38
39    #
40    # Send the mail based upon the defined parameters
41    #
42    proc Mail_Send { mail_server subject sender recipient
43                     content_type contents } {
44
45      # Create and connect the client socket
46      set sock [ socket $mail_server 25 ]
47
48      # Configure for immediate receipt
49      fconfigure $sock -buffering line
50
51      # Catch any errors that occur
52      set err [ catch {
53
54        # Look for the initial e-mail salutation
55        dialog $sock "" "220"
56
57        # Send HELO and await response
58        dialog $sock "HELO thisdomain.com\n" "250"
59
60        # Send "MAIL FROM" command and await response
61        dialog $sock "MAIL FROM: $sender\n" "250"
62
63        # Send "RCPT TO" command and await response
```

```
64       dialog $sock "RCPT TO: $recipient\n" "250"
65
66       # Send "DATA" to start the message body
67       dialog $sock "DATA\n" "354"
68
69       # Send out the mail headers (from/to/subject)
70       dialog $sock "From: $sender\n" ""
71
72       dialog $sock "To: $recipient\n" ""
73
74       dialog $sock "Subject: $subject\n" ""
75
76       # Send the content type
77       dialog $sock "Content-Type: $content_type\n\n" ""
78
79       # Send the actual message body
80       dialog $sock "$contents" ""
81
82       # Send the end-of-email indicator
83       dialog $sock "\n.\n" "250"
84
85       # Finally, close out the session
86       dialog $sock "QUIT\n" "221"
87
88     } ]
89
90     # Emit to standard-out if an error occurred
91     if { $err } {
92
93       puts "Couldn't send mail"
94
95     }
96
97     close $sock
98
99   }
100
101
102   #
103   #  Send a test e-mail
104   #
105   Mail_Send "mtjones.com" "The Subject"
106             "tim@mtjones.com" "mtj@mtjones.com"
107             "text/html"
108             "<HTML><BODY><H1>This is the mail</H1></BODY></HTML>"
109
```

A About the CD-ROM

The CD-ROM included with *BSD Sockets Programming from a Multi-Language Perspective* includes all the code from the examples found in the book.

SYSTEM REQUIREMENTS

The minimum system requirements for the tools and accompanying patterns and examples are:

- Pentium I processor or above
- CD-ROM drive
- Hard drive (350 MB available space for tools, 1 MB for patterns and examples)
- 128 MB of RAM

OPERATING SYSTEM REQUIREMENTS

The tools and code examples provided in this book require *Linux* (tested with *Red Hat 7.2*, but will run any standard distribution) or *Windows 95, Windows 98, Windows NT®, Windows 2000*, or *Windows XP*.

ON THE CD-ROM

The CD-ROM contains the software patterns discussed in Chapter 16 through Chapter 21. Where applicable, software patterns for stream sockets, datagram sockets, multicast sockets, broadcast sockets, simple HTTP servers, and simple SMTP clients are provided. Each language discussed in the book is provided with a sample implementation that provides the basis for more complex applications, and as a basis for comparison to better understand how each language provides a particular network feature. Also provided on the CD-ROM are complete code examples from Chapters 2, 5, 6, and 10–14.

The CD-ROM also contains language interpreters and tools that can be used on either Windows or Linux. C is not provided, as C compilers are common on most operating system distributions. A Java interpreter is also not provided due to licensing issues.

The CD-ROM is made up of two parts, software and tools. In the software directory are the software patterns and examples for each of the languages. In the tools directory are the interpreters and tools used with the examples.

Software Patterns

The ./software directory contains 15 subdirectories covering a number of chapters from the book (./ch16 for C, ./ch17 for Java, ./ch18 for Python, ./ch19 for Perl, ./ch20 for Ruby, and ./ch21 for Tcl). Under each of the language chapters are the pattern chapters, which are labeled ./stream (stream socket client and server), ./dgram (datagram socket client and server), ./mcast (multicast socket client and server), ./bcast (broadcast socket client and server), ./shttp (simple HTTP server), and ./smtpc (simple SMTP client). Subdirectories ./ch2, ./ch5, ./ch6, and ./ch10 through ./ch14 include code examples from those chapters of the book.

Software Tools

The ./tools directory contains software tools and languages of the versions used in this book.

The Perl interpreter is provided in source form for *Linux* or *Cygwin* (./tools/perl/perl-5.8.0.tar.gz) and is provided under the "Artistic license" (provided as ./tools/perl/license.txt).

The Ruby interpreter is in source form for *Linux* or *Cygwin* (./tools/ruby/ruby-1.6.8.tar.gz) and as a *Windows* executable 9x/NT/2000/XP (./tools/ruby/ruby168-8.exe). Ruby is copyright-free software by Yukiro Matsumoto and is distributed under the GPL (./tools/ruby/license.txt).

The Python interpreter is provided in source form for *Linux* or *Cygwin* (./tools/python/Python-2.2.2.tgz) and as a *Windows* executable for 9x/NT/2000/XP (./tools/python/Python-2.2.2.exe). Python is distributed under the GPL license (./tools/python/license.txt).

The Tcl interpreter is provided in source form (./tools/tcl/tcl8.4.3-src.tar.gz). It can be used on *Linux* platforms and *Windows* platforms (9x/NT/2000/XP). Tcl is copyrighted by the Regents of the University of California, Sun Microsystems, Inc., Scriptics Corporation, ActiveState Corporation, and others and is distributed under a BSD-style license (./tools/tcl/license.terms).

SOFTWARE LICENSE

The software patterns (within the ./software directory) provided on the accompanying CD-ROM are covered under the following license agreement. This license is modeled after the BSD license, which means that you can use the software for any purpose so long as the license remains with the code in both source and binary forms.

B Acronyms

This appendix provides a list of the terms and acronyms used within this book along with their meaning.

- ACK — Acknowledge
- ALGOL — ALGOrithmic Language
- ANSI — American National Standards Institute
- API — Application Program Interface
- ARP — Address Resolution Protocol
- ASCII — American Standard Code for Information Interchange
- BASIC — Beginners All-purpose Symbolic Instruction Code
- BBC — British Broadcasting Company
- BCPL — Basic Combined Programming Language
- BSD — Berkeley Source Distribution
- CGI — Common Gateway Interface
- CIDR — Classless Inter-Domain Routing
- CPAN — Comprehensive Perl Archive Network
- CWI — Centrum voor Wiskunde en Informatica
- DF — Don't Fragment
- DNS — Domain Name System
- EOF — End of File
- FAQ — Frequently Asked Questions
- FIFO — First In, First Out
- FQDN — Fully Qualified Domain Name
- FTP — File Transfer Protocol
- GCC — GNU C Compiler
- GNU — GNU's Not Unix
- HBO — Host Byte Order
- HTML — HyperText Markup Language
- htonl — Host To Network Long
- htons — Host To Network Short
- HTTP — HyperText Transport Protocol
- ICMP — Internet Control Message Protocol
- IDE — Integrated Development Environment

- IDLE   Integrated DeveLopment Environment
- IEC   International Electrotechnical Commission
- IEEE   Institute of Electrical and Electronics Engineers
- IGMP   Internet Group Management Protocol
- I/O   Input/Output
- IPC   InterProcess Communication
- IP   Internet Protocol
- IPv4   Internet Protocol Version 4
- IPv6   Internet Protocol Version 6
- ISO   International Standards Organization
- JDK   Java Development Kit
- JRE   Java Runtime Environment
- JVM   Java Virtual Machine
- KB   KiloByte
- LAN   Local Area Network
- MAC   Medium Access Control
- MB   MegaByte
- MIT   Massachusetts Institute of Technology
- MSS   Maximum Segment Size
- MTU   Maximum Transfer Unit
- NAT   Network Address Translation
- NBO   Network Byte Order
- ntohl   Network To Host Long
- ntohs   Network To Host Short
- OOB   Out-of-Band
- PDU   Protocol Data Unit
- PERL   Practical Extraction and Report Language
- PEP   Python Enhancement Proposal
- POP3   Post Office Protocol, Version 3
- PSF   Python Software Foundation
- RAM   Random Access Memory
- RDP   Reliable Datagram Protocol
- RDM   Reliably Delivered Messages
- RDMA   Remote Direct Memory Access
- RFC   Request for Comments
- RMI   Remote Method Invocation
- RTT   Round Trip Time
- SCTP   Stream Control Transmission Protocol
- SED   Stream Editor
- SMTP   Simple Mail Transport Protocol
- STD   Standard
- SYN   Synchronize
- TCL   Tool Command Language

- TCP — Transmission Control Protocol
- TOS — Type of Service
- TTL — Time to Live
- UART — Universal Asynchronous Receiver/Transmitter
- UDP — Universal Datagram Protocol
- URL — Uniform Resource Locator
- USENET — User's Network
- WAN — Wide Area Network

Index